ND | Springer Series in **Nonlinear Dynamics**

M. Lakshmanan (Ed.)

Solitons

Introduction and Applications

Proceedings of the Winter School,
Bharathidasan University, Tiruchirapalli,
South India, January 5–17, 1987

With 29 Figures

Springer-Verlag
Berlin Heidelberg New York Tokyo

Professor Dr. **Muthusamy Lakshmanan**

Department of Physics, Bharathidasan University, Tiruchirapalli–620 024,
Tamil Nadu, India

ISBN-13: 978-3-642-73195-2 e-ISBN-13: 978-3-642-73193-8
DOI: 10.1007/978-3-642-73193-8

Softcover reprint of the hardcover 1st edition 1988

2153/3150-543210

Preface

It is more than 20 years since the concept of the soliton was introduced into nonlinear dynamics by Zabusky and Kruskal in their now famous numerical experiments on the Kortweg-deVries equation. Since then the field has grown in an almost exponential manner and has now entered a period of stability and high respectability. It has attracted the attention of researchers in all modern areas of mathematics, physics, engineering and biology. On the one hand various mathematical concepts such as prolongation structures, jet bundles, space curves and surfaces, gauge-equivalence, Lie-algebraic properties including Kac-Moody and Virasoro algebras, symplectic structures, Lie-Bäcklund symmetries, singularity structures, and so on have been attributed to soliton properties, while on the other hand numerous applications have been found in such wide areas as fluid dynamics, lattice theory, plasma physics, condensed matter physics, superconductivity, magnetism, nonlinear optics, particle theory, general relativity, aerodynamics, meteorology and electrical networks.

The Science and Engineering Research Council of the Government of India, Department of Science and Technology (DST), has formulated a programme of annual summer/winter schools to encourage research by younger scientists in the frontier areas of nonlinear phenomena. A winter school in this series on the topic "Solitons" was held at the Bharathidasan University, Tiruchirapalli, South India, January 5–17, 1987. This book contains the proceedings of this winter school. It includes eighteen articles by the speakers at the winter school (of which two articles were given in absentia) and six contributions by participants.

The book consists of five sections. The first section (Part I) deals with introductory remarks on integrability and dynamics and historical aspects of the solitary wave which eventually led to the concept of the soliton. Part II deals with the mathematical theory of solitons in both 1+1 and 2+1 dimensions. The inverse scattering transform (IST), Lie-Bäcklund symmetry, singularity structure and integrability aspects of nonlinear evolution equations are discussed here. Then in the next section (Part III) lattice solitons are considered. The quantum field theoretical and statistical mechanical aspects of solitons are described in Part IV. Finally in Part V a few selected physical and biological applications are considered.

We are extremely grateful to the DST for its financial support and to Professor A. Gnanam, Vice-Chancellor, Bharathidasan University for his enthusiastic encouragement. The school was organized in collaboration with Professor P.K. Kaw, Institute for Plasma Research, Gandhinagar, and his support is also gratefully acknowledged. My colleagues offered me unflinching cooperation in this endeavour and I am particularly indebted to Dr. R. Sahadevan, Messrs. S. Rajasekar, K. Porsezian, and S. Parthasarathy for their help. Finally, I record my appreciation of the very efficient typing of Mr. S. Venugopal of the entire manuscript.

Tiruchirapalli, India
August, 1987

M. Lakshmanan

Contents

Part I Introduction

Inaugural Address – The Dynamics of Dynamics
By E.C.G. Sudarshan 2

"The Wave" "Par Excellence", the Solitary Progressive Great Wave of Equilibrium of the Fluid: An Early History of the Solitary Wave
By R.K. Bullough (With 8 Figures) 7

Part II Mathematical Theory: IST, Symmetries, Singularity Structure and Integrability

Topics Associated with Nonlinear Evolution Equations and Inverse Scattering in Multidimensions
By M.J. Ablowitz 44

Inverse Problems and a Unified Approach to Integrability in 1, 1+1 and 2+1 Dimensions*
By A.S. Fokas and V. Papageorgiou (With 3 Figures) 66

Gauge Unification of Integrable Nonlinear Systems
By A. Kundu (With 3 Figures) 86

Prolongation Structure in One and Two Dimensions
By A. Roy Chowdhury 105

Integrable Equations in Multi-Dimensions (2+1) are Bi-Hamiltonian Systems
By A.S. Fokas and P.M. Santini 118

Painlevé Analysis and Integrability Aspects of Nonlinear Evolution Equations
By M. Lakshmanan and K.M. Tamizhmani 145

Generalised Burgers Equations and Connection Problems for Euler-Painlevé Transcendents
By P.L. Sachdev 162

Bäcklund Transformations and Soliton Wave Functions*
By J.A. Rao and A.A. Rangwala 176

Comparison of Some Numerical Schemes for the K-dV Equation*
By A. Hasan and M.S. Kalra 184

K-dV Like Equations with Domain Wall Solutions and Their Hamiltonians*
By Bishwajyoti Dey 188

Part III **Lattice Solitons**

Lattice Solitons and Nonlinear Diatomic Models
By P.C. Dash 196

Recent Results in Toda Lattice
By Z. Popowicz 212

Construction of Exact Invariants for One- and Two-Dimensional Classical Systems*
By R.S. Kaushal 226

Nonlinear Chains and Kink-Impurity Interactions*
By Bishwajyoti Dey 233

Part IV **Statistical Mechanics and Quantum Aspects**

Quantum Solitons: An Overview*
By R. Rajaraman 240

Soliton Statistical Mechanics: Statistical Mechanics of the Quantum and Classical Integrable Models
By R.K. Bullough, D.J. Pilling, and J. Timonen 250

Exactly Solvable Models in Statistical Mechanics
By M. Wadati and Y. Akutsu (With 12 Figures) 282

Part V **Applications: Physics and Biology**

Solitons and Some Other Special Solutions in Field Theory
By K. Babu Joseph 308

Solitary Waves of the "2-Dimensional Ferromagnet"*
By R. Rajaraman 321

Soliton Propagation in Optical Fibres
By A. Kumar (With 2 Figures) 328

* Contribution by the Participants

Davydov's Soliton
By A.C. Scott 343

Generalized Nonlinear Schrödinger Equations in Quantum Fluid Dynamics*
By B.M. Deb and P.K. Chattaraj 359

Index of Contributors 367

Part I

Introduction

Inaugural Address – The Dynamics of Dynamics

E.C.G. Sudarshan

The Institute of Mathematical Sciences, Madras 600113, India

A dynamical system is defined by a collection of configurational coordinates and equations of motion obeyed by them. These equations of motion may be generated by a suitable principle or may themselves be postulated. Given such equations of motion we would like to solve them so that the dynamical variables at any time may be determined as a function of the initial variables and time. For a system which is a generalization of a Newtonian system these would be even in number. The central problem of dynamics is the determination and characterization of the solutions. Naturally if we can solve the problem completely then we could consider various aspects of the solutions including the dependence of the solution on the initial data. Except for the really trivial cases, even in relatively elementary examples there are interesting dependences and qualitatively new features emerging. For example if we consider the elementary problem of uniform acceleration, say a projectile moving vertically in terms of the initial position s_o, initial velocity u and the acceleration due to gravity -g, the distance s travelled in time t is:

$$s = s_o + ut - \frac{1}{2} gt^2 . \tag{1}$$

But if we ask the time t_1 at which the distance s_1 is reached we have to solve a quadratic equation in which $s_1 - s_o$, g and u appear as coefficients. We may get no solution, one solution or two solutions depending on the set of initial values.

Returning to the question of a general dynamical system, what constitutes a solution? Most of the time "reducing to quadratures" is considered as having solved the problem, though we may not get the solutions in terms of elementary functions. In most cases singular integrals and singularities may appear. One immediate consequence of this is that the solution may become multiple valued and thus imply an unexpected richness of the dynamics.

If the equations of motion of a system can be reduced to a polynomial in a single dynamical variable and its time derivative equated to zero, the problem is reducible to quadratures in terms of Abelian integrals. These have pole and branch point (including logarithmic) singularities. Correspondingly, the dynamical system develops qualitatively new physical behaviour. For simple harmonic motion the singularity corresponds to the limits of the simple harmonic motion.

In a recent paper M. Lakshmanan and R. Sahadevan [1] have given a succinct exposition of nonlinear dynamics from the point of view of integrability and Painlevé analysis with many standard examples and applied the method to two, three and N-coupled quartic anharmonic oscillators [2].

There is a close connection between integrable systems and Lie groups. Gelfand and Kirillov have shown that under very general conditions the generators of a Lie algebra of rank γ and dimension $N=2n+\gamma$ can be realized in terms of rational functions of n pairs of canonical variables and γ unknowns. If one of these generators turns out to be the Hamiltonian, then the dynamical evolution can be viewed as a one-parameter group of transformations generating an orbit. Naturally this is true not only for polynomial Hamiltonians but also Hamiltonians which are rational functions.

But few systems of practical interest are integrable; and so we must resort to a study of qualitative dynamics. The state of the system can be plotted as a point in phase space; and the evolution in a small but finite time maps this point to another point, unless singularities intervene. These phase space maps can be viewed as an alternate form of (discrete time!) dynamics, and one could ask questions about longtime behaviour and other qualitative aspects. Among dynamics so defined we may identify systems which "mix" (in which the long-time evolute is independent of the initial conditions) and those in which the number of constants of motion are much less than normally expected.

Another class of questions of interest involve stability and secular behaviour. These questions are expected to be complicated for nonlinear systems, but they can become nontrivial even for innocent looking systems. For example, consider a Hamiltonian system with n degrees of freedom whose Hamiltonian is quadratic in the phase space variables. The equations of motion are all linear and finite time solutions all exist. However the diagonalization is nontrivial (though already done by

Williamson [3]). It is only in one of the Williamson classes is the solution properly bounded.

Just because the equations of motion of a dynamical system have been integrated it does not follow that the system behaves in a satisfactory manner. If we consider a system with a central inverse cube force added to an inverse square force, the orbit would be a rosette made from a precessing ellipse. There are more startling examples of integrable chaotic systems. In fact the most acceptable forms of chaos are generated by such mechanisms.

We are thus led to the unexpected and unsettling recognition that the lawfulness and the determinism embodied in the orbits of an integrable dynamical system does not automatically imply our intuitive notions of continuity and stability. Slight disturbances could dramatically alter the orbits; and the conserved quantities of dynamics vary irregularly over the phase space. Long term trends cannot be predicted on ideas abstracted from simple dynamical systems.

When a dynamical system has nonlinear equations of motion, the dynamic inertia of the system becomes dependent on the configuration. If it happens that this dynamic inertia tends to vanish these are points of maximum fluctuation where even a miniscule change in the configuration can cause a substantial change in the outcome. Gone is the smooth dependence of the outcome on the initial conditions.

Such irregular configurations could also result from constraints. Given a system with holonomic constraints we can get rid of the constraints, at least locally, by choosing generalized coordinates. But with nonholonomic constraints we have to evolve a whole new theory of constrained systems. The extraction of the true degrees of freedom and the true equations of motion are nontrivial tasks and the question of uniqueness must be investigated in each case.

The elimination of constraints, even when they are holonomic, may not be possible globally. A simple example is provided by a charged particle moving in a monopole magnetic field. In this case the configuration space is $\mathbb{R}^3$ with the origin deleted, and the equations of motion can be written down and seen to be explicitly rotationally invariant. However, the Lagrangian formulation of the problem requires the introduction of the vector potential. This vector potential cannot be chosen in a rotationally invariant manner; and any choice involves a "line singularity" along which the potential cannot be defined. The proper way to handle this problem is to use fibre bundle formalism and identify the choice of any vector potential as a particular section of the fibre bundle. In this case the structure is quite simple; the group space of SU(2) may be seen as a fibre bundle on S^2, the unit sphere, as base. Such dynamics on manifolds is not any longer a curiosity, but appropriate to many problems of dynamics particularly to gauge field theories considered as dynamical systems with infinite number of degrees of freedom.

The invariance group of the differential equations of motion leads to integrals of motion, and serves to simplify the problem of complete integration of the equations of motion. In the Lagrangian form if the Lagrangian is invariant under a group of transformations, so will the equations of motion be invariant. The reverse is not necessarily true. The equations of motion may exhibit quasi-invariance and change of the Lagrangian by a total derivative which may or may not require a redefinition of the generators of change. For example, the system moving in two dimensions having the Lagrangian

$$\mathcal{L} = \frac{1}{2} m(\dot{x}^2 + \dot{y}^2) + eB(x\dot{y} - \dot{x}y) \tag{2}$$

possesses a modified translation invariance with generators $\partial_x + eB\partial_{p_y}$, $\partial_y - eB\partial_{p_x}$ which commute amongst themselves. On the other hand the system with Lagrangian

$$\mathcal{L} = \frac{1}{2} m(\dot{x} - \dot{y})^2 + e(x\dot{y} - \dot{x}y) + \frac{1}{2}(x - y)^2 \tag{3}$$

is only quasi invariant under the translation $\partial_x + \partial_y$; but by the addition of $\frac{d}{dt}\{\frac{e}{2}(x^2 - y^2)\}$ this becomes strictly invariant under the same translation. Of course, the equations of motion are unchanged by this addition.

In the study of mechanics we have come a long way. From the idealized free particles and the two-body celestial mechanics problems we get the impression that mechanics is an exact strictly casual discipline with solutions which may be difficult to compute but which are generally well behaved. This good behaviour includes smooth variations of the trajectory with regard to specification of initial data, and regular longtime behaviour. For nonlinear systems, even relatively simple ones, neither aspect of good behaviour may obtain. Catastrophes and singularities may vitiate the first tendency; and the discovery of completely integrable chaotic systems puts in evidence unexpected possibilities in longtime behaviour. It is a classical result of celestial mechanics that present knowledge is unable to predict whether the moon would escape or undergo capture in the moon-earth-sun system; but many more new features come to light in simple dynamical systems with nonlinear interactions, including the existence of mixing systems.

The generalization of mechanics to general topological manifolds also introduces a new flavour to dynamics itself; and reminds us that we have not always been dealing with the most natural frameworks in mechanics. We are in a period where older results and problems are restudied with new insights and points of view.

It is therefore appropriate that Bharathidasan University has organized this winter school on 'Solitons'. I am honoured to inaugurate it.

REFERENCES

[1] M. Lakshmanan and R. Sahadevan, Painlevé analysis, Lie-symmetries, and integrability of coupled nonlinear oscillators of polynomial type, Physics Reports, to be published; R. Sahadevan, Painlevé analysis and integrability of certain coupled nonlinear oscillators, Ph.D. Thesis, University of Madras, 1986 (Unpublished).

[2] M. Lakshmanan and R. Sahadevan, Phys. Rev. **A31** (1985) 861; R. Sahadevan and M. Lakshmanan, Phys. Rev. **A33** (1986) 3563.

[3] J. Williamson, Amer. J. Math. **58** (1936) 141.

"The Wave" "Par Excellence", the Solitary Progressive Great Wave of Equilibrium of the Fluid: An Early History of the Solitary Wave

R.K. Bullough

Department of Mathematics, U.M.I.S.T., P.O. Box 88,
Manchester M60 1QD, United Kingdom

It is shown how in 1876 Rayleigh resolved the conflict between Russell on the one hand, and Stokes and Airy on the other, about the nature of the solitary wave and the formula for its velocity of propagation c. However, Boussinesq had already done this in 1872 when he published the Boussinesq equation and gave its solitary wave solution. The fundamental articles by Russell of 1840 and 1844 which first introduced the solitary wave and gave the formula $c = \sqrt{g(h+k)}$ for its velocity are surveyed. They show that he understood the collision properties of solitons in 1835, though these objects and their mathematics emerged only some 130 years later.

1. INTRODUCTION

This lecture is concerned with a few points surrounding the early history of the solitary wave and of the soliton it gave birth to.

Because the 1973 review of solitons [1] made its now well known quotation from John Scott Russell's 1844 paper [2], everybody knows how in the month of August 1834 he rode his horse along the banks of a certain Scottish canal [3] in pursuit of a disturbance of the water which in particular Stokes [4], and more latterly ourselves have called a 'solitary wave'. Recall Russell 'was observing the motion of a boat which was rapidly drawn along a narrow channel by a pair of horses when the boat suddenly stopped--not so the mass of water in the channel which it had put in motion: it accumulated round the prow of the vessel in a state of violent agitation, then suddenly leaving it behind, rolled forward with great velocity, assuming the form of a solitary elevation, a rounded smooth and well defined heap of water, which continued its course along the channel apparently without change of form or diminution of speed. I (He) followed it on horseback and overtook it still rolling at some eight or nine miles an hour, preserving its original figure some thirty feet long and a foot to a foot and a half in height. Its height gradually diminished and after a chase of one or two miles I lost it in the windings of the channel. Such, in the month of August 1834, was my first chance encounter with the singular and beautiful phenomenon which I have called the Wave of Translation, a name which it now generally bears: which I have since found to be an important element in almost every case of fluid resistance, and ascertained to be of the type of that great moving elevation of the sea, which, with the regularity of a planet, ascends our rivers and rolls along our shores". This 'first chance encounter' was actually reported first of all in an earlier paper [5] and in a rather more prosaic fashion (Ref. [5], p. 61).

Notice that Russell is using 'Wave of Translation' not 'solitary wave'. As we shall see it was certainly he who introduced the word 'solitary' to describe it, and on the p. 61 of Ref. [5] he used '*a large solitary progressive wave*' (Russell's italics) propably the first such use. But generally he preferred 'Wave of Translation', or 'The Wave' ([5], p. 61) as it is used in the title of this lecture. It was Wave of Translation he continued to use throughout his life and finally in the posthumous book [6], 'The Wave of Translation in the Oceans of Water, Air and Ether' published in 1885, on which I comment shortly. However, I shall follow Stokes [4], Rayleigh [7], and indeed current practice, and use the term 'solitary wave' for Russell's 'Wave' here; and this lecture is concerned with the history of the solitary wave from August 1834, when Russell first saw it, upto 1876 when Rayleigh [7] calculated its profile.

Of course the history of the solitary wave, or even of this particular solitary wave, does not stop in 1876. In 1895 Korteweg and de Vries published their paper [8] in which they gave their now famous equation. And they also gave the solitary wave solution of that equation. It was probably only then, as we shall see, that the controversy surrounding the whole idea of the solitary wave as Russell conceived it ceased, and during the period 1876-95 a number of papers contributed to its discussion--two by McCowan [9,10] amongst others.

These days we usually quote the KdV equation in a scaled form such as [11]

$$u_t + 6uu_x + u_{xxx} = 0\ , \tag{1.1}$$

where u_t means $\frac{\partial u}{\partial t}$, etc. And its solitary wave solution is then

$$u = 2\xi^2 \operatorname{sech}^2 \xi(x - 4\xi^2 t) \qquad (1.2)$$

in which ξ is a free, real, parameter. Both the equation (1.1) and its solution (1.2) actually refer to a frame translating at the 'sound' speed, but Russell was not concerned with subtleties like that. However he was very much aware of the formula for that sound speed, namely

$$c_o = \sqrt{gh} \qquad (1.3)$$

in which g is the acceleration due to gravity and h, using Russell's terminology [2], is the depth of undisturbed water in the canal. Equation (1.1) is thus in a frame translating at c_o with c_o set equal to unity.

Following the discoveries of Zabusky and Kruskal [12], and Gardner, Greene, Kruskal and Miura [13], we now know that (1.2) is not just a solitary wave: it is a soliton with the remarkable collision properties of solitons [12]. Viewed in these terms, and without any computer power at his disposal, Russell's mathematical capabilities were relatively weak. He knew [2] Euler's 'general formula for the motion of fluids in the Memoirs of the Academy of Sciences of Berlin' [14]. And he seems to have had a useful knowledge of classical texts like Lagrange's [15] and Laplace's [16], and even Poisson's [17], while he particularly appreciated [2] a work [18] by the two brothers Weber, Professors at Leipzig and Halle respectively. But it was on his acuteness of observation and a strong if variable physical insight and understanding on which he relied; and from his experiments on water waves in channels constructed in the laboratory during 1834-40 [2], as well as the earlier experiments he performed on the canal [5] in the period 1834-35, he already knew that the solitary wave he had first seen generated in 1834 had this collision property. Perhaps what prevented him from stressing this particular, and to us so much more remarkable, feature was the more compelling need he faced to justify to his scientific peers the idea of the solitary wave itself. It is with this aspect that this historical note is concerned.

2. JUSTIFICATION OF THE SOLITARY WAVE: BOUSSINESQ'S PAPER

Although, as I shall show, Rayleigh's paper [7] should have settled the whole argument in 1876, it was settled even before this in a remarkable paper by Boussinesq presented in 1871 [19] and published in 1872 [20]. Moreover, and despite the discussion between 1876 and 1895 culminating in the paper by Korteweg and de Vries [8], this paper might in one respect have opened up the problem again rather than closed it. This was because much of Russell's own case for the existence of the solitary wave rested on the exact truth of his formula for the velocity of propagation of that wave

$$c = \sqrt{g(h+k)} \tag{2.1}$$

which he first gave as such in [2]: again the terminology is Russell's, and k is the height of the peak of the solitary wave above the surface of the undisturbed water. Obviously (2.1) is a natural generalisation of (1.3) and I believe this is how Russell found it. However he then went on to demonstrate its truth by experiment [2]. The point now is that the formula (2.1) is not quite the velocity of the solitary wave solution (1.2) of the KdV equation (1.1), even when that solution is placed in the laboratory, rather than the moving, frame. I establish the actual error shortly. Later I show how Korteweg and de Vries [8] really did settle the whole matter, however.

Still the matter was settled by Boussinesq in 1871-72. For in his paper [20] he gave the analysis by which, starting from Euler's equation for the conservation of momentum under pressure gradients ∇p and body forces $\nabla\Omega$ in an incompressible fluid of density ρ

$$\frac{D}{Dt}\rho \underset{\sim}{u} = \rho(\underset{\sim}{u}_t + (\underset{\sim}{u}.\nabla)\underset{\sim}{u}) = -\nabla_p - \nabla\Omega, \quad \underset{\sim}{u} = (u,v,w) \tag{2.2}$$

(in a modern notation), together with the equation for conservation of mass

$$\nabla.\underset{\sim}{u} = 0 , \tag{2.3}$$

he reached the partial differential equation

$$u_{tt} = c_o^2 u_{xx} + c_o^2 \left(\frac{3}{2}\frac{u^2}{h} + \frac{h^2}{3} u_{xx}\right)_{xx} ; \tag{2.4}$$

h is the depth of undisturbed water and c_o is given by (1.3). Boussinesq also gave the solitary wave solution of this equation, namely,

$$u(x,t) = k \operatorname{sech}^2 \left(\frac{1}{2}\sqrt{\frac{3k}{h^3}}(x - ct)\right) . \tag{2.5}$$

The height at peak of this solitary wave is indeed k, while its velocity c proves to be given by (2.1) exactly.

One easily scales (2.4) to

$$u_{tt} = u_{xx} + \left(\frac{3}{2} u^2 + u_{xx}\right)_{xx} , \tag{2.6}$$

called the Boussinesq equation. The solitary wave solution of (2.6) is

$$u = \lambda^2 \operatorname{sech}^2 \tfrac{1}{2}\lambda(x - ct) \tag{2.7}$$

and

$$c^2 = 1 + \lambda^2 . \tag{2.8}$$

The Boussinesq equation is integrable and (2.7) is a soliton; but the spectral problem which solves it is a third order (or 3 × 3 matrix) spectral problem [21], and so it is rather more complicated than the Schrödinger spectral problem $-\Psi_{xx} + u\Psi = k^2\Psi$, already familiar from wave mechanics, which solves the KdV equation (1.2) [13]. It is therefore fascinating to speculate on the history of the soliton itself if Boussinesq's paper of 1872 had then, and subsequently, received the attention it deserved. It is just possible that this meeting at Tiruchirapalli would not have taken place had it done so!

Of course the Boussinesq equation (2.6) 'contains' the KdV equation (1.1). For in reaching (2.4) terms effectively of order $0(k^2/h^2)$ are dropped. And to this order (2.4) can be 'factorised' so that (2.6) is written as

$$\left(\frac{\partial}{\partial t} - \frac{\partial}{\partial x}\right)\left[\frac{\partial}{\partial t} + \frac{\partial}{\partial x} + \frac{1}{2}\frac{\partial}{\partial x}\left(\frac{3}{2} u + \frac{\partial^2}{\partial x^2}\right)\right] u = 0 . \tag{2.9}$$

A solution of

$$u_t + u_x + \frac{1}{2}\left(\frac{3}{2}u^2 + u_{xx}\right)_x = 0 \tag{2.10}$$

solves (2.9), and such a solution is

$$u = \lambda^2 \operatorname{sech}^2 \tfrac{1}{2}\lambda(x - c_1 t) \tag{2.11}$$

with

$$c_1 = 1 + \tfrac{1}{2}\lambda^2 . \tag{2.12}$$

Evidently

$$c_1^2 = 1 + \lambda^2 + \tfrac{1}{4}\lambda^4 \tag{2.13}$$

and c_1^2 differs from c^2, equation (2.8), by $0(\lambda^4) = 0(k^2/h^2)$. On the other hand in a moving frame, so that $\partial/\partial t + \partial/\partial x \to \partial/\partial t$, (2.10) easily scales to the KdV equation (1.1), so that (2.13) makes the point that the KdV soliton moves at a speed slightly different from Russell's formula (2.1)! Still it was at $0(\lambda^2)$ that the argument raged, as we shall see, and $\lambda^2 \approx 1/5$ in Russell's first experiments [5] on his canal. Larger values of λ^2 arise in the more detailed laboratory experiments reported in [2], however: the largest value quoted seems to be $\lambda^2 = 1/3$ (Ref. [2] Table VII, p. 336) but values of $\lambda = 1$ were examined for Russell [2] certainly knew that the waves broke in this case as he points out (e.g. [2], p. 340). Rayleigh [7] quotes Airy [22] as saying the wave always broke in this case and shows why. Rayleigh's actual remarks in this connection are given at the very end of this paper.

In the remainder of this note I survey some aspects of Russell's work, that due to Airy [22] which disagreed with Russell's, some due to Stokes [4] which did likewise, and Rayleigh's paper [7] which confirmed both the solutions (2.5) or (2.7) (upto (k^2/h^2)!) for Russell's solitary wave and his key formula (2.1) for the velocity of that solitary wave exactly.

3. THE BOOK 'THE WAVE OF TRANSLATION' AND RUSSELL'S PAPERS OF 1840 AND 1844

Although almost everybody alive must now know of Russell's 1844 paper [2], few have read it, and only the well-known quotation from it survives. Still fewer people are aware that the 1844 paper was preceded by the longer paper [5] published in 1840 in the Transactions of the Royal Society of Edinburgh. These two papers, [2] and [5], contain Russell's permanent contribution to the evolution of our science. In a working lifetime of some 56 years (he was born in 1808, had studied at three of the four Scottish universities, Edinburgh, St. Andrews and Glasgow, before graduating from the last at the age of 16, and died in 1882) he published some 49 scientific articles [11]--including 21 British Association Reports, 4 Edinburgh Royal Society Transactions or Proceedings, and one Royal Society (of London) Proceedings [23]. (He was elected FRS in 1847.) He also wrote at least three books, namely the posthumous [6], an original, and indeed seminal, contribution to Naval Architecture 'The Modern System of Naval Architecture' [24], and a pedagogical contribution a 'Systematic Technical Education' [25]. But he never had a permanent academic or scientific appointment (Sir W. R. Hamilton, the great Hamilton, wrote of Russell as a 'person of active and inventive genius' in support of his application for the

Chair of Mathematics at Edinburgh University in 1842; but this was not enough to get Russell the job). Russell spent the period upto 1846 on his scientific researches often in his spare time [5]. By 1855 he was certainly building the great 12,000 ton iron ship 'The Great Eastern' at his own shipyard at Millwall on the Thames, for the engineer Isambard Kingdom Brunel; and in 1856 he published his major work on Naval Architecture 'The Modern System'. Further details of a remarkable life are given in the Appendix to [11] and the biography [26]. He seems to have reflected on his solitary wave throughout that life and on June 16, 1881, one year before his death on June 8, 1882, he gave his paper [3] at the Royal Society entitled 'The Wave of translation and the work it does as the carrier wave of sound'. This was followed by the posthumous book [6] of 1885.

In the paper [23] Russell apparently argued, as his book certainly does, that since in general dispersion will obviate any possibility of transmitting information by sound, the solitary wave, which, virtually by definition has permanent shape, must be used instead. The 'Wave of translation' is Russell's solitary wave so the formula (2.1) applies. From $c \sim 1100$ ft sec^{-1}, $g \approx 32$ ft sec^{-2}, one finds $h+k \sim 8$ miles; Russell corrects for varying density by a factor 2/3 so the equivalent depth of the atmosphere is about 5 miles. From there Russell [6] used the velocity of light for c, the same g, and certain apparently arbitrary factors $\sim 5 \times 10^4$, to reach a 'depth of the universe' of 5×10^{17} miles: this is in error by only 5 orders of magnitude--quite a respectable estimate in my view given all the circumstances!

The permanent work in the book [6] is the Appendix. This is simply a reprint of the 1844 paper [2]. That paper is primarily concerned to report the formula (2.1) which it then validates pragmatically by showing that the speeds of the solitary waves actually measured by Russell in the laboratory fit to it: no theory leading to (2.1) is presented anywhere in Russell's work. The paper also contains a categorisation of waves which was very much Russell's concern at the time. He here gives four 'orders': First, the wave of translation--wave of first order and its two species, positive and negative, though Russell well recognised the important difference between the positive and negative cases--see below; second, oscillatory waves, positive and negative and second order; third, surfaces agitated to minute depth, that is capillary waves, third order; and fourth, corpuscular waves, sound waves in fluids reflecting the existence of molecular forces. There appears the Table I "System of Water Waves". First order is as above: it is solitary in character, it can be free or forced, and includes the 'wave of resistance', the 'tide wave', and the aerial sound wave (as explained already). Waves of the second and third orders are gregarious: order two includes stream ripples, wind waves, ocean swell, stationary or progressive, free or forced; third order includes 'dentate waves' and 'zephyral waves', free or forced. The corpuscular wave of fourth order is solitary and includes the water and sound waves.

That Russell knew about soliton break-up is evidenced by the Fig.6 in the Plate 47 of the 1844 paper [2]: this Plate 47 is now reproduced here as the Fig. 1. One sees the first example of break-up is positive and so really is soliton break-up. However, the second example shows break-up into a positive wave leading and a negative one following. Still, Russell certainly knew the negative wave is not really a soliton at all--or even a solitary wave: its speed did not conform to the formula (2.1) as the experiments he reports in the Fig. 5 of Plate 48 of [2] showed, while in the Plate 52 of [2] he shows how the negative wave though at first solitary breaks up into gregarious oscillating waves of order two. The Plates 48 and 52 of [2] are here reproduced as the Figs. 4 and 7.

The 'wave of resistance' of first order referred to in the Table I and appearing already in the quotation from [2], is a reference to Russell's work in the earlier paper [5]. In this long paper he is not so much concerned with the velocity formula (2.1), which he quotes in words only, but with the very practical problem of the resistance of boats towed (or travelling) through water of finite depth. One should recall that Russell was actually on his horse on the banks of canal in the month of August 1834 charged by the Union Canal Company to determine the efficiency of canals for the possible transport of steam driven barges [3]. The conventional wisdom of the time, due to Newton, Bernouilli and d'Alembert to whom he refers, was that the resistance R was simply proportional to the square V^2, of the velocity V of the boat through the water. Being aware as he says [5] 'of the very imperfect state of that part of Theoretical Hydrodynamics which relates to the Resistance of Fluids to the Motion of Floating Bodies, and that there had been found in its application to the solution of practical questions, discrepancies so wide between predicted results and the observed phenomena, as render the principles of the theory exceedingly false guides, when followed as maxims of art' Russell determined to investigate the situation by a series of experiments which he carried out in the summers of 1834 and 1835. He found instead of $R \propto V^2$ that resistance X plotted against velocity Y followed curves like those

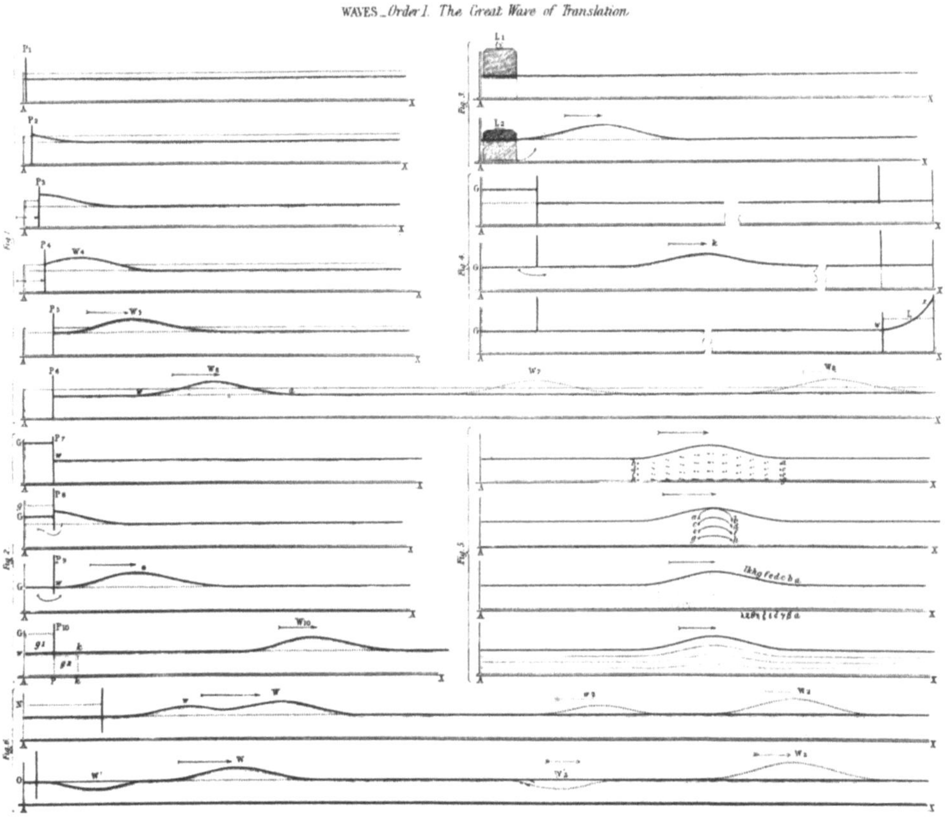

Fig. 1 This Figure is the Plate 47 of Ref. [2]. This plate shows the ways in which Russell created solitary waves in troughs some 6 inches deep in the laboratory. For soliton theory the two most interesting Figures in this plate are the two at the bottom (Fig. 6) as explained in the text. Fig. 5 on the plate shows in its four diagrams the motion of individual water particles during the passage of a solitary wave (the particles were actually material particles suspended in the water). The first diagram describes simultaneous motions at different points; the second, motion of four particles throughout the whole passage of the solitary wave from A to X (the curves are semi-ellipses according to Russell and show that there is a net translation of position which he takes to be the volume of water to create the wave divided by the breadth of the channel); the third shows the motions of vertical planes during the passage (small particles were suspended from stalks in equally spaced vertical planes in the undisturbed water); the fourth shows movement of initially equally spaced horizontal planes of particles during the passage.

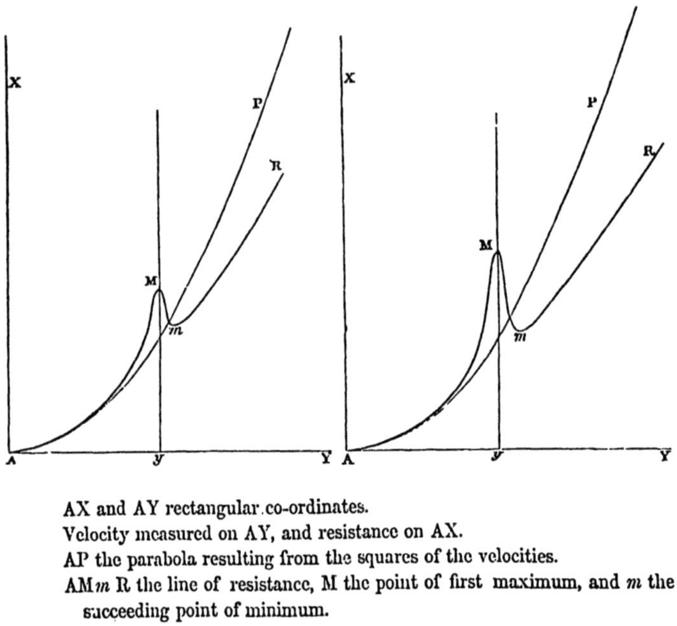

Fig. 2 Two sets of observations made by Russell on his canal of the resistance X of a boat plotted against its speed Y. The curves P are the parabola of the conventional view at the time and the curves R are what Russell found in 1834-35(taken from [5]).

shown in Fig. 2: curves P are $R \propto V^2$, curves R with a maximum at M and a minimum at m, *both depending on the depth h of the fluid*, are followed by the rising curves to R; but, later, for velocities ~29 mph, he argues that these curves too fallaway from a second maximum. In this later discussion Russell gives an explanation of the whole form. Its essentials are these: a crucial feature is the emersion (i.e. emergence) of the whole boat: this explains the second maximum of resistance at $V \approx 29$ m.p.h. (which Russell [5] computes as 4g/3) and its subsequent fall where at 43.8 m.p.h. (V = 2g) the 'floating body emerges wholly from the fluid and skims its surface'. However, the more profound influence arises, as Russell was the first to realise, from the generation and propagation of waves. These waves travelled with speed c given by (2.1), according to Russell, so were unconnected with the form of the vessel. It was this speed which coincided with the first maximum at M. When $V < c$ the effect was to generate Russell's 'Great Anterior Wave' [5], a wholly positive disturbance, near the prow and a depression, the 'Great Posterior Wave' near the stern. Thus the vessel tilted and increased its resistance. But for $V > c$ the vessel could ride horizontally on the water diminishing the resistance to m. Russell noticed that in this region the destructive power due to what he called the 'stern surge', destructive to the banks of the channel and dangerous in navigating shallow water, disappeared. Whilst the great anterior wave of so called displacement was bell shaped, the great posterior wave of so called replacement was oscillatory, breaking into the 'surge' or 'breaking' wave. At this stage he therefore distinguished four species of wave, the 'Ripple or Dentate', the 'Oscillatory', the 'Surge', and '"The Wave" "par excellence", the solitary progressive great wave of equilibrium of the fluid'. For the vessel itself the Great Primary Wave of displacement i.e. the great anterior wave, was the only example of 'The Wave'. 'The Wave' is the solitary wave as I use it here--so thus the title of this lecture!

Russell noticed the independence of 'The Wave' from its mode of genesis and that its speeds (8 m.p.h. for the channel 5.5 ft deep he was concerned with) were independent of the speed of the generating bodies (e.g. 2, 5, 6 and 12 m.p.h.). Moreover 'Another observation equally simple served to show that a large or high wave had a greater velocity than a small one. When a small wave preceded a large one, the latter invariably overtook the other, and when the large wave was before the less, their mutual distance invariably became greater'(Ref. [5], p. 6). This remark certainly supports the formula (2.1), but it also supports the concept of the soliton also. Moreover, reference to Plate II of [5] shows that Russell had already drawn the shapes of different observed examples of his "Wave". And he also draws there three different examples of soliton break-up which he must therefore have known about in 1835, and strictly from his observations on the canal not from the experiments of [2].

Russell was naturally concerned with the problem of moving a boat into the minimum resistance region m and noted some of the possible difficulties. 'It appears', he says, 'that increased force applied gradually to the vessel for the purpose of rendering the velocity of the body equal to or greater than that of the wave, has the effect at the same time of increasing at a more rapid rate the retarding forces, and a limit is soon reached, which it has in many case been found impossible to pass' (later he shows how a singularity may develop at M). 'It is the circumstance of the very rapid increase of the resistance in approximating to the velocity of the wave that has led to the false idea that there is a final and low limit to velocity in shallow water. There are circumstances in which this limit is final, the channel being very shallow, and the boat very bluff in its formation, I have seen in such an extreme case, when the depth of the channel was about five feet, the channel laid bare in the stern hollow behind the wave, so that the stern of the vessel no longer floated but rested on the bottom, while the bow was elevated and buried in a large anterior wave, rising more than two feet above the level, and overflowing the banks, and the posterior wave rushed furiously on behind, roaring and foaming, tearing up the banks of the channel, and threatening the destruction of the vessel, which, indeed, on stopping, it nearly accomplished. In such a case the persons in the vessel were not visible from the shore, being sunk in the hollow between the great anterior and posterior waves.'

He then goes on to discuss the passage beyond the point M, namely to m, where the boat rides horizontally upon the wave. Figures 7 and 8 from [5] appearing in Fig. 3 show exactly what he intends. 'But

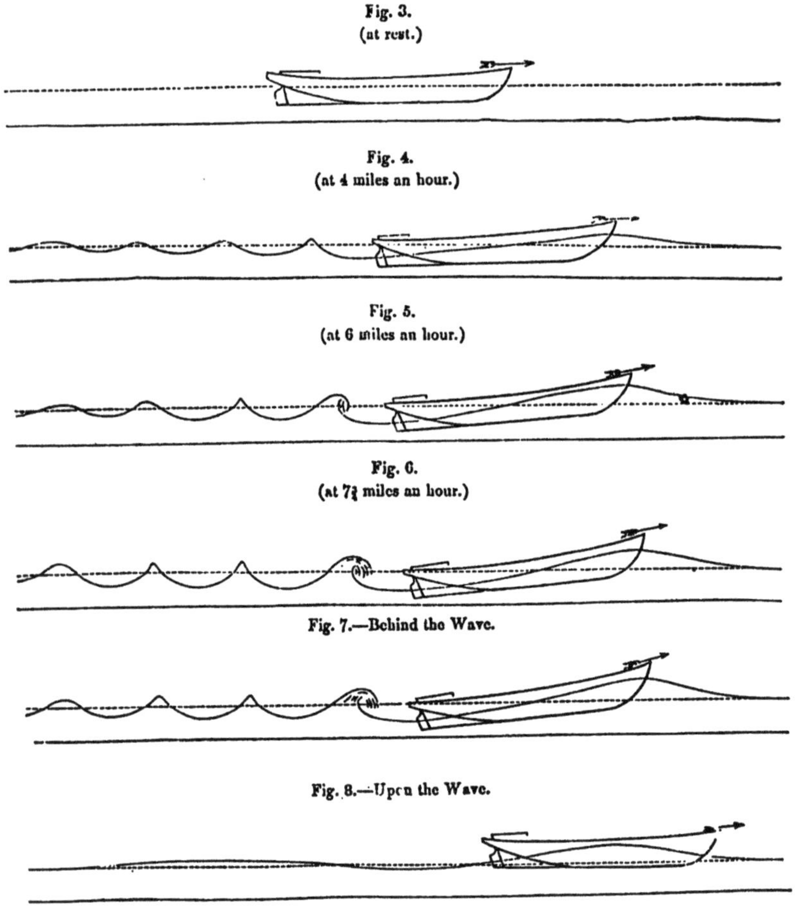

Fig. 3 Figs. 3-8 taken from Russell [5]. Figs. 7 and 8 show how a maximum resistance (Fig. 7) is replaced by a minimum as the vessel rides on the solitary wave.

it will be inquired, how is a vessel to be placed in such circumstances? How is the extreme resistance of the anterior wave to be vanquished, and the vessel planted on its summit? This is admitted to be a practical problem, often of extreme difficulty; sometimes it is impracticable. There are forms of vessel that do not admit of a position of stable equilibrium on the top of a wave. Still, however, it is a practical problem solved everyday on all canals navigated on the Scotch system. Vessels of greater length than the wave, having a fine entrance, built of light materials, and drawn by well trained highly bred horses, and guided by experienced postillions, are raised by a sudden and powerful jerk to the top of the wave (at from 6 to 8 m.p.h.) and are drawn along on the summit of the wave with greater ease at 10 or 12 m.p.h., than at 6 or 7.' (see §IX of [5]).

I have given these various quotations from [5] at length in order to show Russell's grasp of the situation and incidentally to exhibit the lively and imaginative character of his writing: I hope it will stimulate others to search out his articles and read them. They reflect a much more leisured scientific age, and amaze, stimulate, and infuriate all at the same time. In the rest of this paper I want to show why Russell had such difficulty in establishing the existence of the object we now call the solitary wave. Herschel's comment "It is merely half of a common wave that has been cut off" [27] seems typical of his critics. To simplify the account here I shall focus only on the objections raised by Airy [22] and Stokes [4]. This leads to some historical comment on the shape of Russell's wave. Then, in a final section, I shall quickly show how Rayleigh [7], in particular, fully justified Russell's position.

4. THE OBJECTIONS OF AIRY AND STOKES AND SOME COMMENTS ON THE SHAPE OF THE WAVE

In publishing the 1844 paper [2], which just preceded the appearance of Airy's treatise [22], Russell already knew that Airy disagreed with his formula (2.1) for the velocity c of propagation of a wave of permanent shape. Such a wave is necessarily a wave $u(x,t) = u(x - ct)$. Russell believed his 'Wave' was in this category and, by implication was the only such wave being necessarily positive and lump shaped i.e. solitary. Airy believed to the contrary. Rayleigh [7] quotes from Airy [22] "We are not disposed to recognize this wave as deserving the epithets 'great' or 'primary' . . . and we conceive that ever since it was known that the theory of shallow waves of great length was contained in the relation $d^2X/dt^2 = gkd^2X/dx^2$,... the theory of the solitary wave has been well known". And again "Some experiments were made by Mr. Russell on what he calls a negative wave--that is, a wave which is in reality a progressive hollow or depression. But (we know not why) he appears not to have been satisfied with these experiments, and has omitted them in his abstract. All the theories of our IV Section, without exception, apply to these as well as to positive waves, the sign of the coefficient only being changed." That the second comment is unfair in the light of the material published in [2] is clear--though Airy might not have seen that. However, it is immediately obvious that the linear equation given in the first quotation can only have harmonic solutions albeit, since it is not dispersive, they are unique in speed as well as of permanent shape: But they are nowise solitary. More generally Airy [22] found the velocity formula

$$c^2 = \frac{g}{m}\frac{e^{mh} - e^{-mh}}{e^{mh} + e^{-mh}} = \frac{g}{m} \tanh mh \tag{4.1}$$

which reduces to c_o^2, (1.3), for wave number $m \ll h^{-1}$. However he also included the effect of finite disturbance k to reach

$$c^2 = g(h + 3k) \ . \tag{4.2}$$

As we shall see Rayleigh [7] subsequently found an argument for the same result; but, finding it for a *distorting* wave, he then developed the very different argument that arrives at (2.1). Boussinesq [19,20] quotes Bazin as confirming (2.1), for (2.5), and (4.2) with $\frac{3}{2}k$ not 3k, for a step function. But in 1844 Russell [2] was simply very much concerned to show that Airy's formulae did not fit his actual observations. And the Fig. 4, reproduced from Plate 48 of [2] shows in its Figs. 3 and 4 how well he succeeds in doing this. Russell also knew a result by Kelland [28] quoted in [2] p. 333 as [29]

$$c^2 = \frac{g}{\alpha}\frac{e^{ah} - e^{-ah}}{e^{ah} + e^{-ah}} \div \left\{1 - \varepsilon\,\frac{ae^{ah} - e^{-ah}}{e^{ah} + e^{-ah}}\right\} \tag{4.3}$$

in which ε is given as the semi-elevation, h the depth of water in repose, and λ the length of the wave. He suggests (Ref. [2],p.334) that Airy [22] has simply followed Kelland 'over the same ground in an elaborate paper on waves' and concludes 'a theory of the first order accurately representing this characteristic phenomenon is still wanting, a worthy object for the enterprise of a future wave mathematician'. There proved to be at least two of these, Bousinnesq [19,20] and Rayleigh [7].

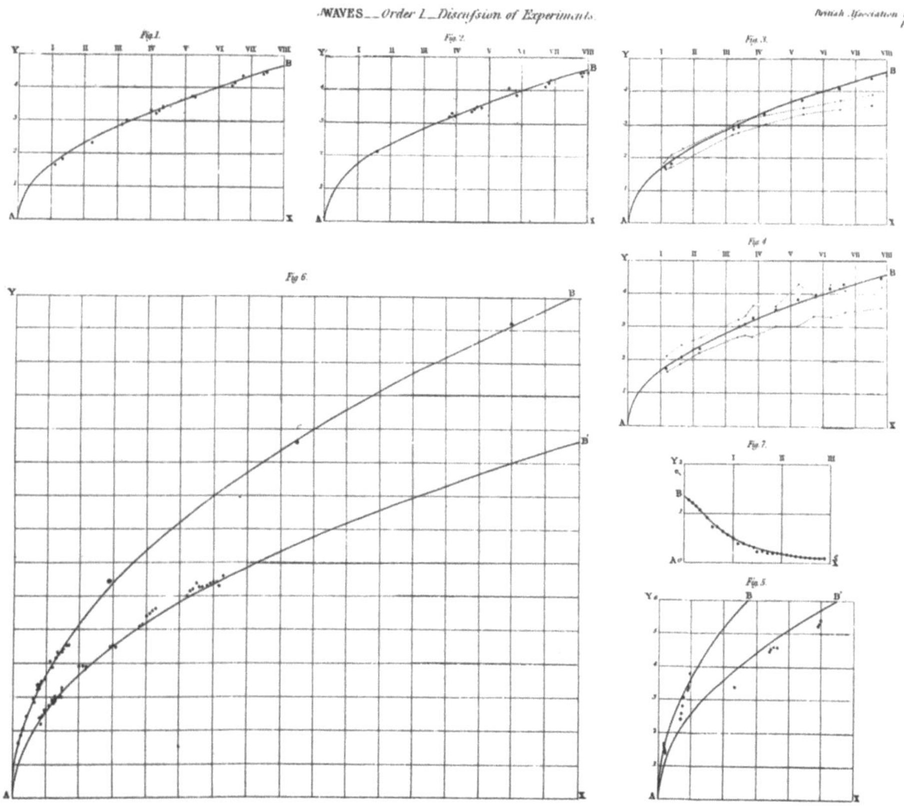

Fig. 4 This Figure is the Plate 48 of [2]. Its Fig.1 is a check of the formula (2.1) against observations for channels 1.0 to 8.0 inches deep and wave heights up to 0.54 inches; Fig. 2 is for depths 1.62-8.0 inches deep and wave heights up to 0.96 inches. Figs. 3 & 4 compare experimental points against (2.1) (bold line) and against four different formulae due to Airy which Russell does not specifically quote (dotted lines). Fig. 5 compares the formula (2.1) (curve AB) against observations for a negative wave in a rectangular channel and formula $c^2 = 1/2\, g(h+k)$ (curve AB') for a negative wave in a triangular channel. Fig. 6 plots general results for c (in units of one foot per second) against (h+k) (in units of five inches): AB is (2.1) for a rectangular channel; AB' is $c^2 = 1/2\, g(h+k)$ for a triangular one.

At this stage it is useful to summarize what Russell apparently knew. Lagrange [15] first integrated the conservation of momentum equation (2.2) in the case of irrotational motion $\underset{\sim}{u} = \nabla\varphi$. This integral is

$$\varphi_t(x,t) + F(t) + \frac{1}{2}\underset{\sim}{u}^2 = -\rho^{-1}p - \rho^{-1}\Omega \qquad (4.4)$$

$(\underset{\sim}{u}^2 = u^2 + v^2 + w^2)$. For steady motion

$$\frac{p}{\rho} + \frac{1}{2}\underset{\sim}{u}^2 + \frac{\Omega}{\rho} = 0 \qquad (4.5)$$

(by choosing the zero for Ω) and this result contains both Bernouilli's theorem [30] and Torricelli's [31]. Russell well understood Torricelli's theorem, which he quotes in [5], as well as (4.5) and its application in the pitot tube. He built a shallow bottomed boat 'The Skiff' [5] in which he placed a series of pitot tubes to measure the velocity profile of water passing a boat in connection with the 1834-35 experiments.

The formula (1.3) quoted by Russell in [2] is apparently due to Lagrange who used the 'method of long waves' to find it [15]. Consider motion in the (x,y)-plane only with an undisturbed surface $y = y_o$, and a disturbed one $y = \eta(x,t) + y_o$. At an arbitrary point (x,y) within the disturbed fluid the pressure p there is p_o the pressure at the disturbed surface enhanced by the potential term $g\rho(y_o + \eta - y)$ so that

$$p - p_o = g\rho(y_o + \eta - y) \ . \tag{4.6}$$

The content of the method of long waves is to identify y and y_o so that p is independent of y: then from (2.2), all particles in a plane perpendicular to x get the same acceleration, and all particles in this plane remain in the same plane and the horizontal velocity u depends on x, t only, $u = u(x,t)$. Equation (2.2) is

$$u_t + uu_x = -\frac{1}{\rho}\frac{\partial p}{\partial x} \tag{4.7}$$

and the nonlinear term is simply dropped as being of the second order. Though this is not a priori a necessary feature of the method of long waves, there is no dispersion term left in the method to balance it. From (4.6) $\partial p/\partial x = g\rho\partial\eta/\partial x$ and so

$$u_t = -g\eta_x \ . \tag{4.8}$$

If $\xi = \int u \, dt$, $\xi_{tt} = -g\eta_x$, and continuity (2.3) means

$$-\frac{\partial}{\partial x}(\xi \ h \ b)\delta x = \eta \ b \ \delta x, \tag{4.9}$$

b being the breadth of the canal--both expressions being the volume of fluid which, up to time t, has entered the space between x, x+δx. From (4.9)

$$-h\xi_x = \eta, \quad \text{so} \quad \xi_{tt} = gh \ \xi_{xx} \quad \text{and}$$

$$\eta_{tt} = c_o^2 \ \eta_{xx} \tag{4.10}$$

with c_o^2 given by (1.3). This is the relation $d^2X/dt^2 = gkd^2X/dx^2$ of Airy [22] quoted by Rayleigh [7].

The argument given here is really that as presented by Lamb [32]. George Green (of Green's Theorem) considered long waves in a rectangular canal with slowly varying breadth β and depth y and found a wave height $\propto \beta^{-\frac{1}{2}} y^{-\frac{1}{4}}$ and the velocity formula (1.3) as $c = \sqrt{gy}$ [33]. For a triangular canal with one side vertical [35] Green found $c = (\frac{1}{2}gy)^{\frac{1}{2}}$, a result Russell [2] also applies to the isosceles triangle and confirms experimentally (see Fig. 6 line AB' of the Plate 48 of [2] in our Fig.4). Kelland [28] found $c = \sqrt{gA/b}$ for a canal of cross-sectional area A and breadth b at the surface. But this result Russell [2] apparently confirmed only for the rectangular and triangular cases disagreeing otherwise (Ref. [2], p. 355). Our Fig. 5, which is Plate 49 of [2], shows the different triangular cross-sections Russell worked with. Nevertheless Stokes [36] says Kelland's formula 'agrees with the experiments of Mr. Russell' simply.

We turn now to the work of Stokes himself. In the British Association Report of 1846 [36] Stokes says "It is the opinion of Mr. Russell that the solitary wave is a phenomenon *sui generis* in nowise deriving its character from the circumstances of the generation of the wave. His experiments seem to render this conclusion probable. Should it be correct the analytical character of the solitary wave remains to be discovered". Note that he expects to agree with Russell, and that already he firmly uses the designation 'solitary wave'. However, in a paper read on 1 March 1847 and published in the Trans. Camb. Phil. Soc. [4] "On the theory of oscillatory waves" he effectively concludes (for axes x horizontal and y vertically downwards) that

$$y = a \cos m(x - ct) - Ka^2 \cos 2m(x - ct) \tag{4.11}$$

(with $a = -m\, Ac\, g^{-1}$ and m arbitrary, but K then fixed and depending on a, m and h, and c given by (4.1))" is the only form of wave which possesses the property of being propagated with a constant velocity and without change of form--so that a solitary wave cannot be propagated in this manner. Thus the degradation observed by Russell is not due to the imperfect fluidity of the fluid and its adhesion to the sides and bottom of the canal but is an essential characteristic of the solitary wave".

Stokes's argumentation is as follows: He starts from (4.4) in the form

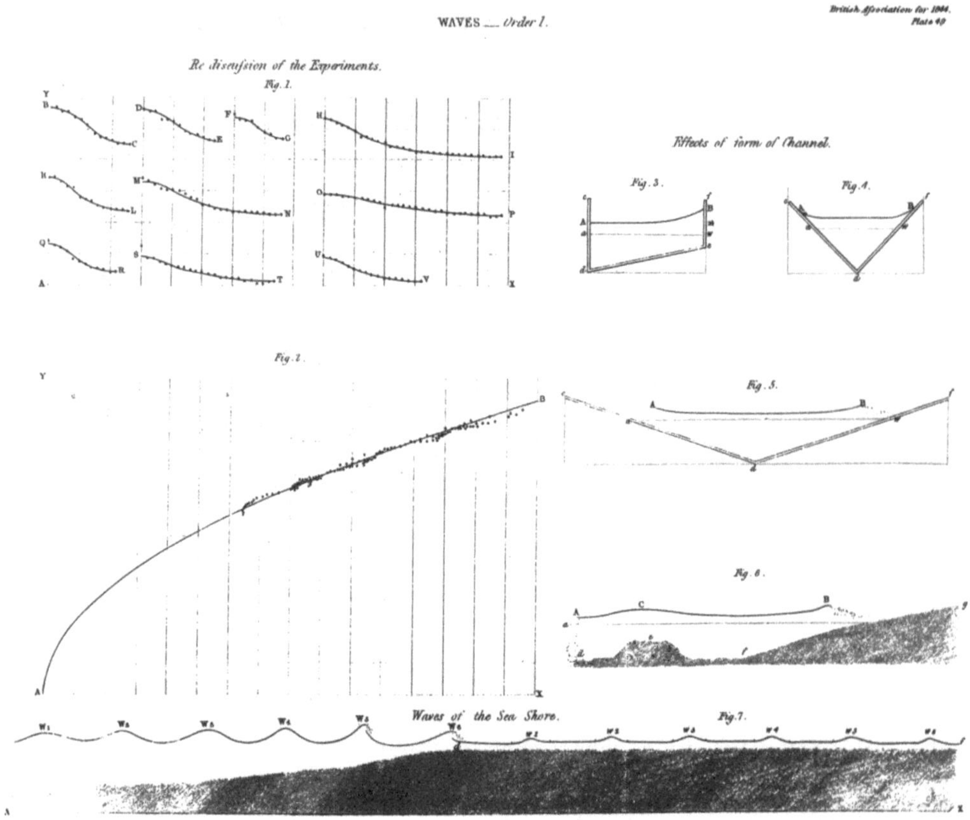

Fig. 5 This Figure is the Plate 49 of [2]. The Figs. 3,4,5 show the triangular profiles used by Russell in his experiments. The curves AB describe the crest of the wave and in Fig. 5 this crest is breaking in the regions aA and Bb. Fig. 1 is the smooth curves drawn through observed wave heights in order to remove observational error, and similar curves were drawn for the velocities. These smoothed results were then plotted in Fig. 2 and compared against (2.1) plotted as the curve AB.

$$p = g\rho y - \rho\frac{d\varphi}{dt} - \frac{1}{2}\rho\left\{\left(\frac{d\varphi}{dx}\right)^2 + \left(\frac{d\varphi}{dy}\right)^2\right\}, \tag{4.12}$$

with continuity

$$\frac{d^2\varphi}{dx^2} + \frac{d^2\varphi}{dy^2} = 0 , \tag{4.13}$$

and with

$$\frac{d\varphi}{dy} = 0 \quad \text{when } y = h , \tag{4.14}$$

and

$$\frac{dp}{dt} + \frac{d\varphi}{dx}\frac{dp}{dx} + \frac{d\varphi}{dy}\frac{dp}{dy} = 0 \text{ when } p = 0 , \tag{4.15}$$

the condition for the free surface remaining the free surface throughout the motion. The velocities $u = d\varphi/dx$ and $v = d\varphi/dy$, and p, depend on x - ct, the condition for permanent profile, together with y. From (4.12) we can admit in φ a term linear in t to fix p so

$$\varphi = f(x - ct) + Ct \; ; \tag{4.16}$$

and (x - ct) is called x, while C is set to -gk. Then (4.12) is

$$p = g\rho(y+k) + c\rho\frac{d\varphi}{dx} - \frac{1}{2}\rho\left[\left(\frac{d\varphi}{dx}\right)^2 + \left(\frac{d\varphi}{dy}\right)^2\right] \tag{4.17}$$

so that for u = v = 0, p = 0 when y = -k. Also, from dependence on x - ct, (4.15) is

$$\left(\frac{d\varphi}{dx} - c\right)\frac{dp}{dx} + \frac{d\varphi}{dy}\frac{dp}{dy} = 0 \text{ when } p = 0 \;, \tag{4.18}$$

and (4.17) into (4.18) gives

$$g\frac{d\varphi}{dy} - c^2\frac{d^2\varphi}{dx^2} + 2c\left(\frac{d\varphi}{dx}\frac{d^2\varphi}{dx^2} + \frac{d\varphi}{dy}\frac{d^2\varphi}{dy^2}\right) - \left(\frac{d\varphi}{dx}\right)^2\frac{d^2\varphi}{dx^2} -$$

$$2\frac{d\varphi}{dx}\frac{d\varphi}{dy}\frac{d^2\varphi}{dxdy} - \left(\frac{d\varphi}{dy}\right)^2\frac{d^2\varphi}{dy^2} = 0 \tag{4.19}$$

when

$$g(y+k) + c\frac{d\varphi}{dx} - \frac{1}{2}\left\{\left(\frac{d\varphi}{dx}\right)^2 + \left(\frac{d\varphi}{dy}\right)^2\right\} = 0 \;. \tag{4.20}$$

At this stage the argument is exact, but Stokes now first of all neglects all the (nonlinear) terms in the second row of (4.19), retaining those in the first row however. He sets

$$\varphi_x \equiv \left(\frac{d\varphi}{dx}\right)_{y=0}, \quad \varphi_y \equiv \left(\frac{d\varphi}{dy}\right)_{y=0} \tag{4.21}$$

(y = 0 is the undisturbed surface) so that (4.20) implies

$$g(y + k) + c(\varphi_x + y\varphi_{xy}) - \frac{1}{2}(\varphi_x^2 + \varphi_y^2) = 0 \tag{4.22a}$$

and by one iteration

$$y = -k - \frac{c}{g}\varphi_x + \frac{c}{g}\varphi_{xy}\left(k + \frac{c}{g}\varphi_x\right) + \frac{1}{2g}(\varphi_x^2 + \varphi_y^2) \tag{4.22b}$$

in which y evidently describes the ordinate of the disturbed surface. He now puts $y = -(k + (c/g)\varphi_x)$ in the first two terms of (4.19) and $y = 0$ in the second two terms so that

$$g\varphi_y - c^2\varphi_{xx} - g(\varphi_{yy} - c^2\varphi_{xxy})\left(k + \frac{c}{g}\varphi_x\right) + 2c(\varphi_x\varphi_{xx} + \varphi_y\varphi_{xy}) = 0. \tag{4.23}$$

Thus φ satisfies (4.13), with (4.14) at $y = h$, and with (4.23) at $y = 0$. When φ is determined (4.22b) gives the ordinate y of the surface and k is fixed by the condition that the mean value of y shall be zero.

The first approximation to (4.23) is

$$g\,\frac{d\varphi}{dy} - c^2\,\frac{d^2\varphi}{dx^2} = 0\ , \quad \text{when } y = 0\ . \tag{4.24}$$

Put $\varphi = \sum\limits_{m^2+n^2=0} A\,e^{(mx+ny)}$ which satisfies (4.13): the boundary conditions are chosen to make m imaginary (so the behaviour along x is matched to a Fourier series and periodic boundary conditions have entered). Then

$$\varphi = \sum\left(e^{m(h-y)} + e^{-m(h-y)}\right)(A \sin mx + B \cos mx) \tag{4.25}$$

since the choice satisfies (4.14). Then one non-trivial contribution to (4.25) satisfies

$$c^2 = \frac{g}{m}\,\frac{e^{mh} - e^{-mh}}{e^{mh} + e^{-mh}}\ , \tag{4.26}$$

which is of course the formula (4.1). Stokes then proves that $d(\log c^2)/d\,(2mh) < 0$ which implies a unique m satisfies (4.26).

By choosing the origin of x

$$\varphi = A\left(e^{m(h-y)} + e^{-m(h-y)}\right)\sin mx\ , \tag{4.27}$$

and this, put into (4.20), yields

$$y = -m\,Ac\,g^{-1}\left(e^{mh} + e^{-mh}\right)\cos mx \tag{4.28}$$

(he sets $k = 0$ since the mean value $y = 0$) and this is the equation for the free surface.

For a second approximation he uses φ from (4.28) in the small terms of (4.23) retained, so that

$$g\varphi_y - c^2\varphi_{xx} - 2cA^2c^2m^2 \sin 2mx = 0 . \tag{4.29}$$

Stokes continues "The general value of φ given by (4.25), which is derived from (4.13) and (4.14), must now be restricted to satisfy (4.29). It is evident that no new terms in φ involving sin mx and cos mx need be introduced, since such terms may be included in the first approximate value, and the only other term which can enter is one of the form

$$B \left(e^{2m(h-y)} + e^{-2m(h-y)}\right) \sin 2mx . \tag{4.30}$$

Substituting this term in (4.29) and simplifying by means of (4.26)

$$B = 3m\ A^2/c^2 \left(e^{mh} - e^{-mh}\right)^2 . \tag{4.31}$$

Moreover, since the term in φ containing sin mx must disappear from (4.29), the equation (4.26) will give c to a second approximation"-- so the velocity formula (4.26) is actually unchanged.

If we now set a for the coefficient of cos mx in the first approximation (4.28),

$$a = -\ mAcg^{-1}\ (e^{mh} + e^{-mh}) \tag{4.32a}$$

and

$$A = -ca/(e^{mh} - e^{-mh}), \tag{4.32b}$$

while

$$\varphi = -ac \frac{e^{m(h-y)} + e^{-m(h-y)}}{e^{m(h-y)} - e^{-m(h-y)}} \sin mx + 3ma^2c \frac{(e^{2m(h-y)} + e^{-2m(h-y)})}{(e^{mh} - e^{-mh})^4} \sin 2mx. \tag{4.33}$$

Thus the ordinate of the surface to the second approximation from (4.22b) is

$$y = a \cos mx - Ka^2 \cos 2mx \tag{4.34}$$

with $x \to x - ct$, c^2 given by (4.26), and

$$K = \frac{1}{2}m \frac{(e^{2mh} + e^{-2mh} + 4)}{(e^{mh} - e^{-mh})^2} \coth mh \tag{4.35}$$

which is the statement (4.11).

To proceed further Stokes simplifies to $h >> \lambda$(depth h >>length λ of the wave). Then (4.33), (4.34) and (4.35) are

$$\varphi = - ace^{-my} \sin mx$$

$$y = a \cos mx - \frac{1}{2} ma^2 \cos 2mx$$

$$K = \frac{1}{2} m = \pi/\lambda \; ; \qquad (4.36)$$

and $c^2 = g\lambda/2\pi$; the parameter $k = 0$. He then finds at next order that (with $h >> \lambda$),

$$\varphi = -ac \left(1 - \frac{5}{8} m^2 a^2\right) e^{-my} \sin mx$$

$$y = a \cos mx - \frac{1}{2} ma^2 \cos 2mx + \frac{3}{8} m^2 a^3 \cos 3mx$$

$$c = \left(\frac{g}{m}\right)^{1/2} \left(1 + \frac{1}{2} m^2 a^2\right) = \left(\frac{g\lambda}{2\pi}\right)^{1/2} \left(1 + \frac{2\pi^2 a^2}{\lambda^2}\right) \qquad (4.37)$$

and only at this order does the wave amplitude a enter the expression for c. Stokes draws the form of y for $a = 7\lambda/80$ (for some reason) but notes the term in a retainer "is almost insensible". His most significant result is that the figure is not symmetrical above and below $y = 0$.

Stokes also notes that "It is remarkable that this expression(for y) coincides with that of the prolate cycloid, if the latter equation be expanded according to ascending powers of the distance of the tracing point from the centre of the rolling circle, and the terms of the fourth order be omitted. The prolate cycloid is the form assigned by Mr. Russell to waves of the kind considered here (Reprints of the British Association Vol. VI p. 448 (Ref. [34], p. 448)". For h/λ not great the form of the surface is not a prolate cycloid even to second approximation.

Before discussing further Stokes's results (4.11) and their extension (4.36) and (4.37), it is interesting to take up his comment on the form of Russell's wave. The reference by Stokes to [34] is correct for 1837 but not for 1847. For in [2] (of 1844) Russell already described the shapes of his waves as curves of versines not cycloids(see below). The paper [34] (of 1837) is the report of a "Committee on Waves" set up by the British Association in 1836. The committee had two members--Russell and Sir John Robinson, Secretary to the Royal Society of Edinburgh. It was charged to determine: first 'What is a Wave'; second what is the nature of the 'Waves of the Sea'; third does the behaviour of the 'Tidal Elevation' obey the same laws as other waves; and is its propagation affected by 'Local Winds'. The report is plainly Russell's report on these questions.

In it he was most concerned to establish what he had already believed to be true--namely that the tidal wave was an example of his solitary wave. In September 1836 he carried out experiments on the River Dee, near Chester, where there is a 5 mile channel of the river straight and uniform in depth and width. Then he carried out experiments on the Firth of Clyde, where conditions were less uniform. From these experiments, and a further series in the laboratory, the Committee concluded: 1. the certain existence of a 'Great Primary Wave' of fluid in its laws different from that of the oscillatory waves treated hitherto; 2. that its velocity c satisfied (2.1); 3. that c is independent of the mode of generation; 4. the motion of the particles of water involves their actual net translation (refer Fig. 5 in our Fig.1); 5. the form of the wave is cycloidal, being first a prolate cycloid, becoming a cusped 'common' cycloid as the height k increases to h, and breaking beyond that height; 6. in channels of arbitrary cross-section c is 'that due to gravity acting through a height equal to the depth of the centre of gravity of the transverse cross-section (so (2.1) applies to a rectangular channel and $c^2 = \frac{2}{3}g(h+k)$ in a triangular one (refer [35])); 7. that the increased height k in a wedge shaped narrowing of the channel may follow $k \propto b^{-1/2}$ (b the breadth (refer [33])); 8. in a uniformly shelving channel the wave breaks when $k = h$; 9. the solitary wave can be normally reflected without any change but reversal of direction; 10. the solitary waves cross each other without change of any kind (solitons again); 11. Waves of the Sea are not of this type but are second order oscillatory; 12. these waves nevertheless become solitary waves as they approach the shore and break for $k \leq h$; 13. waves at the surface of the sea do not move with the velocity due to the total depth h+k; 14. that at sea every 3rd or 7th or 9th wave can be the largest (this is in fact a reference to waves in deep water governed by the nonlinear Schrödinger equation) and these, in particular, will break on the shore; 15. the Tide Wave is the only wave of the ocean which is a wave of first order (is a solitary one); 16. the tide itself is a compound wave; 17. a tidal bore is created when the water is so shallow at low water that 'the first waves of flood tide move with velocity so much less than that due to the succeeding part of the tidal wave, as to be overtaken by the subsequent waves, or wherever the tide rises so rapidly, and the water on the shore or the river so shallow, that the height of the first wave of the tide is greater

than the depth of the fluid at that place. Hence in deep water vessels are safe from the waves of rivers which injure those on the shore';
18. that because the Tidal Wave is solitary (2.1) means that deepening a channel means an advance in the time of observation of high water;
19. that likewise spring tides (large k) travel faster than neap tides (small k) with a consequent effect on tide tables.

This report was followed [37] by a further brief note on the form of the tidal wave in the Firth of Forth. Again the note identifies the Tidal Wave with Russell's solitary wave, while 'It appears that, like the great wave of Translation, tidal waves could not only meet each other without losing their individuality, but they could pass over each other when going in the same direction'.

Thus Russell confirms yet again that he had observed the soliton property. However, he never found the analytical sech^2 formula for the soliton's shape. His shapes are cycloids in [34]; his Wave was a trochoid in [5], p. 85; and it becomes a 'curve of versines' in [2] whilst in that paper his cycloid refers to his gregarious oscillatory waves of order two. Russell shows some of his laboratory observations of wave shapes in the Plates 50 and 51 of [2]. The latter is reproduced here as the Fig. 6 and shows two waves breaking. The Plate 52 of [2], reproduced here as Fig. 7, shows that Russell's 'curve of versines' is a $\frac{1}{2}k(1 + \cos\Theta)$ curve, i.e. it is harmonic but rendered positive. At first sight Russell's insight has really failed him here. But note that for large k (Figs. 4 and 5 in the Plate) he is already correcting this harmonic shape by an effect due to the translation of the particles. The translation in the Fig. 4, AA', is the volume of water used to create the wave divided by the breadth of the channel, and Russell has already demonstrated this is the net translation of planes of water particles (cf. our Fig. 1). The Fig. 6 in our Fig. 7 shows a single particle path in the case of Fig. 5, where $k \approx h$, and three successive positions of the wave in regard to it.

Evidently Russell would have been in a much stronger position in 1844 if he could have given an analytical form to his wave by analytical methods. I now return to Stokes's [4] result of 1847, namely (4.11) and its extension, and indicate what was wrong with it.

The passage to $h \gg \lambda$ (i.e. $m\,h \to \infty$) involved in (4.36) and (4.37) prejudices any discussion with respect to Russell. But Stokes's results at second order are at first quite general and in particular, for the

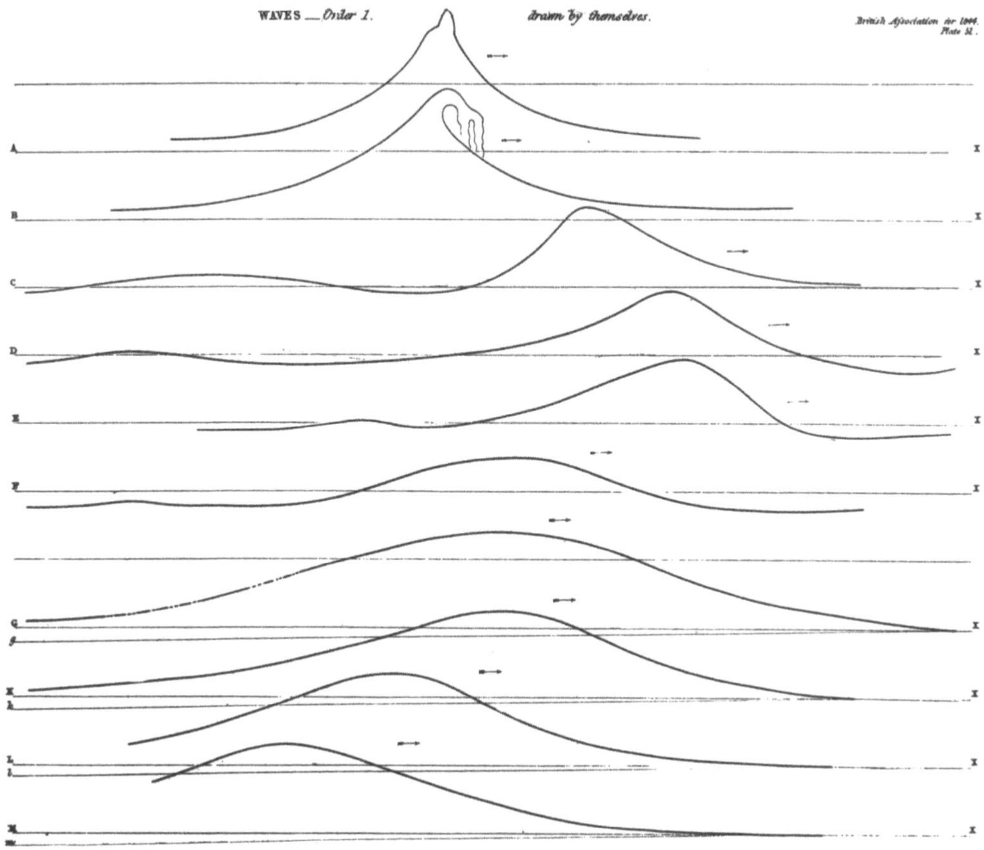

Fig. 6 This Figure is Plate 51 of [2]. The waves are 'drawn by themselves' by using pairs of equal waves travelling in opposite directions and registering their wetting profile in collision. The second four curves are all drawn this way; the first two, which are breaking, are apparently found similarly. The last four curves are also apparently drawn in the same way but the troughs have the sloping bottoms gX, kX, ℓX and mX with gradients one in twelve. Otherwise h = 2 inches for all the curves. The dimensions were reduced by Russell by a factor 2/3 and the parallel lines are really one inch apart (the preceding Plate 50 is similar but is drawn full size).

'long wave' condition mh small, Ka^2 in (4.35) reduces to $3a^2/4m^2h^3$. Since Stokes's Fourier series must converge he then requires $3a^2/4m^2h^3 < a$ and $(a/h) < m^2/h^2$. According to Stokes this resolves an apparent discrepancy with Airy [22] (Art. 198, etc.) who working under the long wave condition $mh \ll 1$ finds the form y of his wave alters at second order as the wave proceeds while c depends on wave height at that order (Airy may even reach his expression (4.2) for c at this point). Evidently Stokes has $(a/h) < m^2h^2$ and Airy has mh small with $a/h > (mh)^2$. Thus the two assume "different physical circumstances".

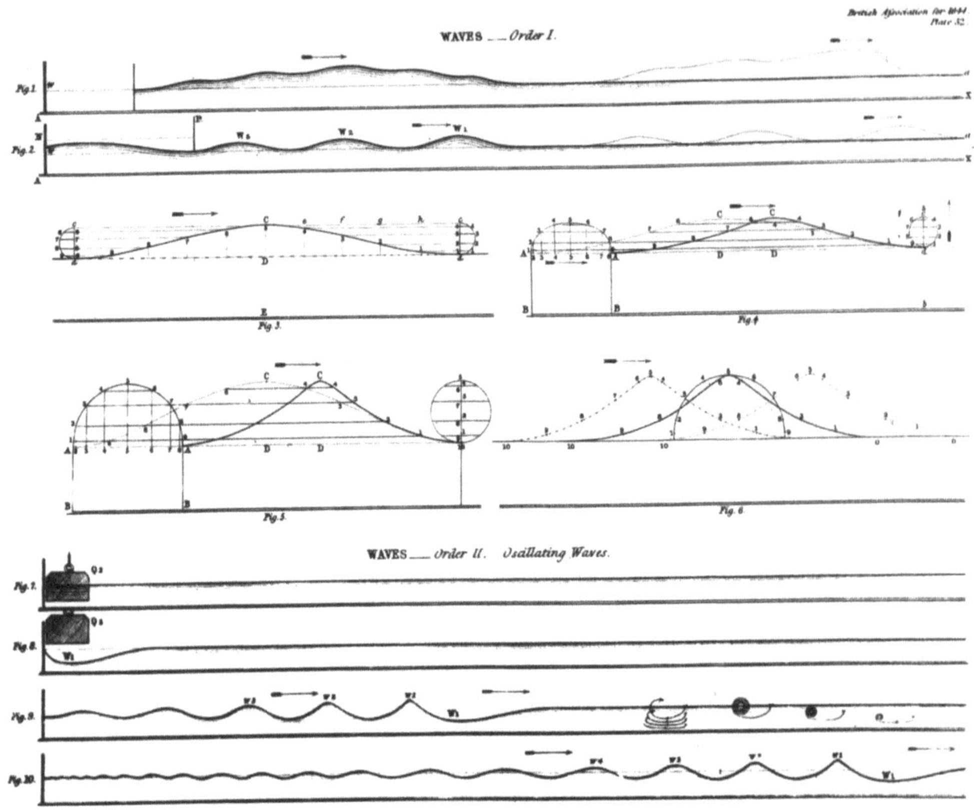

Fig. 7 This Figure is the Plate 52 taken from [2]. Figs. 3,4,5 show Russell's curve of versines description of the shapes of his Waves and Figs. 4,5 show the corrections he made for the translation of the particles. Fig. 6 shows the single particle path. Figs. 1 and 2 are each superb examples of soliton break up: Fig. 2 is generated from a long low column of water, Fig. 1 by a plate with variable force and longitudinal velocity. The lower Figs. show the creation of a solitary negative wave and its break-up into gregarious oscillatory waves of Order II.

This is almost but not quite the nub of the matter. To second approximation Stokes has no error and everything depends on the approach to the long wave limit $\lambda = 2\pi m^{-1} \to \infty$. In this respect both Stokes and Airy, who assume $\lambda = \infty$ *a priori* but then work incorrectly, are wrong.

Certainly Stokes avoids the assumption of p independent of y, characteristic of the method of long waves, and nonlinear terms are retained at sufficient accuracy in (4.19) at his second approximation. For mh small the two terms of (4.11) then prove to be the first two in the Fourier expansion of the square cn^2 of the Jacobian elliptic function cn. And in later work (see Stokes in [39]) Stokes finds the third term also. Since cn^2

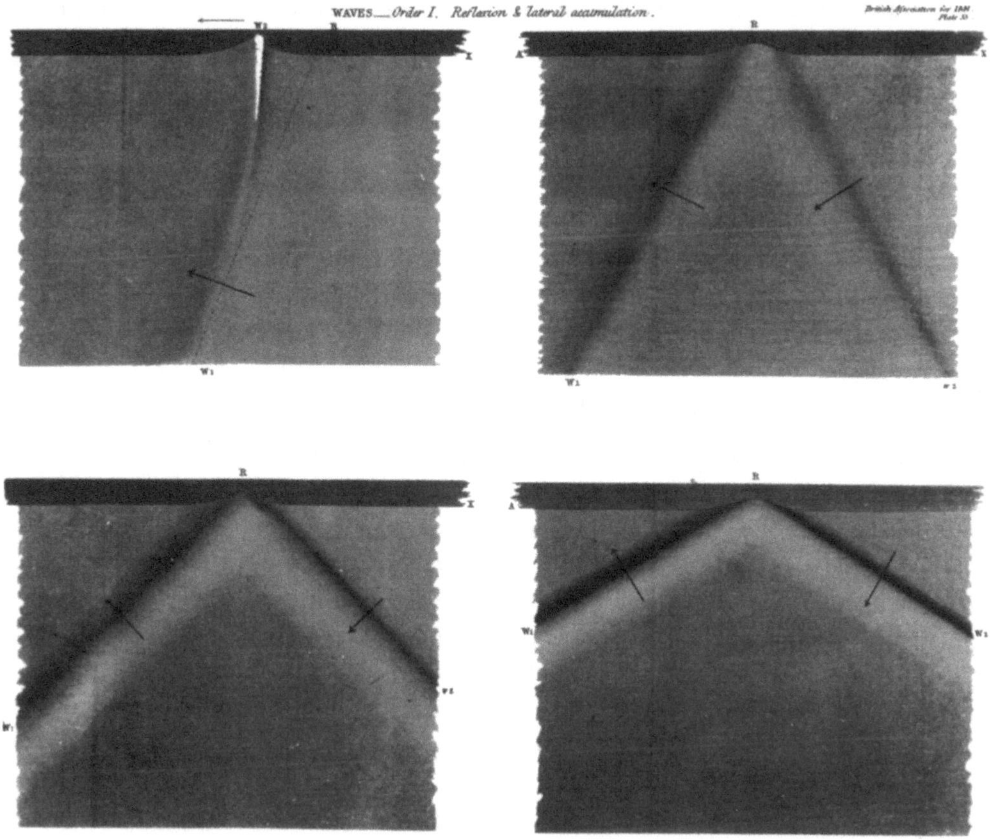

Fig. 8 This Figure is Plate 53 of [2] and is not referred to in the text. It shows reflexion of a solitary wave from the surface R for different angles of incidence measured from R. The reflected wave decreases in amplitude with decreasing angle of incidence until at about 15° incidence it prefers to travel as a single wave by increasing its height in the plane of R.

$\rightarrow$ sech^2 as the period λ finally goes to infinity Stokes's real error has been to impose periodic boundary conditions in all of his analyses. It was Korteweg and de Vries [8] who pointed out that Stokes [4] had found the first two terms of the cn^2 function and it was in this way they finally settled the matter. No such escape seems available to Airy however. The critically different physics described by periodic boundary conditions is still not always appreciated [38].

Later Stokes acknowledged his error [39] but Airy never did. His opinion carried weight well beyond Rayleigh's paper [7] and Korteweg and de Vries [8] seem to have written their paper in 1895 essentially for that reason. They quote Boussinesq [19,20], Rayleigh [7] and St.

Venant [40] as establishing the theory of the solitary wave but noted that (in 1895) treatises by Lamb ([32], 2nd edition 1895) and Basset still assert that Airy was correct in his opinion. The same is true of the Encyclopaedia Brittanica article on Russell of 1886. Moreoever, Korteweg and de Vries say that even Rayleigh [7] and McCowan [9] do not directly refute Lamb's and Basset's assertions and "It is the desire to settle this question definitively which has led us into the somewhat tedious calculations which are to be found at the end of this paper". The KdV equation itself, though noted as "very important" in [8] gets rather less discussion.

5. RAYLEIGH'S PAPER AND HIS RESOLUTION OF THE PROBLEM

These various remarks in [8] seem quite extraordinary when we actually look at what Rayleigh did in his paper [7] in 1876. It is somewhat of a relief to turn to this paper after those of Stokes which are also "somewhat tedious". The paper [7] is 'On Waves', and in it Rayleigh exploits his physical insight as usual rather than any extraordinary manipulative mathematical skills. In these terms his paper [7] should have settled the matter in England at least, and Russell, a Scot by birth should have benefitted for he had long since moved to London [26]. Boussinesq [20] working in Paris could be ignored but Rayleigh scarcely. Yet he seems to have been if Korteweg and de Vries [8], working in Holland however, were right. McCowan's papers [9,10] suggest very much that they were.

Rayleigh uses h for the height of the proposed solitary wave above the undisturbed water, and he uses ℓ for the depth of undisturbed water. At the risk of confusing, I shall at first keep Rayleigh's usage and change it to Scott Russell's only when I turn, with Rayleigh, to his actual analysis for the solitary wave.

Rayleigh first notes (in effect by appeal to (4.5)) that a steady water velocity u_o along x in the undisturbed region will lead to an increase in pressure inside the solitary wave where, by continuity, if $h > 0$, u_o drops to u. On the other hand the pressure will decrease from the $\rho_o^{-1}\Omega = gh$, and the possibility of a balance exists. He first of all assumes the postulate of the theory of long waves, i.e., $\lambda \gg h$, so that the velocity v (in the y-direction) is negligible and u (the velocity along x) is independent of y: u is of course uniform along z, i.e., across the canal. Continuity then means

$$u = \frac{\ell u_o}{\ell + h} \qquad (5.1)$$

and

$$u_o^2 - u^2 = u_o^2 \frac{2\ell h+h^2}{(\ell+h)^2} \quad . \tag{5.2}$$

'The principle of hydrodynamics' (namely (4.5)) now means that the increase in pressure due to the fall from u_o to u is

$$\frac{1}{2}\rho(u_o^2 - u^2) = \frac{1}{2}\rho u_o^2 \frac{2\ell h+h^2}{(\ell+h)^2} \tag{5.3}$$

while the **net** gain is therefore

$$\left\{\frac{\rho u_o^2}{\ell} \frac{1 + h/2\ell}{(1+h/\ell)^2} - g\rho\right\}h \approx \left[\frac{\rho u_o^2}{\ell} - g\rho\right]h \ . \tag{5.4}$$

Implicitly he now moves to a frame where the velocity u_o is reduced to still water so the region where the velocity is u moves up the x-direction (actually he treats a stationary wave and points out this also gives the velocity of the wave in still water). The condition for a free surface (no net gain in pressure) is to the order just realised then $u_o^2 = g\ell$, which is Lagrange's (1.3). To second approximation however it follows that

$$\delta p = g\rho h \left\{\frac{(1+h/2\ell)}{(1+h/\ell)^2} - 1\right\} \approx -\frac{3}{2}\frac{g\rho h^2}{\ell} \tag{5.5}$$

so that p is defective wherever $h \neq 0$, and it is impossible for a long wave of finite height to be propagated in still water 'without change of type'. Evidently (from (5.4)) if $h > 0$ one obtains a better result if u_o is increased, thus making the left hand expression less negative. But if $h < 0$ one obtains a better result if u_o is decreased, making the curly bracket there less positive. The wholly positive wave Russell conceived therefore has a velocity $c > c_o$, while a negative wave has $c < c_o$.

Rayleigh now notes that (5.1) is

$$u' \equiv u - u_o = (g/\ell)^{1/2}h \tag{5.6}$$

(since he has just shown h is infinitesimal). For vanishingly small h, as required, (5.6) is the condition that the disturbance moves up the positive x direction. If the condition is violated the wave emerges in the negative x direction. Rayleigh adopts this condition as a local condition for the profile at any finite distance h above the undisturbed water so that, with velocity and height changing gradually, the local condition for no negative wave is

$$du' = \sqrt{g/(\ell+h)}\ \ dh \tag{5.7}$$

and the resultant net condition at final height h is that the total discrepancy $u' = u - u_o$ will be

$$u' = 2\sqrt{g}\ \{\sqrt{\ell+h} - \sqrt{\ell}\} \ . \tag{5.8}$$

Now the formula (1.3) means that at total depth $(h+\ell)$, $u_o = \sqrt{g(h+\ell)}$ so that u, the velocity of the peak relative to still water is

$$u = 2\sqrt{g}\{\ \sqrt{\ell+h}\ \ -\sqrt{\ell} + \sqrt{g(\ell+h)}\}$$

$$\approx \sqrt{g}\ \sqrt{\ell+3h} \ . \tag{5.9}$$

This is actually Airy's formula (4.2). However, the condition is an always positive moving profile and this profile necessarily distorts (as Rayleigh has proved already).

Rayleigh next uses the same line of reasoning to regain Kelland's formula $c = \sqrt{gA/b}$ as well as those of Green. Let A be the area of the cross section below the undisturbed level, b the breadth at that level. Then continuity means

$$(A + bh)u = Au_o \tag{5.10}$$

for small enough h; so instead of (5.2)

$$u_o^2 - u^2 = (2bh/A)u_o^2 \ . \tag{5.11}$$

But, since 'by dynamics' (namely (4.5) $u_o^2 - u^2 = 2gh$ if the upper surface is free,

$$u_o^2 = gA/b \ , \tag{5.12}$$

which is Kelland's formula.

Rayleigh then goes on to argue as follows: "The energy of a long wave is half potential and half kinetic. If we suppose that initially the surface is displaced, but that the particles have no velocity, we shall evidently obtain (as in the case of sound) two equal waves travelling in opposite directions". Evidently the total energies of each are equal and make up the original energy of displacement, these derived waves have each one half the elevation of the original and potential energies one quarter. The energy of each wave is one half so potential and kinetic energies are equal in the two derived waves. Now apply this result to the case of gradually changing b and A. The potential energy $\propto$ {length $\times$ breadth $\times$ $(\text{height})^2$} of the wave which is then true of the total energy. Since wave-length $\propto$ propagation velocity ($\propto\sqrt{A/b}$ from (5.12)), energy

$$E \propto \sqrt{(A/b)} \times (\text{height})^2 \times b \tag{5.13}$$

and, since this is constant for a slowly varying canal,

$$\text{height} \propto A^{-1/4} b^{-1/4} \ . \tag{5.14}$$

This is $\propto \ell^{-\frac{1}{4}} b^{-\frac{1}{2}}$ for a rectangular canal, namely Green's result [33] for a canal of slowly varying depth and breadth. Both (5.12) and (5.14) are also given by Airy [22] (Art. 260) and Stokes [36] points out that his proof of (5.12) is actually simpler than Kelland's in [28]. However, Rayleigh simply refers them jointly to Green, Kelland and Airy by reference to Stokes [36].

Next Rayleigh proves that h cannot exceed the value due to the velocity u_o by generalising (5.3), a result he notes as 'otherwise obvious' (which it is). He then shows that if u_o is less than the velocity of the free wave ($gA_o > bu_o^2$) a contraction of area A_o to A produces a depression of the surface while an expansion from A_o to A produces an elevation the effects being reversed for $gA_o < bu_o^2$. He notes that a stationary wave can sustain itself "in a stationary position without requiring a variation in the channel; and the effects of such a variation are naturally much intensified" and goes on to show that the situation when $u_o^2 > gA_o b^{-1}$ is unstable.

At this point Rayleigh starts a wholly new section entitled "The Solitary Wave". He says "This is the name given by Mr. Scott Russell to a peculiar wave described by him" (in [2]). Since Russell's wave is 6 or 8 times the canal depth in length it should be treatable by the theory of longwaves. However "there are several circumstances observed by Mr. Russell which indicate it has a character distinct from that of long waves"--notably the different behaviours of positive and negative waves, the former having a "remarkable permanence", the latter being "soon broken up and dissipated".

Rayleigh remarks that Airy "appears not to recognize anything distinctive in the solitary wave" and quotes from Airy [22] "We are not disposed . .." in the form of the two quotes given above. On the other hand he then also quotes the earlier view expressed by Stokes in [36] "It is the opinion of Mr. Russell . . . (also given above). After dismissing a paper by Earnshaw [41], which appeals to experiment in order to validate the postulates of the theory of long waves and thereby runs into a problem of matching rotational motion in the wave to irrotational motion beyond it, Rayleigh goes on to consider whether there can be compensation between pressure variation at the surface from the finiteness of height and variation due to the departure from the law of uniform velocity proper to very long waves. This is the crucial consideration.

To analyse this he introduces stream lines Ψ = const. and a velocity potential Φ setting

$$\Phi + i\Psi = F(x + iy) = e^{iy\frac{d}{dx}} F(x) , \tag{5.15}$$

this choice being motivated by the fact that one stream line, at the bottom of the canal, is straight. Thus

$$\Phi = F - \frac{y^2}{2!} F'' + \frac{y^4}{4!} F^{(iv)}$$

$$\Psi = yF' - \frac{y^3}{3!} F''' + \frac{y^5}{5!} F^{(v)} ; \tag{5.16}$$

$\Psi = 0$ on the bottom of the canal, and at this point by changing back to Russell's notation and choosing h for the depth of undisturbed water, $\Psi = -ch$ at the free surface. On this surface p is uniform so that (from (4.5))

$$u^2 + v^2 = c^2 - 2g(y-h) . \tag{5.17}$$

Then

$$u = F' - \frac{y^2}{2!} F''' + \frac{y^4}{4!} F^{(v)} + \ldots \tag{5.18a}$$

and

$$v = -yF'' + \ldots , \tag{5.18b}$$

so that

$$(F')^2 - y^2F'F''' + y^2(F'')^2 + \ldots = c^2 - 2g(y - h) \tag{5.19}$$

on the free surface. But the free surface is also $\Psi = -ch$ so that

$$yF' - \frac{y^3}{3!} F''' + \ldots = -ch . \tag{5.20}$$

The procedure now is to eliminate F from (5.19) and (5.20) to get a differential equation for y, the ordinate of the free surface. If F varies slowly one can solve these two equations for y by iteration so that

$$F' = -\frac{ch}{y} + \frac{1}{6} y^2 F''' + \ldots$$

$$= -ch \{\frac{1}{y} + \frac{1}{6} y^2 (\frac{1}{y})'' + \ldots\} \tag{5.21}$$

and (5.19) becomes

$$\frac{1}{y^2} - \frac{2}{3} y (\frac{1}{y})'' + y^2 (\frac{1}{y})' = \frac{1}{h^2} - 2g \frac{(y - h)}{c^2 h^2} \tag{5.22}$$

or

$$(y')^2 = \frac{3(y - h)^2}{h^2} \left(1 - \frac{gy}{c^2}\right) \tag{5.23}$$

from which, when $y' = 0$, $y = h$ or $y = c^2/g$. Then if $1 - gy/c^2 \geq 0$, $y = c^2/g$ is maximum, $y = k + h$ (say). Thus

$$c^2 = g(k + h) . \tag{5.24}$$

Also

$$y - h = k \operatorname{sech}^2 \tfrac{1}{2}(x/b) \tag{5.25}$$

and

$$b^2 = h^2(h + k)/3k . \tag{5.26}$$

The formula (5.24) for c^2 is of course precisely Russell's (2.1). Also (5.25) is

$$u(x, t) = y - h = k \operatorname{sech}^2 \left[\frac{1}{2} \left(\frac{3k}{h^3}\right)^{\frac{1}{2}} \left(\frac{h}{h+k}\right)^{\frac{1}{2}} x \right] \tag{5.27}$$

(at $t = 0$) and the maximum height is Russell's k. However the formula differs from Boussinesq's result (2.5) by the scaling $(h/h+k)^{\frac{1}{2}}$ on x. This scaling has the same order of discrepancy as does the KdV solitary wave (2.11), for $[1/(1 + k/h)]^{\frac{1}{2}} \approx 1 - \frac{1}{2}(k/h)$. But now the exact result (5.24) is found for c^2 (to within the order worked). Thus, in this very direct way, Rayleigh further justifies Boussinesq's analysis and wholly justifies (2.1) and therefore, for Russell, his whole position.

It would be tempting now to trace the further developments to modern methods which, for example, identify symplectic manifolds and infinite dimensional Lie algebras within the theory, quantise it and find Yang-Baxter integrability conditions. I leave this fascinating series of developments to other contributions to this meeting.

I therefore finish with Rayleigh's own summing up of his work on the solitary wave. He says [7] "The velocity of propagation is given by (5.24), which is Scott Russell's formula exactly. In words, the velocity of the wave is that due to half the greatest depth of water.

Another of Russell's observations is now readily accounted for:- 'It was always found that the wave broke when its elevation above the general level became equal, or nearly so, to the greatest depth. The application of mathematics to this circumstance is so difficult, that we confine ourselves to the mention of the observed fact [22,42]'. When the wave is treated as stationary it is evident from dynamics that its height can never exceed that due to the velocity of the stream in the undisturbed parts; that is, k is less than $u_o^2/2g$. But $u_o^2 = g(h + k)$, and therefore k is less than $\frac{1}{2}(h + k)$ or k is less than h. When the wave is on the point of breaking, the water at the crest is moving with the velocity of the wave".

Rayleigh then goes on to consider Periodic Waves in Deep Water.

REFERENCES

[1] A. C. Scott, F. Y. F. Chu, D. W. McLaughlin, Proc. IEEE **61** (1973) 1443.

[2] J. S. Russell, "Report on Waves", British Association Reports (1844).

[3] The canal is believed to be the Union Canal which passes below the campus of Heriot-Watt University, Edinburgh, Scotland. The Union Canal Company certainly paid for the boats Russell constructed in 1835 for the experiments described in [5] below. However, Encyclopaedia Brittanica 1886 says "Having been consulted as to the possibility of applying steam-navigation to the Edinburgh and Glasgow Canal, he replied that the question could not be answered without experiment, and that he was willing to undertake such if a portion of the canal were placed at his disposal". Still the 100th anniversary of Russell's death in 1882 was held at Heriot-Watt University. Despite perfect arrangements in all other respects, the organisers were unable to generate a 1½ foot solitary wave on this occasion despite the several powerful boats stopped for this purpose! In the light of such difficulties, Russell's experiments on this canal in 1834-35 described in [5] below are astonishing.

[4] Sir George Stokes, Trans. Camb. Phil. Soc. **8** (1847) 441.

[5] J. S. Russell, "Experimental Researches into the Laws of Certain Hydrodynamical Phenomena that accompany the Motion of Floating Bodies and have not previously been reduced into conformity with the Laws of Resistance of Fluids" Trans. Roy. Soc. Edinburgh **14** (1840) 47-109.

[6] J. S. Russell, **The Wave of Translation in the Oceans of Water, Air and Ether** (Trubner, London, 1895).

[7] Lord Rayleigh (J. W. Strutt), "On Waves", Philos. Mag. **1**(1876)257.

[8] D. J. Korteweg, G. de Vries, "On the Change of Form of Long Waves Advancing in a Rectangular Canal, and on a New Type of Long Stationary Waves", The London, Edinburgh, and Dublin Philosophical Magazine and Journal of Science Series 5, 39, No. 241, June 1895, pp. 422-443 (Philos. Mag. **39** (1895) 422).
[9] J. McCowan, "On the Solitary Wave", Philos. Mag. **31** (1891) 45.
[10] J. McCowan, "On the Highest Wave of Permanent Type", Philos. Mag. **38** (1894) 351.
[11] R. K. Bullough and P. J. Caudrey, eds., **Solitons**, Topics in Current Physics 17 (Springer, Heidelberg, 1980), Chap. 1.
[12] N. Zabusky, M. D. Kruskal, Phys. Rev. Lett. **15** (1965) 240.
[13] C. S. Gardner, J. M. Greene, M. D. Kruskal, R. M. Miura, Phys. Rev. Lett. **19** (1967) 1095.
[14] L. Euler, Memoirs of the Academy of Sciences of Berlin (1755).
[15] J. L. Lagrange, Méchanique Analytique pp. XII, 512 (Chez La Veuve Desaint, Paris, 1788); Nouvelle Edition Augmente Par L'auteur 2 Vols. Paris 1811-15.
[16] P. S. Laplace (Marquis de Laplace): Traite de Méchanique Celeste (Chez, J. B. M. Duprat, Paris an VII, 1798-1823).
[17] S. D. Poisson, "Memoire sur la théorie des ondes" Mem. de l'Acad. Royale des Sciences (i) (1816).
[18] E. H. Weber, W. Weber, **Wellenlehre auf Experimente gegründer,oder über tropfbarer Flüssigkeiten mit Adwendung auf die Schall und Licht-Wellen** (Leipzig, 1825).
[19] Reference [20] was presented on 13 Nov. 1871 and published in 1872. Rayleigh [7] refers to the note by Boussinesq (J. Boussinesq, "Theorie de l'intumesence liquid appelée onde solitaire ou de translation, se propagent dans un canal rectangulaire" Comptes Rendus LXX II, 755(1871)) and acknowledges Boussinesq's priority.
[20] J. Boussinesq, "Theorie des ondes et des remous qui se propagent le long d'un canal rectangulaire horizontal, en communiquant au liquide contenu dans ce canal des vitesses sensiblement pareilles de la surface au fond" J. Math. Pures et Appliquées **2**(1872)55.
[21] P. J. Caudrey, "The Inverse Problem for a General $N \times N$ Spectral Equation", Physica **6D** (1982) 51-66.
[22] Sir G. B. Airy (Astronomer Royal), "Tides and Waves", Encyclopaedia Metropolitana (1845).
[23] J. S. Russell, Proc. Roy. Soc. **32** (1881) 382.
[24] J. S. Russell, **The Modern System of Naval Architecture** (Day & Son, London, 1865).
[25] J. S. Russell, **Systematic Technical Education for the English People** (Day & Son, 1869).
[26] G. S. Emerson, **John Scott Russell: A Great Victorian Engineer and Naval Architect** (John Murray, London, 1977).
[27] Sir John Herschel, quoted by Russell [24], p. 208.
[28] The Reverend P. Kelland, "On the Theory of Waves Part I" Trans. Roy. Soc. Edinburgh **14** (1839) 497-545. The result $c^2 = gAb^{-1}$ quoted below (4.10) appears on the p. 530.
[29] Reference to Kelland [28] shows that Kelland's formula in his notation is, first of all (p. 514)

$$\frac{2\pi}{\lambda} c^2 = g \frac{e^{\alpha h} - e^{-\alpha h}}{e^{\alpha h} + e^{-\alpha h}} \frac{1}{1 - \left(\frac{2\pi a}{\lambda}\right)^2 [e^{\alpha h} - e^{-\alpha h}]^2} \tag{A.1}$$

with a the amplitude of the wave given by $a = (\lambda/\pi)(e^{\alpha h}+e^{-\alpha h})^{-1}$. However in a Sec. III devoted to "Solitary Wave Motion" Kelland defines this motion to have a velocity along x positive at all points along the length of the wave and reaches the formula (4.3) with the semi-elevation $\varepsilon \equiv (b/c\,\alpha)(e^{\alpha h} - e^{-\alpha h})$ where u (the velocity along x) defines b through $u = b(e^{\alpha y} - e^{-\alpha y})(1 + \sin(x - ct))$.

(Evidently Russell mixes α and a in (4.3)). Kelland [28] acknowledges that his formula seems to mean that larger wave height k means longer wavelength λ contrary to Russell. Russell [2] p. 334 plainly appreciates the candour of Kelland's "my solution can only be regarded as an approximation, nor does it very accurately agree with observation", and in contrast with his response to Airy on the same page.

[30] D. Bernouilli, **Hydrodynamica** (Argentorati, 1738).

[31] E. Torricelli, **De motu gravium naturaliter accelerato** (Firenze, 1643).

[32] Sir Horace Lamb, **Hydrodynamics**, 6th ed. (Cambridge University Press, 1952).

[33] George Green, Trans. Camb. Phil. Soc. **6** (1837) 457. Russell [34] supported the breadth dependence $\beta^{-\frac{1}{2}}$ of the wave height (see [35] as well as 7 in our remarks on [34] in the text). He observed that the height of the wave increased as the depth of the fluid decreased, but that variation in wave height was very slow compared with variation in depth across the canal.

[34] Sir John Robinson, K. H., John Scott Russell, "Report on the Committee on Waves" British Association Reports **6** (1837) 417-496.

[35] George Green, Trans. Camb. Phil. Soc. **7** (1839) 87. Russell [34] gave the formula $c = \left(\frac{2}{3}gy\right)^{\frac{1}{2}}$, but Green shows that Russell's experimental data better fit $c = \left(\frac{1}{2}gy\right)^{\frac{1}{2}}$.

[36] Sir George Stokes, British Association Reports (1846).

[37] Sir John Robinson, J. S. Russell, "Report, 1840" British Association Reports, pp. 441-443 (1840).

[38] R. K. Bullough, D. J. Pilling and J. Timonen, Soliton Statistical Mechanics, p. 276 in this volume.

[39] I cannot now find actual reference to this. But Rayleigh [7] in a footnote added in 1899 (to his collected papers) quotes Stokes's supplement to [4] in Stokes **Mathematical and Physical Papers** Vol. I (Cambridge University Press, Cambridge, 1880) indicating that like the work of KdV [8] Stokes's later work confirms the existence of absolutely permanent waves of finite height. In this supplement to [4] Stokes develops the analysis at finite depth to 3rd order. He finds

$$\frac{mc^2}{y} = \frac{D_1}{S_1} + \frac{1}{S_1 D_1}(S_4 + 2S_2 + 12)b^2 \qquad (A.2)$$

where $S_i = e^{imk/c} + e^{-imkc}$ $(i = 1,2,\ldots)$, $D_i = e^{imk/c} - e^{-imkc}$, $k = ch$ and $bD_1 = a$. In this paper he introduces the stream function Ψ as well as the velocity potential Φ = -k (a different k) at the canal bottom.

[40] B. De St. Venant, Comptes Rendus, Vol. ci (1885) reference taken from [8].

[41] S. Earnshaw, The Mathematical Theory of the Two Great Solitary Waves of the First Order, Trans. Camb. Phil. Soc. **VIII** (1849) 326-341.

[42] Beyond the analysis which Rayleigh gives in [7] and despite the paper [10] by McCowan this quotation from Airy [22] is still surely right: the problem is of current interest, but the broken wave is dramatically turbulent, and deterministic chaos theory (for example) does not handle such a case yet. (For a picture see e.g., front cover to 'Physics Bulletin' **38**, No. 5, May 1987, and the article by I. Grant "Flow Measurement by Pulsed Laser (Speckle)" ibid. pp. 175-177 which describes the methods used to observe it.)

Part II

Mathematical Theory: IST, Symmetries, Singularity Structure and Integrability

Topics Associated with Nonlinear Evolution Equations and Inverse Scattering in Multidimensions

M.J. Ablowitz

Department of Mathematics and Computer Science, Clarkson University, Potsdam, NY 13676, USA

In recent years the basic structure required to implement the inverse scattering transform in 1+1 and 2+1 dimensions has been clarified and extended. Aspects involved with fully multidimensional problems have also been treated. In particular the inverse scattering associated with various multidimensional operators and generalizations of the sine-Gordon and self-dual Yang-Mills equations have been studied. A review of some of this work will be discussed in this review.

The Inverse Scattering Transform (I.S.T.) is a method to solve certain nonlinear evolution equations. There has been wide ranging interest in this method for many reasons. A review of earlier work can be found in [1]. A surprisingly large number of physically interesting nonlinear equations can be solved via IST; there are many applications in physics including: surface waves, internal waves, lattice dynamics, plasma physics, nonlinear optics, particle physics and relativity. Mathematically speaking the field is also quite rich, with nontrivial results in the areas of analysis, group theory, algebra, differential and algebraic geometry being used by various researchers. From our point of view, IST allows us to solve the Cauchy problem for these nonlinear systems. We shall concentrate on questions in infinite space. All of the nonlinear equations discussed below arise as the compatibility condition of certain linear equations, one of which is identified as a scattering (direct and inverse scattering is required) problem and the other(s) serves to fix the "time evolution" of the scattering data.

In one spatial dimension the prototype problem is the (KdV) equation

$$u_t + 6uu_x + u_{xxx} = 0. \tag{1}$$

The KdV equation is compatible with

$$v_{xx} + u(x,t)v = \lambda v \tag{2}$$

$$v_t = (\gamma+u_x)v - (4\lambda+2u)v_x \tag{3}$$

i.e., $v_{xxt} = v_{txx}$ implies (1). Equation (2) is the time independent Schrödinger scattering problem, λ the eigenvalue (γ = const. in (3)). The solution of (1) on the line: $-\infty<x<\infty$ for initial values $u(x,t=0)$ vanishing sufficiently rapidly at infinity is obtained by studying the associated direct and inverse scattering problem of (2) and using (3) to fix the time evolution of the scattering data. It turns out that the inverse problem accounts to solving a matrix Riemann-Hilbert boundary value problem (RHBVP) whose jump discontinuity depends explicitly on the scattering data. Calling $\lambda = -k^2$, $v(x,k) = \mu(x,k)e^{-ikx}$ the RHBVP takes the following form,

$$(\mu_+ - \mu_-)(x,t,k) = \mu_-(x,t,\alpha(k))V(x,t,k) \text{ on } \Sigma,\ \mu\pm \to 1,\ |k|\to\infty, \qquad (4)$$

where $V(x,t,k) = r(k,t)e^{2ikx}$, $\alpha(k) = -k$, $\Sigma = \{k:k\in\mathbb{R}\}$, and $\mu_\pm$ are the limiting boundary values, as $\mathrm{Im}k\to 0\pm$, of meromorphic functions in the upper (+) lower (-) half plane. (4) may be converted into a linear integral equation by taking a minus projection and the potential is reconstructed via

$$u(x,t) = -\frac{1}{\pi}\frac{\partial}{\partial x}\int_C \mu(x,t,-k)V(x,t,k)dk, \qquad (5)$$

where the contour is taken above all poles of $r(k,t)$; of which there is at most a finite number, $k_j = i\kappa_j$, $\kappa_j > 0$, $j = 1,\dots,N$. The scattering data: the reflection coefficient, $r(k,t)$ evolves simply in time

$$r(k,t) = r(k,0)\, e^{8ik^2t} \qquad (6)$$

The above scheme may be extended so as to solve a surprisingly large number of interesting nonlinear evolution equations. There are two scattering problems of particular interest in one dimension:

(i) Scalar scattering problems:

$$\frac{d^nv}{dx^n} + \sum_{j=2}^{n} u_j(x)\frac{d^{n-j}v}{dx^{n-j}} = \lambda v,$$

$$v(x,k),\quad u_j \in \mathbb{C}$$

(ii) First order systems - generalized AKNS

$$\frac{dv}{dx} = i\,k\,J\,v + q\,v$$

$$v(x,k),\ q(x) \in \mathbb{C}^{N\times N},\quad J = \mathrm{diag}(J^1,\dots,J^n)$$

$$J^i \neq J^j,\ i \neq j$$

$$q^{ii} = 0.$$

Via an appropriate transformation the inverse problem associated with (i), (ii) can be expressed as a matrix RHBVP of the form (4). The potentials u_j, q can be shown to satisfy nonlinear evolution equations by appending to (i) and (ii), suitable linear time evolution equations. One then finds that the scattering data V(x,t,k) evolves simply in time. Well-known solvable nonlinear equations include the Boussinesq, modified KdV, sine-Gordon, nonlinear Schrödinger, and three wave interaction equations. The reader may wish to consult for example [2a-e] for a detailed discussion of some of this material.

It is most significant that these concepts can be generalized to two spatial plus one time dimension. Here the prototype equation is the Kadomtsev-Petviashvili (K-P) equation:

$$(u_t + 6uu_x + u_{xxx})_x = -3\sigma^2 u_{yy} \tag{7}$$

which is the compatibility equation between the following linear problems:

$$\sigma v_y + v_{xx} + u(x,y,t)v = 0 \tag{8}$$

$$v_t + 4v_{xxx} + 6uv_x + 3(u_x - \sigma \int_{-\infty}^{x} u_y dx')v + \gamma v = 0 \tag{9}$$

(γ = const.). We shall consider the question of solving (7) for u(x,y,0) decaying sufficiently rapidly in the plane $r^2 = x^2 + y^2 \to \infty$. Physically speaking, both cases $\sigma^2 = -1$ (KPI) $\sigma^2 = +1$ (KPII) are of interest. Whereas KPI can be related to a RHBVP of a certain type (nonlocal; see ref. [3]), KPII turns out to require new ideas. Letting

$$v = \mu(x,y,k) \exp(ikx + k^2 y/\sigma)$$

$\sigma = \sigma_R + i\sigma_1$, $\sigma_R \neq 0$. Then there exist functions μ bounded for all x,y satisfying $\mu \to 1$ as $|k| \to \infty$. However such a function turns out to be nowhere analytic in k, rather it depends nontrivially on both the real and imaginary parts of $k(k=k_R + ik_I)$. $\mu = \mu(x,y,k_R,k_I)$.

In fact μ satisfies a generalization of a RHBVP--namely a $\bar{\partial}$(DBAR) problem where μ satisfies,

$$\frac{\partial \mu}{\partial \bar{k}} = \mu(x,y,\xi_0,k_I)V(x,y,k_R,k_I) \tag{10}$$

where $\frac{\partial}{\partial \bar{k}} = \frac{1}{2}(\frac{\partial}{\partial k_R} + i\frac{\partial}{\partial k_I})$ and V has the structure

$$V(x,y,k_R,k_I) = \frac{\operatorname{sgn}(k_0)\exp[i\beta(x,y,k_R,k_I,\xi_0)]}{2\pi|\sigma_R|} T(k_R,k_I),$$

$$\beta(x,y,k_R,k_I,\xi_0) = (x + 2y\frac{k_I}{\sigma_R})(\xi_0 - k_R) = -2(x + 2y\frac{k_I}{\sigma_R})k_0,$$

$$\xi_0 = -k_R - \frac{2\sigma_I}{\sigma_R}k_I, \quad k_0 = k_R + \frac{\sigma_I}{\sigma_R}k_I \tag{11}$$

(10-11) may be converted into a linear integral equation by employing the generalized Cauchy formula. $T(k_R,k_I)$ is viewed as the "nonphysical" data (i.e., inverse scattering data or inverse data) and the potential is reconstructed via

$$u(x,y) = \frac{2i}{\pi}\frac{\partial}{\partial x}\iint \mu(x,y,\xi_0,k_I)\, V(x,y,k_R,k_I)\, dk_R dk_I. \tag{12}$$

The basic ideas used in order to derive these equations is as follows. We convert the equation for $\mu = \mu(x,y,k)$:

$$\sigma\mu_y + \mu_{xx} + 2ik\mu_x - u(x,y)\mu = 0 \tag{13}$$

into an integral equation

$$\mu(x,y,k) = 1 + \tilde{G}(u,\mu) \tag{14}$$

where

$$\tilde{G}(f) = G*f = \iint G(x-x', y-y',k)\, f(x',y')dx'dy', \tag{15}$$

the Green's function kernel being given by ($k=k_R+ik_I$):

$$G(x,y,k_R,k_I) = \frac{1}{(2\pi)^2}\iint\frac{\exp[i(\xi x+\gamma y)]}{(i\sigma\eta-\xi^2-2k\xi)}d\xi\, dy$$

$$= \frac{\operatorname{sgn}(y)}{2\pi\sigma}\int d\xi\ \exp[ix\xi+\xi(\xi+2k)y/\sigma]$$

$$\cdot\ \Theta(-y\sigma_R(\xi^2+2\xi k_0))d\xi \tag{16}$$

where

$$k_0 = k_R + \frac{\sigma_I}{\sigma_R}k_I \text{ and } \Theta(x) = \{1: x > 0, 0: x < 0\}.$$

The $\bar{\partial}$ derivative of the Green's function is especially simple,

$$\frac{\partial G}{\partial \bar{k}}(x,y,k_R,k_I) = \frac{\operatorname{sgn}(k_0)}{2\pi|\sigma_R|} \exp[i\beta(x,y,k_R,k_I)] \tag{17}$$

where

$$\frac{\partial}{\partial \bar{k}} = \frac{1}{2}\left(\frac{\partial}{\partial k_R} + i\frac{\partial}{\partial k_I}\right), \text{ and}$$

$$\beta(x,y,k\ ,k\ ) = -2\left(x + 2y\frac{k_I}{\sigma_R}\right)k_0.$$

Taking the $\bar{\partial}$ derivative of (14)

$$\frac{\partial \mu}{\partial \bar{k}}(x,y,k_R,k_I) = \iint \frac{\partial G}{\partial \bar{k}}(x-x',y-y',k_R,k_I)u(x',y')\mu(x',y',k_R,k_I)dx'dy'$$

$$+ \iint G(x-x',y-y',k_R,k_I)u(x',y')\frac{\partial \mu}{\partial \bar{k}}(x',y',k_R,k_I)dx'dy' \tag{18}$$

and using (17) shows that

$$\frac{\partial \mu}{\partial \bar{k}} = \frac{\operatorname{sgn}(k_0)}{\pi\sigma} T(k_R,k_I)w(x,y,k_R,k_I) \tag{19}$$

where $T(k_R,k_I) = \iint \exp[-i\beta(x,y,k_R,k_I)]u(x,y)\mu(x,y,k_R,k_I)dxdy$ and

$w(x,y,k_R,k_I)$ satisfies:

$$w(x,y,k_R,k_I) = \exp[i\beta(x,y,k_R,k_I) + \iint G(x-x',y-y',k_R,k_I)$$

$$u(x',y')w(x',y',k_R,k_I)dx'dy']. \tag{20}$$

Multiplying (20) by $\exp[-i\beta(x,y,k_R,k_I)]$ and employing the following symmetry condition on the Green's function

$$\exp[-i\beta(x,y,k_R,k_I)]\ G(x,y,k_R,k_I) = G(x,y,\xi_0,k_I) \tag{21}$$

where $\xi_0 = -k_0 - \frac{\sigma_I}{\sigma_R}k_I$, yields

$$w(x,y,k_R,k_I) = \exp[i\beta(x,y,k_R,k_I)]\ \mu(x,y,\xi_0,k_I) \tag{22}$$

whereupon (10-11) follow. The eigenfunction μ is recovered with the generalized Cauchy formula

$$\mu(x,y,k_R,k_I) = 1 + \frac{1}{\pi} \iint \frac{\frac{\partial\mu}{\partial\bar{k}}(x,y,k_R',k_I')}{k-k'} \, dk_R', dk_I' \qquad (23)$$

noting that using (10-11), (23) becomes a linear integral equation for μ. The potential $u(x,y)$ is recovered by taking $k \to \infty$ in (13) or (14) and (23). For the K-P the evolution of the data obeys ($\gamma = 4ik^3$ in (9))

$$\frac{\partial T}{\partial t} = (8ik_0)(6kk_0 - 4k_0^2 - 3k^2)T \qquad (24)$$

where $k_0 = k_R + \frac{\sigma_I k_I}{\sigma_R}$, $k = k_R + ik_I$.

Special cases include $\sigma = \sigma_R + i\sigma_I$:

(a) KP_{II}; $\sigma = -1$: $\sigma_R = -1$, $\sigma_I = 0$

$$\frac{\partial T}{\partial t} = 8ik_R(3k_I^2 - k_R^2)T \qquad (25)$$

(b) KP_I; $\sigma = i$: $\sigma_R \to 0-$, $\sigma_I = 1$, $\hat{k}_I = k_I/\sigma_R$

$$\frac{\partial T}{\partial t} = -8i(k_R + \hat{k}_I)(k_R^2 + 2k_R\hat{k}_I + 4\hat{k}_I^2)T \qquad (26)$$

These formulae allow us in principle to solve the Cauchy problem for K-P and in particular the limit (ii) discussed above allows us to give an alternative solution for KP_I via $\bar{\partial}$ and not via a nonlocal RHBVP.

Similar ideas apply to higher order scalar problems

(iii) $$\sigma \frac{\partial v}{\partial y} + \frac{\partial^n v}{\partial x^n} + \sum_{j=2}^{n} u_j(x) \frac{\partial^{n-j} v}{\partial x^{n-j}} = 0$$

where $v, u_j \in \mathbb{C}$ and to first order systems

(iv) $$\sigma \frac{\partial v}{\partial y} + J \frac{\partial v}{\partial x} + q(x,y)v = 0$$

where $v, q \in \mathbb{C}^{N\times N}$, $J = \mathrm{diag}(J^1,\ldots,J^N)$, $J^i \neq J^j$, $i \neq j$ with $q^{ii} = 0$.

Interested readers may consult reference 4a, and review 4b for more details.

The notion of $\bar{\partial}$ extends to higher dimensional scattering and inverse scattering problems. However as we shall mention, despite the fact that the inverse scattering problem is essentially tractable there does not appear to be any local nonlinear evolution equations in dimensions greater than 2+1 associated with multidimensional generalizations of (iii) or (iv).

Our prototype scattering problem will be

$$\sigma v_y + \Delta v + u(x,y)v = 0$$

$$\Delta = \sum_{\ell=1}^{n} \frac{\partial^2}{\partial x_\ell^2}, \quad x \in \mathbb{R}^n,\ y \in \mathbb{R}. \tag{27}$$

Letting

$$v = \mu(x,y,k)\exp(ikx + k^2 y/\sigma)$$

$$k = k_R + ik_I, \quad k \in \mathbb{C}^n$$

$$k.x = \sum_1^n k_j x_j, \quad \sigma = \sigma_R + i\sigma_I.$$

Then there exist functions μ bounded for all x, y satisfying $\mu \to 1$, as $|k_j| \to \infty$, $j = 1,\ldots,n$. When $\sigma_R \neq 0$, μ turns out to be nonanalytic in each of the variables k, i.e., $\mu = \mu(x,y,k_{R_1},\ldots,k_{R_n}, k_{I_1},\ldots,k_{I_n})$ and satisfies a $\bar\partial$ problem linear in μ, in each of the variables k_j; i.e., we shall show that μ satisfies an equation of the form,

$$\frac{\partial\mu}{\partial\bar{k}_j} = \tilde{T}_j(\mu); \quad j = 1,\ldots,n \tag{28}$$

where $\tilde{T}_j$ is an appropriate linear integral operator.

The basic idea in order to derive (28) follows a similar format to the two dimensional case described earlier. From the definition of $\mu(x,y,k)$ below (27) we see that it satisfies

$$\sigma\mu_y + \Delta\mu + 2ik.\nabla\mu + u\mu = 0. \tag{29}$$

We convert to an integral equation

$$\mu = 1 + \tilde{G}(u,\mu) \tag{30}$$

where the Green's function kernel is given by

$$G(x,y,k_R,k_I) = \frac{1}{(2\pi)^{n+1}} \int\int \frac{\exp[i(x.\xi+y\eta)]}{i\sigma y-\xi^2-2k.\xi} d\xi dy \tag{31}$$

$$= \frac{\mathrm{sgn}(y)}{\sigma} \frac{1}{(2\pi)^n} \int \exp[ix.\xi+ \frac{y}{\sigma}(\xi^2+2k.\xi).$$

$$.\Theta(-y\sigma_R(\xi^2+2(k_R + \frac{\sigma_I k_I}{\sigma_R}).\xi)d\xi. \tag{32}$$

Taking the $\bar{\partial}$ derivative of (30)

$$\frac{\partial\mu}{\partial\bar{k}_j} = \frac{\partial\tilde{G}}{\partial\bar{k}}(u\mu) + \tilde{G}(u \frac{\partial\mu}{\partial\bar{k}_j}), \tag{33}$$

and using

$$\frac{\partial G}{\partial\bar{k}_j}(x,y,k_R,k_I) = - \frac{1}{(2\pi)^n} |\sigma_R| \int \exp[i\beta(x,y,k_R,k_I,\xi)]$$

$$.(\xi_j-k_{Rj})\delta(\rho(\xi))d\xi \tag{34}$$

where

$$\beta(x,y,k_R,k_I,\xi) = (x + 2y \frac{k_I}{\sigma_R}).(\xi-k_R)$$

$$\rho(\xi) = (\xi + \frac{\sigma_I}{\sigma_R}k_I)^2 - (k_R + \frac{\sigma_I}{\sigma_R} k_I)^2 \tag{35}$$

shows that

$$\frac{\partial\mu}{\partial\bar{k}_j} = - \frac{1}{(2\pi)^n} \frac{1}{|\sigma_R|} T(k_R,k_I\xi)(\xi_j-k_{Rj})\delta(\rho(\xi))$$

$$.w(x,y,k_R,k_I,\xi)d\xi \tag{36}$$

where

$$T(k_R,k_I,\xi) = \int \exp[-i\beta(x,y,k_R,k_I,\xi)]u(x,y)\mu(x,y,k_R,k_I)dxdy \tag{37}$$

and w satisfies

$$w(x,y,k_R,k_I,\xi) = \exp[i\beta(x,y,k_R,k_I\xi)] + \tilde{G}(uw). \tag{38}$$

Multiplying (37) by $\exp(-i\beta)$ and using the symmetry condition

$$\exp[-i\beta(x,y,k_R,k_T,\xi)] G(x,y,k_R,k_I) = G(x,y,\zeta,k_I) \tag{39}$$

yields

$$w(x,y,k_R,k_I,\xi) = \exp[-i\beta(x,y,k_R,k_I,\xi)]\ \mu(x,y,\xi,k_I) \tag{40}$$

and hence (36) gives

$$\frac{\partial\mu}{\partial\bar{k}_j} = \tilde{T}_j(\mu) = -\frac{1}{(2\pi)^n}\frac{1}{|\sigma_R|}\int T(k_R,k_I,\xi)(\xi_j-k_{Rj})$$

$$\cdot\delta(\rho(\xi))\exp[i\beta(x,y,k_R,k_I,\xi)\ \mu(x,y,\xi,k_I)d\xi. \tag{41}$$

We see that $\tilde{T}_j$ is an integral operator which depends on a scalar scattering function $T = T(k_R,k_I,\xi)\,\xi$ being effectively (n-1) integration parameters (due to the delta function in (41) in the nonlocal operator $\tilde{T}_j$).

One can use a generalized Cauchy formula such as (23) in order to obtain a linear integral equation to reconstruct μ. However due to the redundancy of the data discussed below, we find that an alternative method is more useful. The inverse problem is redundant, i.e., we are given $T(k_R,k_I,\xi)$ (3n-1 parameters) and we must reconstruct a local potential $u(x,y)$ (n+1 parameters). A serious issue is how to characterize admissible inverse data T, i.e., data that really arises from a local potential (small generic changes in $T(k_R,k_I,\xi)$ cannot be expected to arise from a local potential $u(x,y)$). Insight into this question is obtained by noting that T must satisfy a nonlinear constraint, one which is obtained by requiring $\partial^2\mu/\partial\bar{k}_i\partial\bar{k}_j = \partial^2\mu/\partial\bar{k}_j\partial\bar{k}_i$ $(i\neq j)$. The form of this constraint is given by

$$\mathcal{L}_{ij}(T) = \tilde{N}_{ij}[T] \tag{42}$$

where $\mathcal{L}_{ij}$ is a linear operator and $\tilde{N}_{ij}$ a nonlinear (quadratic) nonlocal operator. These operators are given by

$$\mathcal{L}_{ij} = (\xi_j-k_{jR})\left(\frac{\partial}{\partial\bar{k}_i}+\frac{1}{2}\frac{\partial}{\partial\xi_i}\right) - (\xi_i-k_{iR})\left(\frac{\partial}{\partial\bar{k}_j}+\frac{1}{2}\frac{\partial}{\partial\xi_j}\right) \tag{43}$$

$$\tilde{N}_{ij}(T) = \int[(\xi'_j-k_{jR})(\xi_i-\xi'_i) - (\xi'_i-k_{iR})(\xi_j-\xi'_j)]$$

$$\cdot\delta(\rho(\xi')T(k_R,k_I,\xi')T(\xi',k_I,\xi)d\xi'. \tag{44}$$

There is, in fact, an explicit transformation of variables

$$(k_R,k_I,\xi) \to (\chi,w_0,w) \in \mathbb{C}^{n-1}\times\mathbb{R}\times\mathbb{R}^n$$

which simplifies (42). Namely,

$$k_{R1} = \sum_{j=2}^{n} w_j\chi_{Rj} - \frac{w_1}{2} - \frac{\sigma_I w_0 w_1}{2w^2},$$

$$k_{Rj} = -w_1\chi_{Rj} - \frac{w_j}{2} - \frac{\sigma_I w_0 w_j}{2w^2}, \quad (j\geq 2)$$

$$k_{i1} = \sum_{j=2}^{n} w_j\chi_{Ij} + \frac{\sigma_R w_0 w_1}{2w^2},$$

$$k_{ij} = -w_1\chi_{Ij} + \frac{\sigma_R w_0 w_j}{2w^2}, \quad (j\geq 2)$$

$$\xi_1 = \sum_{j=2}^{n} w_j\chi_{Rj} + \frac{w_1}{2} - \frac{\sigma_I w_0 w_1}{2w^2},$$

$$\xi_j = -w_1\chi_{Rj} + \frac{w_j}{2} - \frac{\sigma_I w_0 w_j}{2w^2} . \quad (j\geq 2) \tag{45}$$

transforms (42) into

$$\frac{\partial T}{\partial \bar{x}_j} = \tilde{N}_{ij}(T)(\chi,w_0,w). \quad j=2,\dots,n \tag{46}$$

using the generalized Cauchy formula (23) we have

$$I_j[T](\chi,w,w_0) = T(\chi,w,w_0) - \frac{1}{\pi}\iint \frac{\tilde{N}_{ij}(T)(\tilde{\chi},w,w_0)}{\chi - \tilde{\chi}} d\chi_R' d\chi_I'$$

$$= \hat{u}(w,w_0) , \tag{47}$$

where

$$\tilde{\chi} = (\chi_2,\chi_3,\dots,\chi_j',\dots,\chi_n)$$

$$\hat{u}(w_0,w) = \iint \exp[-i(yw_0+x.w)]u(x,y)dxdy. \tag{48}$$

We have used the fact that when $w_0 = 2k_I.(\xi-k_R)/\sigma_R$ and $w = \xi-k_R$ are kept fixed, $T(\chi,w,w_0) \to \hat{u}(w,w_0)$ (The Fourier Transform of u(x,y)) for large $\chi_j (w_1 \neq 0)$; this is the analogue of the Born approximation.

We expect that for suitably "small" u (i.e., no homogeneous solutions to the relevant integral equations) if I is independent of χ,j and decays sufficiently fast for $|w|,|w_0|\to\infty$, then $T(k_R,k_I,\xi)$ is admissible. Moreover (47) gives a formula to reconstruct the potential by quadratures. Limits to case $\sigma = i$ and reductions to stationary potentials $u(x,y) = u(x)$ can be carried out. Details can be found in Ref. [5a,b]. It should also be noted that in recent work Nachman and Lavine [5c] have extended their ideas to situations where there are homogeneous solutions to the relevant integral equations. (42) also suggests why simple local nonlinear evolution equations have not been associated with equation (27). Namely, in the previous lower dimensional (2+1 and 1+1) problems the time evolution of the scattering data obeyed a particularly simple equation (e.g., $\frac{\partial T}{\partial t} = \omega(k_R,k_I)T$). However in this case such a simple flow will not be maintained due to the nonlinear constraint (42).

These ideas can be generalized to first order systems:

$$\text{(v)} \qquad \frac{\partial v}{\partial y} + \sigma \sum_{j=1}^{n} J_j \frac{\partial v}{\partial x_j} = qv$$

$$v,q \in \mathbb{C}^{N\times N}, \quad J_j = \mathrm{diag}(J_j^1,\dots,J_j^N), \qquad J_j^k \neq J_j^\ell,\ k \neq \ell$$

with many similar results obtained [6a,b,c]; though there are some important differences as well: see ref. [6c]. Again the scattering data satisfies a nonlinear constraint. In general, there is no compatible local nonlinear evolution equation associated with (v). However when certain restrictions are put on J_j then the constraint equation becomes linear and the so-called N wave interaction equations are compatible with the system (v). Nachman and Ablowitz [6a] showed that at most, the system would be 3+1 dimensional, and Fokas [6b] showed that indeed the system is reducible to 2+1 dimensions by a transformation of independent variables (characteristic variables). In [6c] Fokas studies the inverse scattering of (v). For $\sigma = i$ he finds an equation similar to (42). However its integrated form shows that in order for the potential to be reconstructed one must solve a reduced system of equations of the form (v): i.e., for N = 2. This is in contrast to the scalar problem where reconstruction is via quadratures.

Beals and Coifman have an alternative but similar formulation [7a,b] for multidimensional scalar problems.

There is an n-dimensional problem which also fits within the framework of IST: The so-called generalized wave and generalized sine-Gordon equation (GWE and GSGE). These equations arise in the context of differential geometry and serve to extend the classical results of Bäcklund for the sine-Gordon equation to n-dimensions [8]. The n-dimensional Bäcklund transformation is given by:

$$d\chi + \chi A^t\chi = A - \chi B, \tag{49}$$

where

$$d\chi = \sum_{j=1}^{n} \frac{\partial\chi}{\partial\chi_j} d\chi_j,$$

$$A_{ij} = \beta_i(Z) a_{ij} dx_j,$$

$$B_{ij} = \frac{1}{a_{1i}} \frac{\partial a_{1j}}{\partial x_i} dx_j - \frac{1}{a_{1j}} \frac{\partial a_{1i}}{\partial x_j} dx_i, \quad 1 \leq i,j \leq n, \tag{50}$$

and $a = \{a_{ij}\} \in \mathbb{R}^{n\times n}$. Equations (49-50) reduce to the Bäcklund transformation for the generalized sine-Gordon equation (GSGE) when

$$\beta_i(z) = (z^2 + (2\delta_{i1} - 1))/2z, \tag{51}$$

and for the generalized wave equation (GWE) when

$$\beta_i(z) = -(1-z^2)/2z \equiv \lambda(z). \tag{52}$$

The compatibility condition required for the existence of solutions to these Bäcklund transformations results in a system of second-order partial differential equations for an orthogonal $n\times n$ matrix $a = \{a_{ij}\}$ in (49) which is a function of n independent variables $a = a(x_1, x_2, \ldots, x_n)$. The equation has the form

$$\frac{\partial}{\partial x_i}\left(\frac{1}{a_{1i}} \frac{\partial a_{1j}}{\partial x_i}\right) + \frac{\partial}{\partial x_j}\left(\frac{1}{a_{1k}} \frac{\partial a_{1i}}{\partial x_j}\right)$$

$$+ \sum_{k\neq i,j} \frac{1}{a_{1k}^2} \frac{\partial a_{1i}}{\partial x_k} \frac{\partial a_{1j}}{\partial x_k} = \varepsilon\, a_{1i}a_{1j}, \quad i \neq j,$$

$$\frac{\partial}{\partial x_k}\left(\frac{1}{a_{1j}} \frac{\partial a_{1i}}{\partial x_j}\right) = \frac{1}{a_{1k}a_{1j}} \frac{\partial a_{1i}}{\partial x_k} \frac{\partial a_{1k}}{\partial x_j}, \quad i,j,k \text{ distinct},$$

$$\frac{\partial a_{jk}}{\partial x_k} = \frac{\partial a_{ji}}{a_{1i}} \frac{\partial a_{1k}}{\partial x_i}, \quad i \neq k, \tag{53}$$

where $\varepsilon = 1$ for the GSGE and $\varepsilon = 0$ for the GWE.

We observe that when n = 2 and $\epsilon = 1$ (GSGE), the orthogonal matrix $a = \{a_{ij}\}$ given by

$$a = \begin{pmatrix} \cos\frac{1}{2}u & \sin\frac{1}{2}u \\ -\sin\frac{1}{2}u & \cos\frac{1}{2}u \end{pmatrix} \tag{54}$$

for the function $u = u(x,t)$ reduces the GSGE to the classical sine-Gordon equation ($\kappa=-1$),

$$u_{tt} - u_{xx} - \kappa \sin u = 0. \tag{55}$$

On the other hand when n = 2 and $\kappa = 0$, then with (54) the GWE reduces to the wave equation (55). When $n \geq 3$ the generalization of the wave equations discussed here is nonlinear.

The Bäcklund transformations (49) described above are in fact matrix Riccati equations. Linearizations of such a system can be performed in a straightforward manner. Introducing the transformation

$$\chi = UV^{-1}, \tag{56}$$

where U, V and n×n matrix functions of $x_1,\ldots,x_n$, the following linear system is deduced:

$$\begin{pmatrix} dU \\ dV \end{pmatrix} = \begin{pmatrix} 0 & A \\ A^t & B \end{pmatrix} \begin{pmatrix} U \\ V \end{pmatrix} \tag{57}$$

with the components of A, B given by (50). Compatibility ensures that the orthogonal matrix $a = \{a_{ij}\}$ satisfies the GSGE with (51) and GWE with (52). Alternatively, if we call

$$\begin{pmatrix} U \\ V \end{pmatrix} = \psi , \tag{58}$$

the following linear system of 2n o.d.e.'s are obtained:

$$\frac{\partial \psi}{\partial x_j} = \lambda \tilde{A}_j \psi + C_j\psi, \tag{59}$$

where $\tilde{A}_j$, C_j are 2n×2n matrices with the block structure

$$\tilde{A}_j = \begin{pmatrix} 0 & \tilde{a}_j \\ \tilde{a}_j^t & 0 \end{pmatrix}, \quad C_j = \begin{pmatrix} 0 & 0 \\ 0 & \gamma_j \end{pmatrix}. \tag{60}$$

Here $\tilde{a}_j$, $\tilde{\gamma}_j$ are nxn matrices having the following structure:

$$\tilde{a}_j = (\delta/\lambda - 1)e_1 a_j + a_j,$$

$$a_j = ae_j \tag{61}$$

where $e_j = \{e_j\}_{ik}$ is the unit matrix

$$\{e_j\}_{ik} = \begin{cases} 1 & i = k = j \\ 0 & \text{otherwise,} \end{cases} \tag{62}$$

and in component form γ_j takes the form

$$(\gamma_j)_{k\ell} = (1 - \delta_{kj}) \frac{1}{a_{\ell k}} \frac{\partial a_{\ell j}}{\partial x_k} \delta_{\ell j} - (1 - \delta_{\ell j}) \frac{1}{a_{1\ell}} \frac{\partial a_{1j}}{\partial x_\ell} \delta_{kj}. \tag{63}$$

In (61) a is the orthogonal matrix $\mathbb{R}^n \to SO(n)$ associated with the GWE when $\delta = \lambda$ and with the GSGE when $\delta = \frac{1}{2}(z + 1/z)$, $\lambda = \frac{1}{2}(z - 1/z)$, and γ_j is the matrix (63): $\mathbb{R}_n \to M_n(\mathbb{R})$, $\gamma_j^t + \gamma_j = 0$. Equations (53) arise as the compatibility condition associated with (58). More explicitly, for the GWE the scattering problem takes the form $[\psi = \psi(x,\lambda)]$

$$\frac{\partial \psi}{\partial x_j} = \lambda A_j \psi + C_j \psi \tag{64}$$

with

$$A_j = \begin{pmatrix} 0 & a_j \\ a_j^t & 0 \end{pmatrix}, \tag{65}$$

and C_j given by (60,63).

For the GSGE the scattering problem for $\psi = \psi(x,z)$ takes the form

$$\frac{\partial \psi}{\partial x_j} = \delta(z) \begin{pmatrix} 0 & e_1 a_j \\ a_j^t e_1 & 0 \end{pmatrix} \psi$$

$$+ \lambda(z) \begin{pmatrix} 0 & (I-e_1)a_j \\ a_j^t(I-e_1) & 0 \end{pmatrix} \psi + C_j \psi, \tag{66}$$

$\delta(Z)$, $\lambda(Z)$, C_j given above, or equivalently

$$\frac{\partial \psi}{\partial x_j} = \frac{Z}{2} A_j \psi + \frac{Z}{2} B_j \psi + C_j \psi, \tag{67}$$

where

$$B_j = \begin{pmatrix} 0 & ua_j \\ a_j^t u & 0 \end{pmatrix}, \quad u = \mathrm{diag}(+1,-1,\dots,-1). \tag{68}$$

In [8] it is shown how these linear problems may be viewed as a direct and inverse scattering problem for the GWE and GSGE. Namely, the direct and inverse problem may be solved for matrix potentials, depending on the orthogonal matrix a, tending to the identity sufficiently fast in certain "generic" directions. It should be noted that solving the n-dimensional GWE and GSGE reduces to the study of the scattering and inverse scattering associated with a coupled system of n one-dimensional o.d.e.'s. This is in marked contrast to other attempts described earlier to isolate solvable (local) multidimensional nonlinear evolution equation which are compatibility conditions of two Lax-type operators, e.g.,

$$L\psi = \lambda\psi \tag{69}$$

$$\psi_t = M\psi \tag{70}$$

where L is a partial differential operator with the variable t entering only parametrically. Although as we have seen nonlinear evolution equations in three independent variables can be associated with such Lax pairs (e.g., the K-P, Davey-Stewartson, three wave interaction equations, etc.) little progress via this route has been made in more than three dimensions. As discussed earlier one has to overcome a serious constraint inherent in the scattering/inverse scattering theory for higher dimensional partial differential operators in order to be able to isolate associated solvable nonlinear equations, i.e., the scattering data generally satisfies a nonlinear equation (eq. (42)). The analysis associated with the GWE and GSGE avoids these difficulties since the GWE and GSGE problems are simply a compatible set of nonlinear one-dimensional o.e.e.'s. The results in ref. [8] demonstrate that the initial value problem is posed with given data along lines and not on (n-1) dimensional manifolds.

Similar ideas apply to certain n-dimensional extensions of the so-called anti-self-dual Yang-Mills equations (SDYM) [9]. In two complex variables the self-dual Yang Mills equations take the form (see [10])

$$\frac{\partial}{\partial \bar{x}_1}\left(\Omega^{-1}\frac{\partial\Omega}{\partial x_1}\right) + \frac{\partial}{\partial \bar{x}_2}\left(\Omega^{-1}\frac{\partial\Omega}{\partial x_2}\right) = 0, \tag{71}$$

where Ω is a positive matrix valued function of $(x_1,x_2)\in \mathbb{C}^2$. Alternatively SDYM takes the form

$$\frac{\partial A_1}{\partial \bar{x}_1} + \frac{\partial A_2}{\partial \bar{x}_2} = 0 \tag{72}$$

$$\frac{\partial A_1}{\partial x_2} - \frac{\partial A_2}{\partial x_1} + [A_1, A_2] = 0, \tag{73}$$

where

$$A_j = -\Omega^{-1} \frac{\partial \Omega}{\partial x_j}. \tag{74}$$

The SDYM may be obtained via the compatibility condition of the following linear system

$$\frac{\partial m}{\partial x_1} - z \frac{\partial m}{\partial \bar{x}_2} = A_1 m$$

$$\frac{\partial m}{\partial x_2} + z \frac{\partial m}{\partial \bar{x}_1} = A_2 m. \tag{75}$$

Multidimensional extensions may be obtained. For example, consider the linear system

$$D_z^j m(x,z) = A_j(x) m(x,z), \quad j = 1,\dots,n \tag{76}$$

$$D_z^j = \frac{\partial}{\partial x_j} + z s_j \frac{\partial}{\partial \bar{x}_{j+1}} \tag{77}$$

and

$$x_{n+1} = x_1, \quad s_j = (-1)^j.$$

Compatibility (commutativity) implies:

$$D_z^i A_j - D_z^j A_i + [A_i, A_j] = 0 \tag{78}$$

$$\frac{\partial A_i}{\partial x_j} - \frac{\partial A_j}{\partial x_i} + [A_i, A_j] = 0 \tag{79}$$

$$s_j \frac{\partial A_i}{\partial x_{j+1}} - s_i \frac{\partial A_j}{\partial \bar{x}_{i+1}} = 0. \tag{80}$$

A potential Ω may be introduced as before:

$$A_j = \Omega^{-1} \frac{\partial \Omega}{\partial x_j} \tag{81}$$

to obtain

$$s_j \frac{\partial}{\partial \bar{x}_{j+1}} (\Omega^{-1} \frac{\partial \Omega}{\partial x_i}) - s_i \frac{\partial}{\partial \bar{x}_{i+1}} (\Omega^{-1} \frac{\partial \Omega}{\partial x_j}) = 0. \tag{82}$$

Clearly when n = 2 this system reduces to the classical SDYM equation.

Solutions to these equations may be constructed via the $\bar{\partial}$ method. Define

$$D_z^j = L_1^j + zL_2^j \tag{83}$$

with

$$L_1^j = \frac{\partial}{\partial x_j}, \quad L_2^j = s_j \frac{\partial}{\partial \bar{x}_{j+1}}.$$

We shall show that the $\bar{\partial}$ integral equation

$$m(x,z) = I + \frac{1}{2\pi i} \iint \frac{(mV)(x,\zeta)}{\zeta - z} d\zeta \wedge d\bar{\zeta} \tag{84}$$

satisfies (76) ($d\zeta \wedge d\bar{\zeta} = -z_i d\zeta_R d\zeta_I$). Operating on (84) with D_z^j yields,

$$D_z^j m = \frac{1}{2\pi i} \int \frac{(L_1^j m)V + m(L_1^j V)}{\zeta - z} d\zeta \wedge d\bar{\zeta} + J \tag{85}$$

where

$$J = \frac{1}{2\pi i} \int \frac{zL_2^j(mV)}{\zeta - z} d\zeta \wedge d\bar{\zeta}$$

$$= - \frac{1}{2\pi i} \int L_2^j(mV) d\zeta \wedge d\bar{\zeta} + \frac{1}{2\pi i} \int \frac{\zeta L_2^j(mV)}{\zeta - z} d\zeta \wedge d\bar{\zeta}. \tag{86}$$

Putting (85), (86) together gives

$$D_z^j m = \tilde{A}_j + \frac{1}{2\pi i} \int \frac{(D_\zeta^j m)V + m(D_\zeta^j V)}{\zeta - z} d\zeta \wedge d\bar{\zeta} \tag{87}$$

where

$$\tilde{A}_j(x) = - \frac{1}{2\pi i} \int L_2^j(mV) d\zeta \wedge d\bar{\zeta} = - \frac{1}{2\pi i} s_j \frac{\partial}{\partial \bar{x}_{j+1}} \int (mV) d\zeta \wedge d\bar{\zeta}. \tag{88}$$

We shall require V(x,z) to satisfy

$$D_z^j V = 0 \tag{89}$$

in which case using (84) in (87) by writing

$$\tilde{A}_j = \tilde{A}_j\left(m - \frac{1}{2\pi i}\int \frac{mV}{\zeta - z}\, d\zeta \wedge d\bar{\zeta}\right) \tag{90}$$

we find

$$(D_z^j m - \tilde{A}_j m) = \frac{1}{2\pi i}\int \frac{((D_\zeta^j m) - \tilde{A}_j(x)m)V}{\zeta - z}\, d\zeta \wedge d\bar{\zeta}. \tag{91}$$

For V suitably chosen (84) has a unique solution in which case

$$D_z^j m - \tilde{A}_j m = 0. \tag{92}$$

Thus $\tilde{A}_j = A_j$ and solutions of the extended SDYM are obtained.

The condition (89) is satisfied if we take V(x,z) = V(u(x),z), with $u_j(x) = zx_j + s_{j+1}\bar{x}_{j+1}$ and V holomorphic in the u_j. Then

$$D_z^j V = \left(\frac{\partial}{\partial x_j} + zs_j \frac{\partial}{\partial \bar{x}_{j+1}}\right) V(u_i,\ldots,u_n,z)$$

$$= \sum_{\ell=1}^{n} V'(u_\ell,z)(z\delta_{j\ell} + s_j s_{j+1} z\delta_{j\ell}) = 0 \tag{93}$$

by virtue of $s_j = (-1)^j$. In ref. [9] other examples of multidimensional extensions of SDYM and a rigorous derivation of the foregoing is given.

ACKNOWLEDGEMENTS

I am most pleased to acknowledge the many crucial contributions of my colleqgues: A. S. Fokas, D. Bar Yaacov, A. I. Nachman, R. Beals and K. Tenenblat. This work was supported in part by the National Science Foundation under grant number DMS-8202117, the Office of Naval Research under grant number N00014-76-C-0867, and the Air Force Office of Scientific Research under grant number AFOSR-84-0005.

REFERENCES

[1] M. J. Ablowitz and H. Segur, **Solitons and the Inverse Scattering Transform** (SIAM, Philadelphia, 1981).

[2a] M. J. Ablowitz and A. S. Fokas,"Comments in the Inverse Scattering Transform and Related Nonlinear Evolutions Equations," in **Lecture Notes in Physics, 189**, ed. K. B. Wolf (Springer, Berlin, 1982).

[2b] R. Beals and R. Coifman, Commun. Pure Appl. Math., 1984, pp.39-90.

[2c] R. Beals, "The Inverse Problems for Ordinary Differential Operators on the Line," to appear, Amer. J. Math.

[2d] A. B. Shabat, Func. Annal and Appl. **9** (1975); Diff. Eq. **15** (1979) 1824.

[2e] P. Caudrey, Physica **6D** (1982) 51.

[3a] S. V. Manakov, Physica **3D** (1981) 420.

[3b] A. S. Fokas and M. J. Ablowitz, Stud. Appl. Math. **69** (1983) 211.

[4a] M. J. Ablowitz, D. Bar Yaacov and A. S. Fokas, Stud. Appl. Math. **69** (1983) 135.

[4b] A. S. Fokas and M. J. Ablowitz, "The Inverse Scattering Transform for Multidimensional (2+1) Problems," in **Lecture Notes in Physics, 189**, ed. K. B. Wolf (Springer, Berlin, 1982).

[5a] A. I. Nachman and M. J. Ablowitz, Stud. Appl. Math. **71** (1984) 243.

[5b] M. J. Ablowitz and A. I. Nachman, Physica **18D** (1986) 223.

[5c] A. I. Nachman and R. Lavine, "On the Inverse Scattering Transform for the n-Dimensional Schrödinger Operator," to be published as Proceedings of Conference on Nonlinear Evolution Equations, Solitons and the Inverse Scattering Transform, held at Mathematische Forschungsistitut, Oberwolfach, W. Germany.

[6a] A. I. Nachman and M. J. Ablowitz, Stud. Appl. Math. **71** (1984) 251.

[6b] A. S. Fokas, Phys. Rev. Lett. **57** (1986) 159.

[6c] A. S. Fokas, J. Math. Phys. **27** (1986) 1737.

[7a] R. Beals and R. Coifman, "Multidimensional Inverse Scattering on Nonlinear PDE," Proc. Symposium Pure Math. **43** (1985) 45.

[7b] R. Beals and R. Coifman, Physica **18D** (1986) 242.

[8] M. J. Ablowitz, R. Beals and K. Tenenblat, Stud. Appl. Math. **74** (1986) 177.

[9] M. J. Ablowitz, D. G. Costa and K. Tenenblat, Solutions of Multidimensional Extensions of the Anti-Self-Dual Yang-Mills Equations, preprint, INS #66, 1986.

[10] K. Pohlmeyer, Commun. Math. Phys. **72** (1980) 37.

Inverse Problems and a Unified Approach to Integrability in 1, 1+1 and 2+1 Dimensions*

A.S. Fokas and V. Papageorgiou

Department of Mathematics and Computer Science and Institute for Nonlinear Studies, Clarkson University, Potsdam, NY 13676, USA

A unified approach for solving initial value problems for equations in 0+1, 1+1 (one spatial and one temporal), and 2+1 (two spatial and one temporal) dimensions is given. Illustrative examples in each of these cases are provided. Some remarks on inverse problems in higher than two spatial dimensions are made in the context of inverse scattering.

1. INTRODUCTION

The aim of this paper is to emphasize that there exists a unified approach for solving initial value problems for equations in 1, 1+1 (i.e., one spatial and one temporal), and 2+1 (i.e., two spatial and one temporal) dimensions. Furthermore it remarks on inverse problems in higher than two spatial dimensions. Although these inverse problems are not related to physically significant nonlinear evolution equations, they are useful in the context of inverse scattering. In this presentation we emphasize the main ideas. The detailed formalisms can be found in the cited papers.

It turns out that solving the initial value problem for some equations for $q(t)$, or $q(x,t)$, or $q(x,y,t)$ is equivalent to solving an inverse problem for some related eigenfunction $\Psi(z;t)$, or $\Psi(z;x,t)$, or $\Psi(z;x,y,t)$. The inverse problem takes the form of a Riemann-Hilbert (RH) problem for equations in 1 and 1+1, and the form of a nonlocal RH problem or of a $\bar{\partial}$(DBAR) problem for equations in 2+1 (a DBAR problem is generalization of a RH problem). To define the relevant RH or DBAR problems one needs to study the **analyticity properties of Ψ with respect to z.** Furthermore those problems are uniquely defined in terms of certain asymptotic data of the underlying linear system satisfied by Ψ: **Monodromy data** in 1, **scattering data** in 1+1 and some cases of

*This article consists of expanded material of six lectures presented by one of us (A. S. Fokas) at this Winter School on "Solitons".

2+1, and **inverse data** in some cases of 2+1. We use the Painlevé IV(PIV), modified KdV (mKdV) and the Davey-Stewartson (DS) as illustrative examples of equations in 1, 1+1, and 2+1 respectively.

The above inverse problems can be naturally generalized to higher than two spatial dimensions. For example, the generalization of the inverse problem associated with the DS equation leads to an inverse problem for a matrix valued function $\Psi(z;x_0,x)$, $z \in \mathbb{C}^n$, $x_0 \in \mathbb{R}^1$, $x \in \mathbb{R}^n$, $n > 1$. However, while the associated potential $q(x_0,x)$ depends on n+1 variables, the inverse data $T(z_R,z_I,m_2,\ldots,m_n)$, $z_R \in \mathbb{R}^n$, $z_I \in \mathbb{R}^n$, $m_\ell \in \mathbb{R}$, depends on 3n-1 variables. This has important implications: (a) The inverse data must be appropriately constrained. This "characterization" of the inverse data is conceptually analogous to the characterization of the inverse scattering data in the multidimensional Schrödinger equation [1]. (b) The existence of "redundant" scattering parameters can be used to reduce the above problem to one in two spatial dimensions. This is in contrast to the case of the multidimensional Schrödinger equation where the inverse problem can be solved in closed form. (c) Since the inverse problem for Ψ is reduced to one in two spatial dimensions, it follows that, if one allows Ψ, q to depend parametrically on t, $q(x_0,x,t)$ satisfies an evolution equation reducible to two spatial dimensions. In particular, the N-wave interaction equation in n+1 spatial dimensions can always be reduced to two spatial dimensions. Thus a genuine three-spatial-dimensional nonlinear evolution equation, related to an inverse problem, remains to be found. (It should be noted that several other "multidimensional" problems can be reduced to one or two spatial dimensions, see M. J. Ablowitz's contribution in these proceedings.)

We first define the standard RH and DBAR problems.

2. RH AND DBAR PROBLEMS

Let C be a simple, smooth closed contour dividing the complex z-plane into two regions D^+ and D^- (the positive direction of C will be taken as that for which D^+ is on the left).

A function $\Phi(z)$ defined in the entire plane, except for points on C, will be called sectionally holomorphic if: (i) the function $\Phi(z)$ is holomorphic in each of regions D^+ and D^- except, perhaps, at $z = \infty$; (ii) the function $\Phi(z)$ is sectionally continuous with respect to C, approaching the definite limiting values $\Phi^+(\zeta)$, $\Phi^-(\zeta)$ as z approaches

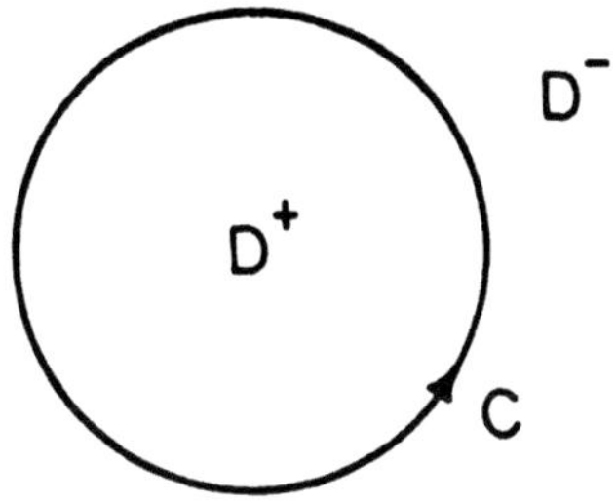

Fig. 1

a point ζ on C from D^+, or D^-, respectively. The classical homogeneous RH problem is defined as follows [2]. Given a contour C, and a function $G(\zeta)$ which is Hölder on C and det $G(\zeta) \neq 0$ on C, find a sectionally holomorphic function $\Phi(z)$, with finite degree at ∞, such that

$$\Phi^+(\zeta) = G(\zeta)\Phi^-(\zeta), \quad \text{on } C, \tag{2.1}$$

where $\Phi^\pm(\zeta)$ are the boundary values of $\Phi(z)$ on C. If $G(\zeta)$ is scalar, (2.1) is solvable in closed form. If $G(\zeta)$ is a matrix valued function, then (2.1) is in general solvable in terms of a system of Fredholm integral equations. Various generalizations of the above RH problem are possible. For example: (i) The contour C may be replaced by a union of intersecting contours. (ii) $G(\zeta)$ may have simple discontinuities at a finite number of points; in this case one allows $\Phi(z)$ to have integrable singularities in the neighbourhood of these points. (iii) RH problems may be considered in other than Hölder spaces (e.g. [3]): (iv) One may consider inhomogeneous RH problems $\Phi^+(\zeta) = G(\zeta)\Phi^-(\zeta) + F(\zeta)$ on C.

The DBAR problem can be defined as follows: Given $\partial\Phi/\partial\bar{z}$, find Φ. If $\partial\Phi/\partial\bar{z} = 0$ everywhere except on a curve, then the DBAR problem reduces to a RH problem (since $\partial\Phi/\partial\bar{z} = \Phi^+ - \Phi^-$, in a distribution sense). The DBAR problem can be solved via the following extension of Cauchy's formula [4]

$$\Psi(z) = \frac{1}{2\pi i}\int_{\mathbb{R}^2} d\bar{\zeta} \wedge d\zeta \, \frac{1}{\zeta - z}\, \frac{\partial\Psi(\zeta)}{\partial\bar{\zeta}} + \frac{1}{2\pi i}\int_C d\zeta \, \frac{\Psi(\zeta)}{\zeta - z}. \tag{2.2}$$

It is interesting that the first RH problem was formulated in connection with an inverse problem (see [12] for references). Actually RH problems are intimately related to solutions of inverse problems in 1+1, 2+1, and 1 dimensions:

3. INVERSE PROBLEMS IN 1+1

We recall that a necessary condition for a given nonlinear equation for $q(x,t)$ to be solvable via IST is that this equation is the compatibility condition of a Lax pair of linear equations. Let us consider the modified KdV equation

$$q_t + q_{xxx} - 6q^2 q_x = 0, \tag{3.1}$$

as an illustrative example. Equation (3.1) is the compatibility condition of

$$\Psi_x(z;x,t) = iz[J_2, \Psi(z;x,t)] + Q\Psi(z;x,t);$$

$$J_2 \doteq \begin{pmatrix} -1 & 0 \\ 0 & 1 \end{pmatrix}, \quad Q \doteq \begin{pmatrix} 0 & q \\ q & 0 \end{pmatrix} \tag{3.2a}$$

$$\Psi_t(z;x,t) = [U_0, \Psi(z;x,t)] + \tilde{Q}\Psi(z;x,t) \tag{3.2b}$$

$$U_0 = \begin{pmatrix} -4iz^3 & 0 \\ 0 & 4iz^3 \end{pmatrix}, \quad \tilde{Q} = \begin{pmatrix} -2izq^2 & 4qz^2+2iq_x z+2q^3-q_{xx} \\ 4qz^2-2iq_x z+2q^3-q_{xx} & 2iq^2 z \end{pmatrix}.$$

We first note that the above Lax pair is **isospectral**, i.e., $\frac{dz}{dt} = 0$. Also it turns out that equation (3.2a) is of primary importance; equation (3.2b) plays only an auxiliary role. To solve the initial value problem for initial data decaying as $|x|\to\infty$, one first formulates an inverse problem for $\Psi(z;x,t)$: **Given appropriate scattering data reconstruct Ψ.**

By studying the analytic properties of ψ with respect to z, where ψ satisfies (3.2a) one establishes the existence of a Ψ which is a sectionally meromorphic function of z, with a jump along the Re z axis. This jump as well as the residues of the poles, are given in terms of appropriate scattering data. **Thus the inverse problem is equivalent to a matrix, regular, continuous RH problem defined along the Re z axis and uniquely specified in terms of scattering data.**

Since in the above discussion we have only used (3.2a), it is evident that one may pose an inverse problem for any function $q(x)$. However, this result is useful for solving the initial value problem for $q(x,t)$

only if q evolves in such a way in t, that the scattering data is known for all t. If Ψ evolves in t according to (3.2b) (i.e., if q solves (3.1)) then it turns out that the evolution of the scattering data with respect to t is simple. Hence, the above RH problem is specified in terms of initial scattering data; its solution yields $\Psi(z;x,t)$ and then (3.2a) gives $q(x,t)$.

We summarize the results of[5,13] concerning mKdV in the case of solitonless potentials.

Proposition 3.1 (Bounded eigenfunctions). A solution of (3.2a) bounded for all complex values of $z = z_R + iz_I$ and tending to I as $z \to \infty$ is given by

$$\Psi(z;x) = \begin{cases} \Psi^+(z;x), & z_I > 0 \\ \Psi^-(z;x), & z_I < 0 \end{cases} \tag{3.3}$$

where $\Psi^{\pm}(z;x)$ satisfy the following integral equations:

$$\Psi^{\pm}(z;x) = I + \int_{-\infty}^{x} d\xi e^{iz(x-\xi)\hat{J}} \pi_{\pm} Q(\xi)\Psi^{\pm}(z;\xi) - \int_{x}^{+\infty} d\xi \; e^{iz(x-\xi)\hat{J}}(\pi_{\mp} + \pi_0)Q\Psi^{\pm} \tag{3.4}$$

where if F and Y are 2 × 2 matrices then

$$e^{\hat{Y}}F = e^{Y}Fe^{-Y}, \quad \pi_+F \doteq \begin{pmatrix} 0 & F_{12} \\ 0 & 0 \end{pmatrix}, \; \pi_-F = \begin{pmatrix} 0 & 0 \\ F_{21} & 0 \end{pmatrix} \tag{3.5}$$

$$\pi_0 F = \mathrm{Diag}(F_{11}, F_{22}).$$

Proposition 3.2 (Departure from Holomorphicity-Scattering Equation). Ψ^+, Ψ^- are holomorphic functions of z for $z_I > 0$, $z_I < 0$ respectively. The departure from holomorphicity for $z = z_R$ is given by

$$\Psi^+(z;x) - \Psi^-(z;x) = \Psi^+ e^{izx\hat{J}}(I - B^{-1}(z)b(z)) \tag{3.6}$$

where

$$B(z) \doteq I + \int_{-\infty}^{\infty} d\xi e^{-iz\xi\hat{J}} \pi_+(Q\Psi^+), \quad b(z) \doteq I + \int_{-\infty}^{\infty} d\xi e^{-iz\xi\hat{J}} \pi_-(Q\Psi^-)$$

so,

$$\Psi^+(z;x)e^{iz\xi\hat{J}}(B^{-1}(z)b(z)) = \Psi^-(z;x). \tag{3.7}$$

Proposition 3.3 (Inverse Problem-Reconstruction of Q)

Q(x) is obtained from

$$Q(x) = [J, \frac{1}{2\pi} \int_{-\infty}^{\infty} dz'\Psi(z';x)e^{iz'x\hat{J}}(I-B^{-1}(z')b(z'))] , \tag{3.8}$$

where $\Psi(z;x)$ solves the following Riemann-Hilbert boundary value problem:

$$\Psi(z;x) = I + \frac{1}{2\pi i} \int_{-\infty}^{\infty} \frac{dz'\Psi(z';x)e^{iz'x\hat{J}}(I-B^{-1}(z')b(z'))}{z' - (z - i0)} . \tag{3.9}$$

Using equation (3.2b) we obtain:

Proposition 3.4 (Evolution of Scattering Data). The evolution of the scattering data from B(z;0), b(z;0) is given by

$$B(z;t) = e^{\hat{U}_0 t} B(z;0), \qquad b(z;t) = e^{\hat{U}_0 t} b(z;0).$$

Since B (resp. b) is a strictly upper (resp. lower) triangular matrix the evolution of the scattering data is given by

$$B_{12}(z;t) = e^{-8iz^3t}B_{12}(z;0), \quad b_{21}(z;t) = e^{8iz^3t}b_{21}(z;0). \tag{3.10}$$

4. INVERSE PROBLEMS IN 2+1

Let us consider the Davey-Stewartson equation (a two dimensional analogue of the nonlinear Schrödinger equation)

$$iQ_t + \frac{1}{2}(\sigma^2 Q_{xx}+Q_{yy}) = -\sigma^2\lambda|Q|^2Q + \Phi Q, \ \Phi_{xx} - \sigma^2\Phi_{yy} = 2\lambda\sigma^2(|Q|^2)_{xx};$$

$$\lambda = \pm 1 \tag{4.1}$$

as an illustrative example. A Lax pair for (4.1) is given by

$$\Psi_x = iz(J\Psi-\Psi J) + q\Psi + \sigma J\Psi_y, \quad J \doteq \begin{pmatrix} 1 & 0 \\ 0 & -1 \end{pmatrix}, \ q \doteq \begin{pmatrix} 0 & Q \\ \bar{Q}\lambda & 0 \end{pmatrix} \tag{4.2a}$$

$$\Psi_t = A_3\Psi_{yy} + A_2\Psi_y + A_1\Psi - z^2(A_3\Psi-\Psi A_{30}) + 2izA_3\Psi_y + izA_2\Psi, \tag{4.2b}$$

where A_1, A_2, A_3, A_{30} are appropriate matrix functions of Q, $\bar{Q}$ (The bar denotes complex conjugate).

The situation is conceptually similar to the case of 1+1: To solve the initial value problem for q(x,y,t) one first formulates an inverse problem for $\Psi(z;x,y,t)$. Depending on the value of σ there exist two different cases (for brevity of presentation we assume non-existence of poles, i.e., non-existence of lumps): (i) $\sigma = 1$. There exists a Ψ which is a sectionally holomorphic function of z and which has a jump along the Re z axis. This jump is also given in terms of scattering data but it depends on them in a non-local way. Thus the inverse problem is equivalent to a **non-local**, matrix continuous RH problem defined along the Re z axis and uniquely specified in terms of scattering data. (ii) $\sigma = -i$. There exists a Ψ which is bounded for all complex z, but which is analytic nowhere in the complex z plane. However, its departure from holomorphicity $\partial\Psi/\partial\bar{z}$ can be expressed in terms of appropriate **inverse data**. Thus, now the inverse problem is equivalent to a $\bar{\partial}$ (DBAR) problem: Given $\partial\Psi/\partial\bar{z}$ reconstruct Ψ .

Using (4.2b), again one shows that the inverse scattering and the inverse data evolve simply in time. Hence, the above RH and $\bar{\partial}$ problems are specified in terms of initial data; their solutions yield $\Psi(z;x,y,t)$ and then (4.2a) gives $q(x,y,t)$.

We summarize the results of [6] concerning DSI ($\sigma = 1$, Proposition 4.1.-4.4) and DSII ($\sigma = -i$, Propositions 4.5-4.8).

Proposition 4.1 (Bounded Eigenfunctions) A solution of (4.2a) with $\sigma = 1$ bounded for all complex values of $z = z_R + iz_I$ and tending to I as $z \to \infty$ is given by

$$\Psi(z;x,y) = \begin{cases} \Psi^+(z;x,y), & z_I > 0 \\ \Psi^-(z;x,y), & z_I < 0 \end{cases} \tag{4.3}$$

where $\Psi^\pm(z;x,y)$ satisfy the following integral equations:

$$\Psi^\pm(z;x,y) = I + \frac{1}{2\pi}\int_{-\infty}^{x} d\xi\, e^{iz(x-\xi)\hat{J}}\int_{-\infty}^{\infty} d\eta \int_{-\infty}^{\infty} dm\, e^{im[(y-\eta)I+(x-\xi)J]}\cdot (\pi_0+\pi_\pm)(q(\xi,\eta)\Psi^\pm(z;\xi,\eta))$$

$$- \frac{1}{2\pi}\int_{x}^{\infty} d\xi\, e^{iz(x-\xi)\hat{J}}\int_{-\infty}^{\infty} d\eta \int_{-\infty}^{\infty} dm\, e^{im[(y-\eta)I + (x-\xi)J]}\cdot \pi_\mp(q(\xi,\eta)\Psi^\pm(z;\xi,\eta)) \tag{4.4$^\pm$}$$

(cf. (3.5) for notation).

Assuming that the linear integral equations $(4.4)^\pm$ have no homogeneous solutions, it follows that:

Proposition 4.2 (Departure from Holomorphicity). Ψ^+, Ψ^- are holomorphic functions of z for $z_I > 0$, $z_I < 0$, respectively. Hence the function $\Psi(z;x,y)$ defined by (4.3) is a sectionally holomorphic function of z. In particular, $\frac{\partial\Psi}{\partial\bar{z}} = 0$ for all z, with $z_I \neq 0$ and $\frac{\partial\Psi}{\partial\bar{z}} = \Psi^+(z;x,y) - \Psi^-(z;x,y)$ for $z = z_R$. The departure from holomorphicity is given by:

$$\Psi^+(z;x,y) - \Psi^-(z;x,y) = \int_{-\infty}^{\infty} dz'\Psi^-(z';x,y)e^{iz'Jx+iz'y} f(z',z)e^{-izJx-izy}, \tag{4.5}$$

for $z = z_R$, where the scattering data $f(z',z)$ are given by:

$$f_{11}(z',z) = -\int_{-\infty}^{\infty} dm\ f_{12}(z',m)f_{21}(m,z),$$

$$f_{12}(z',z) = \frac{1}{2\pi}\int_{-\infty}^{\infty} d\xi\int_{-\infty}^{\infty} d\eta\ Q\Psi^-_{22}\ e^{-(z+z')\xi+i(z-z')\eta} \tag{4.6}$$

$$f_{21}(z',z) = -\frac{1}{2\pi}\int_{-\infty}^{\infty} d\xi\int_{-\infty}^{\infty} d\eta\lambda\ \bar{Q}\Psi^+_{11}\ e^{i(z+z')\xi+i(z-z')\eta},\quad f_{22}=0. \tag{4.7}$$

Proposition 4.3 (Inverse Problem-Reconstruction of the potential q) $q(x,y)$ is obtained from:

$$q(x,y) = -\frac{1}{2\pi}[J, \int_{-\infty}^{\infty} dz' \int_{-\infty}^{\infty} dz\Psi^-(z';x,y)e^{iz'Jx}f(z',z)e^{-izJx+i(z'-z)y}], \tag{4.8}$$

where $\Psi^-(z;x,y)$ solves the following integral equation:

$$\Psi^-(z;x,y) + \frac{1}{2\pi i}\int_{-\infty}^{\infty} dz'' \int_{-\infty}^{\infty} dz' \frac{\Psi'(z'';x,y)e^{iz''Jx}f(z'',z')e^{-iz'Jx+i(z''-z')y}}{z' - z + i0} = I. \tag{4.9}$$

Finally from (4.2b) we obtain the following:

Proposition 4.4 (Evolution of the Scattering Data). The evolution of the scattering data from $t = 0$, $f(z',z;0)$ is given by:

$$f(z',z;t) = e^{-z'^2tA_{30}}\ f(z',z;0)e^{z^2tA_{30}}, \tag{4.10}$$

where

$f(z',z;0)$ is given by (4.7) and $A_{30} = \text{diag}(i,-i)$.

Proposition 4.5 (Bounded Eigenfunctions). A solution of (4.2a) with $\sigma = -i$ bounded for all complex values of $z = z_R + iz_I$ and tending to I as $z \to \infty$ satisfies the following Fredholm linear integral equation

$$\Psi(z;x,y) = I + (G_{z_R,z_I,q}\Psi(z;.,.))(x,y) , \tag{4.11}$$

where

$$\{(G_{z_R,z_I,q}\Psi(z;.,.))\}_{1j} \doteq \frac{1}{2\pi}\left(\int_{-\infty}^{x} d\xi\int_{-\infty}^{c_{1j}z_I} dm\int_{-\infty}^{\infty} d\eta - \int_{x}^{\infty} d\xi\int_{c_{1j}z_I}^{\infty} dm\int_{-\infty}^{\infty} d\eta\right) \times$$

$$\{\exp[(m+i(1-J_j)z)(x-\xi)+im(y-\eta)][q(\xi,\eta)\Psi(z;\xi,\eta)]\}_{1j}, \tag*{(4.12)$_{1j}$}$$

and

$$\{(G_{z_Rz_I,q}\Psi(z;.,.)\}_{2j} \doteq \frac{1}{2\pi}\left(\int_{-\infty}^{x} d\xi\int_{c_{2j}z_I}^{\infty} dm\int_{-\infty}^{\infty} d\eta - \int_{x}^{\infty} d\xi\int_{-\infty}^{c_{2j}z_I} dm\int_{-\infty}^{\infty} d\eta\right)\times$$

$$\{\exp[-(m+i(1+J_j)z)(x-\xi)+im(y-\eta)][q(\xi,\eta)\Psi(z;\xi,\eta)]\}_{2j}, \tag*{(4.12)$_{2j}$}$$

$$c_{1j} = 1-J_j,\ c_{2j} = 1+J_j,\quad j = 1,2.$$

Proposition 4.6 (Departure from Holomorphicity). For every $z \in \mathbf{C}$

$$\frac{\partial\Psi}{\partial\bar z}(z;x,y) = \Psi(\bar z;x,y)\Omega(z_R,z_I;x,y) , \tag{4.13}$$

where the matrix Ω is defined by: $\Omega_{11} = \Omega_{22} = 0$

$$\Omega_{ij} \doteq T_{ij}(z)\exp\Theta_{ij}(z;x,y),\ i \neq j \tag{4.14}$$

$$T_{ij}(z) \doteq \frac{i}{2\pi}\int_{-\infty}^{\infty} d\xi\int_{-\infty}^{\infty} d\eta\{q(\xi,\eta)\Psi(z;\xi,\eta)\}_{ij}\ \exp\{-\Theta_{ij}(z;\xi,\eta)\},\ i \neq j$$

$$\Theta_{12}(z;x,y) \doteq 2i(xz_R + yz_I),\ \Theta_{21}(z;x,y) = 2i(-xz_R+yz_I).$$

Propostion 4.7 (Inverse Problem-Reconstruction of q). q(x,y) is obtained from

$$q(x,y) = [J, \frac{1}{2\pi}\iint_{\mathbf{C}}\Psi(\bar z;x,y)\ \Omega(z_R,z_I;x,y)dz\wedge d\bar z] , \tag{4.15}$$

where $\Psi(z;x,y)$ satisfies:

$$\Psi(z;x,y) - \frac{1}{2\pi i}\iint_{\mathbf{C}}\Psi(\bar z;x,y)\ \Omega(z_R',z_I';x,y)\ \frac{dz'\wedge d\bar z'}{z'-z} = I. \tag{4.16}$$

Finally from equation (4.2b) we obtain:

Proposition 4.8 (Evolution of the Inverse Data). The inverse data at time t, $\Omega(z_R,z_I;x,y,t)$, is given by

$$\Omega(z_R,z_I;x,y,t) = \exp(\bar{z}^2 A_{30}t)\ \Omega(z_R,z_I;x,y,0)\exp(-z^2 A_{30}t) \tag{4.17}$$

where $\Omega(z_R,z_I;x,y,0)$ is given by (4.14) using the initial condition $q(x,y,0)$ and $A_{30} = \mathrm{diag}(i,-i)$.

5. INVERSE PROBLEMS IN 0+1

The Lax pair associated with the PIV equation

$$\frac{d^2y}{dt^2} = \frac{1}{2y}\left(\frac{dy}{dt}\right)^2 + \frac{3}{2}y^3 + 4ty^2 + 2(t^2+\alpha)y + \frac{\beta}{y}, \tag{5.1}$$

is given by

$$Y_z(z) = \left[\begin{pmatrix}1 & 0\\ 0 & -1\end{pmatrix}z + \begin{pmatrix}t & u\\ \frac{2}{u}(v-\Theta_0-\Theta_\infty) & -t\end{pmatrix} + \begin{pmatrix}\Theta_0-v & -\frac{uy}{2}\\ \frac{2v}{uy}(v-2\Theta_0) & -(\Theta_0-v)\end{pmatrix}\frac{1}{z}\right]Y(z), \tag{5.2a}$$

$$Y_t(z) = \begin{pmatrix}1 & 0\\ 0 & -1\end{pmatrix}z + \begin{pmatrix}0 & u\\ \frac{2}{u}(v-\Theta_0-\Theta_\infty) & 0\end{pmatrix}Y(z). \tag{5.2b}$$

Indeed $Y_{zt} = Y_{tz}$ implies

$$\begin{aligned}\frac{dy}{dt} &= -4v + y^2 + 2ty + 4\Theta_0, \quad \frac{du}{dt} = -u(y+2t),\\ \frac{dv}{dt} &= -\frac{2}{y}v^2 + \left(\frac{4\Theta_0}{y} - y\right)v + (\Theta_0+\Theta_\infty)y,\end{aligned} \tag{5.3}$$

where,

$$\alpha = 2\Theta_\infty - 1, \quad \beta = -8\Theta_0^2.$$

As in the cases of 1+1 and 2+1, solving the initial value problem of PIV reduces to solving an inverse problem for Y: **Reconstruct Y(z;t) in terms of appropriate monodromy data.** Again this inverse problem will be solved in terms of a RH problem. Thus it is essential to study the analytic properties of Y with respect to z. However, in contrast to the analogous problem in IST for 1+1 and 2+1, this task here is straightforward: Equation (5.2a) is a linear ODE in z, therefore its analytic structure is completely determined by its singular points. In this particular case $z = 0$ is a regular singular point and $z = \infty$ is an irregular singular point of rank 2. Complete information about $z = \infty$ is provided by the monodromy matrix M_∞ and by the Stokes multipliers a, b, c, d. Solutions of (5.2a), Y_0 and Y_1, normalized at zero and infinity respectively are related via a connection matrix E_0 with entries α_0, β_0, γ_0, δ_0. Taking into consideration the above singularities, there exists a sectionally holomorphic function Y, with jumps across the four rays, $\arg z = -\frac{\pi}{4}, \frac{\pi}{4}, \frac{3\pi}{4}, \frac{5\pi}{4}$ and with singularities at $z = 0$, $z = \infty$. The jumps are specified by a, b, c, d and the nature of singularities by M_0, M_∞. This leads to a matrix, singular, discontinuous RH problem, defined on the above rays and specified in terms of the monodromy data

$$\text{Monodromy Data (MD)} = \{a,b,c,d, \alpha_0,\beta_0,\gamma_0,\delta_0\}.$$

A consistency condition of the above RH problem yields

$$\left(\prod_{j=1}^{4} G_j\right) M_\infty = E_0^{-1} M_0^{-1} E_0,$$

where G_j are the Stokes matrices uniquely defined in terms of the Stokes multipliers. Using (5.5) and certain similarity arguments it can be shown that all MD can be expressed in terms of two of them. Furthermore, equation (5.2b) implies that the MD are time invariant. Hence the above basic RH is specified in terms of two initial parameters (these two initial parameters are obtained from the two initial data of PIV). The solution of this RH problem yields $Y(z;t)$ and hence (5.2a) yields $y(t)$.

The above basic RH problem can be simplified considerably: (i) Assume $0 \le \Theta_0 < 1$, $0 \le \Theta_\infty < 1$, $\Theta_0 \neq \frac{1}{2}$; then the above RH problem is regular. It is interesting that the basic RH problem can be used to obtain Schlessinger transformations which shift Θ_0 and Θ_∞ by a half-integer. By using these transformations the general case is reduced to the regular case. (ii) The basic RH problem can be mapped to a sequence of two RH problems, one on the line $\arg z = \frac{\pi}{4}$ and the other on the line $\arg z = -\frac{\pi}{4}$. The first one is continuous (both at $x = 0$ and $x = \infty$); furthermore, it can be solved in closed form. The second one is discontinuous both at $x = 0$ and $x = \infty$. By using standard auxiliary functions one maps the discontinuous problem to a continuous one. Then the theory of continuous RH problems on simple contours can be used to establish uniqueness and existence of solutions. Elementary solutions of PIV, expressible in terms of Weber-Hermite functions are natually obtained within the above formalism. We summarize the results of [7] concerning PIV.

Proposition 5.1 (Direct Problem). Let Y_0 be the solution of (5.2a) analytic in the neighbourhood of $z = 0$ and normalized by the requirements that $\det Y_0 = 1$ and that Y_0 also solves (5.2b). Let Y_j, $j = 1,\dots,4$ be solutions of (5.2a) analytic in the neighbourhood of infinity such that $\det Y_j = 1$ and $Y_j \sim Y_\infty$ as $|x| \to \infty$ in S_j, where $\sim$ denotes asymptotics, Y_∞ is the formal solution matrix of (5.2a) in the neighbourhood of infinity, and the sectors S_j are given by

$$S_1:\ -\frac{\pi}{4} \le \arg z < \frac{\pi}{4},\quad S_2:\ \frac{\pi}{4} \le \arg z < \frac{3\pi}{4},$$

$$S_3:\ \frac{3\pi}{4} \le \arg z < \frac{5\pi}{4},\quad S_4:\ \frac{5\pi}{4} \le \arg z < \frac{7\pi}{4}.$$

The rays $C_1,\dots,C_4$ are defined by $\arg z = \frac{\pi}{4}, \frac{\pi}{4}, \frac{3\pi}{4}, \frac{5\pi}{4}$ respectively.

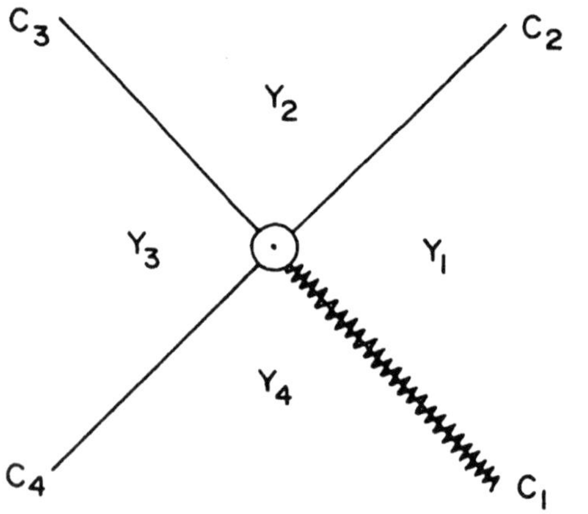

Fig. 5.1

Then the analytic functions $Y_0, Y_1,\dots,Y_4$ satisfy:

(i) $Y_0(z) \sim \hat{Y}_0(z) z^{D_0}$ as $z \to 0$; $D_0 \doteqdot \mathrm{Diag}(\Theta_0, -\Theta_0)$, $\Theta_0 \neq \frac{n}{2}$, $n \in \mathbb{Z}$,

where $\hat{Y}_0(z)$ is holomorphic at $z = 0$. (If $\Theta_0 = n/2$, $Y_0(z)$ has a logarithmic singularity.)

(ii) $Y_j(z) \sim \hat{Y}_\infty(z) e^{Q(z)} (1/z)^{D_\infty}$ as $|z| \to \infty$, z in S_j, $D_\infty \doteqdot \mathrm{Diag}(\Theta_\infty, -\Theta_\infty)$,

$Q(z) \doteqdot \mathrm{Diag}(q, -q)$, $q(z,t) \doteqdot \frac{z^2}{2} + zt$, $\hat{Y}_\infty(z)$ is holomorphic at $z = \infty$.

(iii) $Y_0(ze^{2i\pi}) = Y_0(z)M_0, \quad M_0 \doteq \begin{pmatrix} \exp(2i\pi\Theta_0) & 2i\pi J\exp(2i\pi\Theta_0) \\ 0 & \exp(-2i\pi\Theta_0) \end{pmatrix}$, (5.4)

$J = 0$ if $\Theta_0 \neq \frac{n}{2}$, $J = 1$ if $\Theta_0 = \frac{n}{2}$.

(iv) $Y_2(z) = Y_1(z)G_1, \quad Y_3(z) = Y_2(z)G_2, \quad Y_4(z) = Y_3(z)G_3,$

$$Y_1(z) = Y_4(ze^{2i\pi})G_4M_\infty, \tag{5.5}$$

where

$$G_1 \doteq \begin{pmatrix} 1 & 0 \\ a & 1 \end{pmatrix}, \quad G_2 \doteq \begin{pmatrix} 1 & b \\ 0 & 1 \end{pmatrix}, \quad G_3 \doteq \begin{pmatrix} 1 & 0 \\ c & 1 \end{pmatrix},$$

$$G_4 \doteq \begin{pmatrix} 1 & d \\ 0 & 1 \end{pmatrix}, \quad M_\infty \doteq \exp(2i\pi D). \tag{5.6}$$

(iv) $Y_1(z) = Y_0(z)E_0, \quad E_0 \doteq \begin{pmatrix} \alpha_0 & \beta_0 \\ \gamma_0 & \delta_0 \end{pmatrix}, \quad \det E_0 = 1.$ (5.7)

Furthermore, the parameters

$$MD \doteq \{a,b,c,d, \alpha_0, \beta_0, \gamma_0, \delta_0\} \tag{5.8}$$

satisfy the following consistency condition.

(vi) $\left(\prod_{j=1}^{4} G_j\right)M_\infty = E_0^{-1}M_0^{-1}E_0.$ (5.9)

Proposition 5.2 (Properties of Monodromy Data)

(i) The monodromy data, MD, given by (5.8) and defined in Proposition 5.1, are time-invariant.

(ii) All of the MD can be expressed in terms of two of them.

(iii) $(1+bc)\exp(2i\pi\Theta_\infty) + [ad + (1+cd)(1+ab)]\exp(-2i\pi\Theta_\infty) = 2\cos 2\pi\Theta_\infty.$ (5.10)

In what follows we formulate a RH problem for the case that $0 \leq \Theta_0 < 1$, $0 < \Theta_\infty < 1$. This assumption leads to a **regular** RH problem. The general case follows by considering this result and Schlessinger transformations.

Theorem 5.1 (Inverse Problem). Consider the following matrix, regular homogeneous RH problem along the four rays $C_1,\dots,C_4$ (Figure 5.1): Determine the sectionally holomorphic function $\Psi(z)$, $\Psi(z) = \Psi_j(z)$ if z is in S_j, $j = 1,\dots,4$, from the following conditions:

1. Ψ_j satisfy the jump conditions

$$\Psi_2(\zeta) = \Psi_1(\zeta)g_1(\zeta),\ \Psi_3(\zeta) = \Psi_2(\zeta)g_2(\zeta),\ \Psi_4(\zeta) = \Psi_3(\zeta)g_3(\zeta),$$

$$\Psi_1(\zeta) = \Psi_4(\zeta\, e^{2i\pi})g_4(\zeta) \tag{5.11}$$

along the rays C_2, C_3, C_4, C_1 respectively, where

$$g_j \doteq e^Q G_j e^{-Q},\quad j = 1,2,3,\quad g_4 \doteq e^Q G_4 e^{-Q} M_\infty. \tag{5.12}$$

2. $$\Psi(z) \sim \left(\frac{1}{z}\right)^{D_\infty}\left(I + O\left(\frac{1}{z}\right)\right) \text{ as } |z| \to \infty. \tag{5.13}$$

3. $\Psi(z)$ has at most an integrable singularity at the origin with a monodromy matrix given by

$$\Psi_1(ze^{2i\pi}) = \Psi_1(z)E_0^{-1}M_0E_0,\quad z \to 0. \tag{5.14}$$

In the above, G_j, Q, M_∞, D_∞, M_0 are defined in Proposition 5.1.

4. The monodromy data MD, given by (5.8), satisfy the properties given in Proposition 5.2(ii). Then:

(i) The above RH problem is discontinuous both at the origin and at infinity. Actually

$$\prod_{j=1}^{4} g_j \sim E_0^{-1}M_0^{-1}E_0,\ z \to 0;\qquad \prod_{j=1}^{4} g_j \sim M_\infty,\quad z \to \infty. \tag{5.15}$$

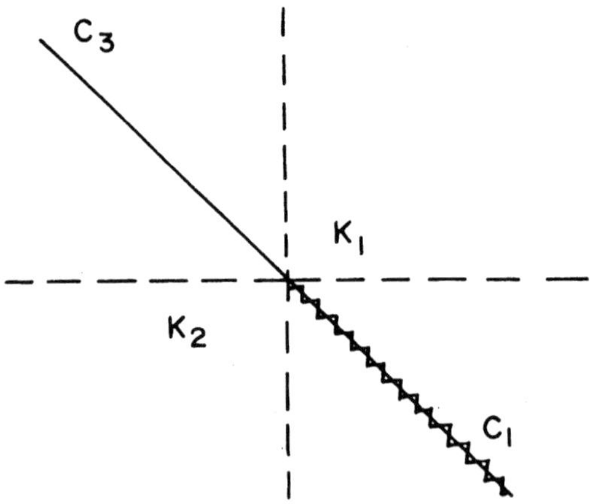

Fig. 5.2

(ii) To obtain the solution of the above RH problem consider the following RH problem along the contour $C_1 + C_3$: Determine the sectionally holomorphic function $K(z)$, $K(z) = K_1(z)$ if z in $S_1 + S_2$, $K(z) = K_2(z)$ if z in $S_3 + S_4$, from the following conditions:

1. K_j satisfy the jump condition

$$K_1 = K_2 \begin{cases} h \begin{pmatrix} 1 & de^{2q} \\ 0 & -a/c \end{pmatrix} M_\infty h^{-1} \text{ on } C_1, \\ h \begin{pmatrix} 1 & -be^{2q} \\ 0 & -a/c \end{pmatrix} h^{-1} \text{ on } C_3 \end{cases} \qquad h(z) \doteqdot \begin{pmatrix} 1 & 0 \\ a\rho(z) & 1 \end{pmatrix},$$

$$\rho(z) \doteqdot -\frac{1}{2\pi i} \int_{C_2+C_4} \frac{d\zeta\, e^{-2q(\zeta)}}{\zeta - z}. \tag{5.16}$$

(If h_1, h_2 denote h in $S_2 + S_3$ and $S_4 + S_1$ respectively then $h=h_1$ on C_1, $h = h_2$ on C_3.)

2. $K(z) \sim \left(\frac{1}{z}\right)^{D_\infty}(I + 0(\frac{1}{z}))$ as $|z| \to \infty$. (5.17)

3. $K(z)$ has at most an integrable singularity at the origin with a monodromy matrix given by

$$K(ze^{2i\pi}) = K(z)h_1(0)E_0^{-1}M_0E_0h_1^{-1}(0), \quad z \to 0. \tag{5.18}$$

The above RH is discontinuous both at the origin and at infinity. Actually if g_{K_1}, g_{K_3} denote the jump matrices along C_1, C_3 respectively then

$$g_{K_3}^{-1}g_{K_1} \sim h_1(0)E_0^{-1}M_0^{-1}E_0h_1^{-1}(0),\ z \to 0; \quad g_{K_3}^{-1}g_{K_1} \sim M_\infty,\ z \to \infty. \tag{5.19}$$

However, the above RH problem can be mapped to a continuous one using the auxiliary functions

$$\left(\frac{z}{z\pm 1}\right)^{\pm\theta_0}, \quad \left(\frac{1}{z\pm 1}\right)^{\pm\theta_\infty}, \tag{5.20}$$

to remove the above singularities.

Ψ is related to K via:

$$\Psi = Kh \text{ if } z \text{ in } S_1+S_2; \quad \Psi = KhM, \quad M \doteq \text{Diag}(1,-a/c), \tag{5.21}$$

if z in S_3+S_4

(i.e., $\Psi_1 = K_1h_1$, $\Psi_2 = K_1h_2$, $\Psi_3 = K_2h_1M$, $\Psi_4 = K_2h_1M$).

Proposition 5.3 (The Solution of PIV). Let $\Psi(z)$ be the solution matrix of the inverse problem formulated in Theorem 5.1. Then y(t),

$$y(t) = -\left(\frac{1}{u}\frac{du}{dt} + 2t\right), \quad u \doteq -2\lim_{|z|\to\infty} \Psi_{21}(z)e^{-2q(z)}, \tag{5.22}$$

solves PIV.

6. INVERSE PROBLEMS IN n SPATIAL DIMENSIONS, $n > 2$

Consider the inverse problem associated with the following system of N first-order equations in n+1 dimensions:

$$\Psi_{x_0} + \sigma \sum_{\ell=1}^{n} J_\ell \Psi_{x_\ell} = q\Psi, \ \sigma = \sigma_R + i\sigma_I, \quad \sigma_I \neq 0, \ n > 1, \tag{6.1}$$

where $q(x_0,x)$ is an $N \times N$ matrix-valued off-diagonal function in $\mathbb{R}^{n+1}$, decaying suitably fast for large x_0, x, and the J_ℓ are constant real diagonal $N \times N$ matrices (we denote the diagonal entries of J_ℓ by $J_\ell^1, \ldots, J_\ell^N$). Alternatively, using the transformation

$$\Psi(z,x_0,x) = \mu(z,x_0,x)\exp\left[i \sum_{\ell=1}^{n} z_\ell(x_\ell - \sigma x_0 J_\ell)\right], \quad z \in \mathbb{C}^n, \tag{6.2}$$

equation (2.13) becomes

$$\mu_{x_0} + \sigma \sum_{\ell=1}^{n} (J_\ell \mu_x + iz_\ell[J_\ell,\mu]) = q\mu. \tag{6.3}$$

We assume that $n \leq N$, otherwise the entries of the J_ℓ matrices will be linearly related and one can always reduce n by a change of coordinates. An inverse problem in this case is defined as follows: **Given appropriate inverse data T, where T is an $N \times N$ matrix-valued off-diagonal function of suitable inverse parameters, reconstruct the potential q.** Again there exists a μ which is bounded for all complex z, $z \in \mathbb{C}^n$. $\partial\mu/\partial\bar{z}$ depends on appropriate inverse data $T(z_R,z_I,m_2,\ldots,m_n)$, $z_R \in \mathbb{R}^n$, $z_I \in \mathbb{R}^n$, $m_\ell \in R$. T satisfies $\frac{\partial^2 T}{\partial\bar{z}_i \partial\bar{z}_j} = \frac{\partial^2 T}{\partial\bar{z}_j \partial\bar{z}_i}$. Using this equation and introducing Born variables,

$$z, m \Leftrightarrow w_0, w, \chi ; \quad w_0 \in \mathbb{R}, \quad w \in \mathbb{R}^n, \quad \chi \in \mathbb{C}^{n-1}, \tag{6.4}$$

one obtains a characterization equation for the inverse data:

$$\hat{T}^{ij}(w_0,w) \doteq T^{ij}(w_0,w,\chi) - \frac{1}{\pi}\int_{\mathbb{R}^2} \frac{d\chi'_{P_R}\, d\chi'_{P_I}\, N^{ij}_{IP}[T](w_0,w,\chi'_P)}{\chi_P - \chi'_P}, \tag{6.5}$$

where N is a quadratic function of T. That is, $T^{ij}(z,m)$ is appropriate inverse data iff the right-hand side of (6.5) is independent of χ. Hence, equation (6.5) serves as both characterizing T^{ij} and defining $\hat{T}^{ij}$. This equation was first introduced by Nachman and Ablowitz [8]. Using equation (6.5) and taking the limit of μ as $|\chi| \to \infty$ we show that the general problem of reconstructing an $N \times N$ potential q in n+1 spatial dimensions, is reduced to one of reconstructing a 2×2 potential with entries q^{ij}, q^{ji} in two dimensions. The inverse data needed for this reconstruction is precisely $\hat{T}^{ij}, \hat{T}^{ji}$. This reduction makes crucial use of the existence of redundant scattering parameters. In this sense it is the analog of the Born approximation. However, the crucial difference is that while in the inverse scattering of the multidimensional Schrödinger equation one can reconstruct the potential in closed form, here one can only reduce the general problem to one for 2×2 matrices in two dimensions. This reduced problem was solved in [6]. In the following, we summarize the results of [9].

Proposition 6.1 (Bounded Eigenfunctions). The function $\mu\ (x_0,x,z)$ defined below, solves equation (6.3), is bounded for all complex values of z and tends to I for large z:

$$\mu^{ij}(x_0,x,z) = \delta^{ij} + \frac{\operatorname{sgn}(\sigma_I J^i_1)}{2\pi i}\int_{\mathbb{R}^2} d\xi_0 d\xi_1 \frac{\exp[i\beta^{ij}(x_0-\xi_0,x_1-\xi_1,z)]}{(x_1-\xi_1) - \sigma J^i_1(x_0-\xi_0)}$$

$$(q\mu)^{ij}(\xi_0,\xi_1,x_2-(x_1-\xi_1)J^i_2/J^i_1,\ldots,x_n-(x_1-\xi_1)J^i_n/J^i_1,z), z \in \mathbb{C}^n, \tag{6.6}$$

where β^{ij} is defined by

$$\beta^{ij}(x_0,x_1,z) \doteq \sum_{\ell=1}^{n} \frac{J^i_\ell - J^j_\ell}{\sigma_I}\left[x_0|\sigma|^2 z_{\ell_I} - \frac{x_1(\sigma z_\ell)_I}{J^i_1}\right], z_\ell = z_{\ell_R} + i z_{\ell_I}.$$

Equivalently μ_{ij} satisfies

$$\mu^{ij}(x_0,x,z) = \delta^{ij} + \frac{\operatorname{sgn}(\sigma_I J_1^i)}{2\pi i} \int_{\mathbb{R}^{n+1}} d\xi_0 d\xi [c_{n-1}\int_{\mathbb{R}^{n-1}} dm^2 e^{i\alpha^i(x-\xi,m)}]$$

$$\frac{\exp[i\beta^{ij}(x_0-\xi_0,x_1-\xi_1,z)](q\mu)^{ij}(\xi_0,\xi,z)}{x_1-\xi_1-\sigma J_1^i(x_0-\xi_0)}, \tag{6.7}$$

where

$$dm^2 \doteq dm_2 \ldots dm_n,\ \alpha^i(x,m) \doteq \sum_{\ell=2}^{n} m_\ell (x_\ell - x_1 \frac{J_\ell^i}{J_1^i}),\ c_n \doteq \frac{1}{(2\pi)^n}. \tag{6.8}$$

Proposition 6.2 (Departure from Holomorphicity). Let μ^{ij} be defined by eq. (6.5). Then

$$\frac{\partial\mu}{\partial\bar{z}_p}(x_0,x,z) = \sum_{i,j} \gamma^i (J_p^i - J_p^j)\exp[i\beta^{ij}(x_0,x_1,z)]$$

$$\times\ c_{n-1}\int_{\mathbb{R}^{n-1}} dm^2 \exp[i\alpha^i(x,m)]T^{ij}(z,m)\mu(x_0,x,\lambda^{ij}(z,m))E_{ij}, \tag{6.9}$$

where $\beta^{ij}(x_0,x_1,z)$, $\alpha^i(x,m)$ are defined by (6.6),(6.8) respectively; E_{ij} is an $N \times N$ matrix with zeros in all its entries except the ij^{th}, which equals 1; and λ^{ij} and T^{ij} are given by

$$\lambda_1^{ij}(z,m) \doteq (z_{1_R}^{ij} - \sum_{\ell=2}^{n} m_\ell \frac{J_\ell^i}{J_1^i}, z_{1_I}),\quad \lambda_r^{ij}(z,m) = (z_{r_R}+m_r, z_{r_1});\ r=2,..,n.$$

$$\gamma^i \doteq \bar{\sigma}/4\pi i|J_1^i\sigma_I|,$$

$$T^{ij}(z,m) \doteq \int_{\mathbb{R}^{n+1}} d\xi_0 d\xi \exp[-i\beta^{ij}(\xi_0,\xi_1,z)-i\alpha^i(\xi,m)](q\mu)^{ij}(\xi_0,\xi,z). \tag{6.10}$$

Proposition 6.3 (Characterization of T)

(a) Assume that $\partial\mu/\partial\bar{z}_p$ is given by Eq. (6.9) and the $T^{ij}(z,m)$ is given by (6.10). Then

$$L_{rp}^{ij}T^{ij}(z,m) = -\sum_{\ell=1}^{n} c_{n-1}\int_{\mathbb{R}^{n-1}} dM^2 T^{i\ell}(\lambda^{\ell j}(z,M),m-M)T^{\ell j}(z,m)$$

$$\times[(J_p^\ell - J_p^j)(J_r^i - J_r^\ell) - (J_r^\ell - J_r^j)(J_p^i - J_p^\ell)] \doteq N_{rp}^{ij}[T](z,m), \tag{6.11}$$

where

$$L^{ij}_{rp} \doteq (J^i_p - J^j_p)\frac{\partial}{\partial \bar{z}_r} - (J^i_r - J^j_r)\frac{\partial}{\partial \bar{z}_p}. \tag{6.12}$$

(b) Assume that $\partial\mu/\partial\bar{z}_p$ is given by Eq. (6.9) and that $\partial^2\mu/\partial\bar{z}_r\partial\bar{z}_p$ is symmetric with respect to r, p. Then $T^{ij}(z,m)$ solves (6.11).

Following A. Nachman and M. J. Ablowitz we introduce appropriate Born variables. Then equation (6.11) can be integrated. Furthermore, we can compute the limit of T^{ij} in the new coordinates as $|\chi_p| \to \infty$ (see below):

Let w_0^{ij}, w_1^{ij}, w_ℓ, $\ell = 2,\ldots,n \in \mathbb{R}^1$ and $\chi_\ell \in \mathbb{C}^1$, $\ell = 2,\ldots,n$, be defined by

$$w_0^{ij} \doteq \sum_{r=1}^{n} \frac{J^i_r - J^j_r}{\sigma_I}|\sigma|^2 z_{r_I}, \quad w_1^{ij} \doteq -\sum_{r=1}^{n} \frac{J^i_r - J^j_r}{\sigma_I J^i_1}(\sigma z_r)_I - \sum_{r=2}^{n} m_r \frac{J^i_r}{J^i_1},$$

$$w_\ell \doteq m_\ell, \qquad \chi^{ij}_\ell \doteq \frac{z_\ell}{J^j_1 - J^i_1}, \; \ell = 2,\ldots,n. \tag{6.13}$$

Assume that

$$(J^r_1 - J^j_1)(J^i_p - J^j_p) \neq (J^i_1 - J^j_1)(J^r_p - J^j_p), \text{ for all distinct } i,j,r \text{ and } p \neq 1. \tag{6.14}$$

For convenience of writing we usually suppress the superscripts, i,j in w_0, w_1, χ. Let z denote $z_1,\ldots,z_n$, m denote $m_2,\ldots,m_n$, χ denote $\chi_2,\ldots,\chi_n$, w denote $w_1,\ldots,w_n$. Then we have the following.

(a) The inverse of the transformation $z,m \to w_0$, w, χ is given by

$$z_\ell = \chi_\ell(J^j_1 - J^i_1), \quad m_\ell = w_\ell, \; \ell = 2,\ldots,n, \quad z_1 = -\sum_{r=2}^{n}(J^j_r - J^i_r)\chi_r + \frac{(\bar{\sigma}/|\sigma|^2)w_0 + \sum_{r=1}^{n} w_r J^i_r}{J^j_1 - J^i_1}. \tag{6.15}$$

(b) In the new coordinates, Eq. (6.11) with r = 1 becomes

$$\frac{\partial T^{ij}}{\partial \bar{\chi}_p}(w_0,w,\chi) = N^{ij}_{1p}[T](w_0,w,\chi), \; p = 2,\ldots,n. \tag{6.16}$$

(c) In the new coordinates,

$$T^{ij}(w_0,w,\chi) = \int_{\mathbb{R}^{n+1}} d\xi_0 d\xi \, \exp[-i(w_0\xi_0+w\xi)](q\mu)^{ij}(\xi_0,\xi,w_0,w,\chi),$$

where

$$w\xi = \sum_{r=1}^{n} w_r \xi_r. \tag{6.17}$$

(d) Let

$$\mu_i^{\ell j} \doteq \mu^{\ell j}(x_0,x,w_0^{ij},w^{ij}, \chi^{ij}), \qquad \hat{\mu}_i^{\ell j} = \lim_{|\chi_p|\to\infty} \mu_i^{\ell j}. \tag{6.18}$$

Then the $\hat{\mu}_i^{\ell j}$ satisfy

$$\hat{\mu}_i^{ij}(x_0,x,w_0,w) = \frac{\operatorname{sgn}(\sigma_I J_1^i)}{2\pi i} c_{n-1} \int_{\mathbb{R}^{2n}} \times$$

$$\frac{dx_0' dx' dw^2 \exp[i\{(x_0-x_0')w_0+w_0+(x-x')w\}]}{x_1-x_1'-\sigma J_1^i(x_0-x_0')} q^{ij}(x_0',x')\hat{\mu}_i^{ij}(x_0',x',w_0,w),$$

$$\hat{\mu}_i^{jj}(x_0,x,w_0,w) = 1 + \frac{\operatorname{sgn}(\sigma_I J_1^j)}{2\pi i} c_{n-1} \int_{\mathbb{R}^{2n}} \times$$

$$\frac{dx_0' dx' dw^2 q^{ji}(x_0',x')\hat{\mu}_i^{ij}(x_0',x',w_0,w)}{x_1-x_1' - \sigma J_1^j(x_0-x_0')}, \quad \hat{\mu}_i^{\ell j} = 0, \text{ for all } \ell,\ \ell\neq i, \ell\neq j. \tag{6.19}$$

(e) $$\lim_{|\chi_p|\to\infty} T^{ij}(w_0,w,\chi) = \int_{\mathbb{R}^{n+1}} d\xi_0 d\xi \, \exp[-i(w_0\xi_0+w\xi)] \times$$

$$q^{ij}(\xi_0,\xi)\hat{\mu}_i^{ij}(\xi_0,\xi,w_0,w) \doteq \hat{T}^{ij}(w_0,w). \tag{6.20}$$

(f) The basic characterization equation is given by

$$\hat{T}^{ij}(w_0,w) = T^{ij}(w_0,w,\chi) - \frac{1}{\pi} \int_{\mathbb{R}^2} \frac{d\chi_{p_R}' d\chi_{p_I}' N_{1p}^{ij}[T](w_0,w,\chi^{p'})}{\chi_p - \chi_p'}, \tag{6.21}$$

where $\chi^{p'}$ denotes $\chi_2,\ldots,\chi_{p-1},\chi_p',\chi_{p+1},\ldots,\chi_n$.

It follows from the above that as $|\chi_p|\to\infty$, the μ^{ij}'s decouple. Furthermore, the $\hat{\mu}_i^{ij}$, $\hat{\mu}_i^{jj}$ satisfy a system of two equations depending on q^{ij}, q^{ji}. It turns out that: (a) By introducing appropriate spatial variables ξ, the $\hat{\mu}_i^{ij}$, $\hat{\mu}_i^{jj}$ satisfy equations in two spatial dimensions. (b) The inverse data needed to reconstruct $\hat{\mu}_i^{ij}$, μ_i^{jj} (and hence q^{ij}, q^{ji}) can be obtained from $\hat{T}^{ij}$:

Proposition 6.4 (Reconstruction of q). Let

$$\alpha_r \doteq \frac{J_2^j J_r^i - J_r^j J_2^i}{J_1^i J_2^j - J_1^j J_2^i}, \quad \beta_r = \frac{J_1^i J_r^j - J_1^j J_r^i}{J_1^i J_2^j - J_1^j J_2^i}, \quad r = 1,\ldots,n, \tag{6.22}$$

where for convenience of writing we have suppressed the dependence of

α_r, β_r on i,j. Let $\xi_0 \in \mathbb{R}$, $\xi \in \mathbb{R}^n$,

$$x_0 = \xi_0, \quad x_1 = \xi_1, \quad x_2 = \xi_2, \tag{6.23}$$

$$x_\ell = \xi_\ell + \alpha_\ell \xi_1 + \beta_\ell \xi_2, \quad \ell = 3,\ldots,n.$$

Then we have the following:

(a) The system (6.19) becomes

$$\hat{\mu}_i^{ij}(\xi_0,\xi,\hat{z}) = \frac{\operatorname{sgn}(\sigma_I J_1^i)}{2\pi i} \int_{\mathbb{R}^2} d\xi_0' d\xi_1' [\xi_1 - \xi_1' - \sigma J_1^i(\xi_0 - \xi_0')]^{-1}$$

$$\times \exp[i\hat{\beta}^{ij}(\xi_0-\xi_0',\xi_1-\xi_1',\hat{z})] q^{ij} \hat{\mu}_i^{jj}(\xi_0',\xi_1',\xi_2 - (\xi_1-\xi_1')\frac{J_2^i}{J_1^i},\xi_3,\ldots,\xi_n,z),$$

$$\hat{\mu}_i^{jj}(\xi_0,\xi,\hat{z}) = 1 + \frac{\operatorname{sgn}(\sigma_I J_1^j)}{2\pi i} \int_{\mathbb{R}^2} d\xi_0' d\xi_1' [\xi_1 - \xi_1' - \sigma J_1^j(\xi_0-\xi_0')]^{-1}$$

$$\times q^{ji} \hat{\mu}_i^{ij}(\xi_0',\xi_1',\xi_2 - (\xi_1-\xi_1')\frac{J_2^j}{J_1^j}, \xi_3,\ldots,\xi_n,\hat{z}), \tag{6.24}$$

where

$$\hat{z} \doteq \sum_{r=1}^{n} (z_r \alpha_r + \frac{J_2^j - J_1^j}{J_1^j - J_1^i} z_r \beta_r),$$

$$\hat{\beta}^{ij}(x_0,x_1,\hat{z}) \doteq \frac{J_1^i - J_1^j}{\sigma_I}[x_0|\sigma|^2 z_I - x_1 \frac{(\sigma z)_I}{J_1^i}]. \tag{6.25}$$

(b) $\hat{T}^{ij}$ in the new coordinates becomes

$$\hat{T}^{ij}(\hat{z},\hat{m}) = \int_{\mathbb{R}^{n+1}} d\xi_0' d\xi' \exp[-i\hat{\beta}^{ij}(\xi_0',\xi_1',\hat{z}) +$$

$$+ \hat{m}_2(\xi_2' - \xi_1' \frac{J_2^i}{J_1^i}) + \sum_{r=3}^{n} \hat{m}_r \xi_r'] q^{ij}\hat{\mu}_i^{jj}(\xi_0',\xi',\hat{z}), \qquad (6.26)$$

where

$$\hat{m}_2 \doteq m_2 + \sum_{r=3}^{n} m_r \beta_r, \ \hat{m}_\ell = m_\ell, \quad \ell = 3,\ldots,n. \qquad (6.27)$$

(c) The inverse data associated with (6.24) and the analogous problem for $\hat{\mu}_j^{ji}$, $\hat{\mu}_j^{ii}$ are given by T^{ij}, T^{ji}. Let

$$T^{ij}(z,\xi_2-\xi_1 \frac{J_2^i}{J_1^i}, \xi_3,\ldots,\xi_n) \doteq c_{n-1} \int_{\mathbb{R}^{n-1}} d\hat{m} \exp[i\hat{m}_2(\xi_2-\xi_1 \frac{J_2^i}{J_1^i})$$

$$+ i \sum_{r=3}^{n} \hat{m}_r \xi_r \hat{T}^{ij}(\hat{z},\hat{m}). \qquad (6.28)$$

Then

$$T^{ij}(\hat{z},\xi_2-\xi_1 \frac{J_2^i}{J_1^i},\xi_3,\ldots,\xi_n) \doteq \int_{\mathbb{R}^2} d\xi_0' d\xi_1' \exp[-i\hat{\beta}^{ij}(\xi_0',\xi_1',z)]$$

$$\times (q^{ij}\hat{\mu}_i^{jj})(\xi_0',\xi_1',\xi_2 - (\xi_1-\xi_1') \frac{J_2^i}{J_1^i}, \xi_3,\ldots,\xi_n,\hat{z}). \qquad (6.29)$$

Equations (6.1)-(6.3) with $\sigma = -1$ lead to a system which appears to be physically more interesting: (a) Since the system is hyperbolic one may consider the physically important question of inverse scattering (IS); i.e., given a scattering amplitude function $S(\lambda,k)$ find the potential $q(x_0,x)$. (b) A special case of the above system, namely if the J_ℓ's are constrained by

$$\frac{J_p^\ell - J_p^j}{J_r^\ell - J_r^j} = \frac{J_p^i - J_p^j}{J_r^i - J_r^j}, \quad p,\ r = 1,\ldots,n, \quad 1,j,\ell = 1,\ldots,N, \qquad (6.30)$$

is associated with the N-wave interaction in n+1 spatial and one temporal dimensions [10]. The above system can be considered as a limiting case of (6.1)-(6.3) [8]. Alternatively, it can be considered on its own right [11]; the problem of reconstruction can be reduced to one for a 2 × 2 matrix problem in two spatial dimensions.

ACKNOWLEDGEMENTS

This work was partially supported by the Office of Naval Research under Grant Number N00014-76-C-0867 and the National Science Foundation under Grant Number DMS-8501325. A. S. Fokas thanks Professor Lakshmanan and his school for their warm hospitality.

REFERENCES

[1] R. G. Newton, **Scattering Theory of Waves and Particles** (Springer, Berlin, 1982).
[2] F. D. Gakhov, **Boundary Value Problems** (Pergamon, New York, 1966); N. P. Vekua, **Systems of Singular Integral Equations** (Gordon and Breach, New York, 1967).
[3] S. Prossdorf, **Some Classes of Singular Equations** (North-Holland, Amsterdam, 1978).
[4] L. Hörmander, **Complex Analysis in Several Variables** (Van Nostrand, New York, 1966).
[5] R. Beals and R. R. Coifman, Comm. Pure Appl. Math. **37** (1984) 39.
[6] A. S. Fokas, Phys. Rev. Lett. **51** (1983); A. S. Fokas and M. J. Ablowitz, J. Math. Phys. **25** (1984) 2505.
[7] A. S. Fokas, U. Mugan, M. J. Ablowitz, A Method of Solution for Painlevé Equations: PIV, PV, preprint INS#73 (1987).
[8] A. I. Nachman, M. J. Ablowitz, Stud. Appl. Math. **71** (1984) 251.
[9] A. S. Fokas, J. Math. Phys. **27** (1986) 1737.
[10] M. J. Ablowitz and R. Haberman, Phys. Rev. Lett. **35** (1975) 1125.
[11] A. S. Fokas, Phys. Rev. Lett. **57** (1986) 159.
[12] A. S. Fokas and M. J. Ablowitz, Comm. Math. Phys. **91** (1983) 381.
[13] A. V. Mikhailov, Physica **3D** (1981) 1&2, 73-117.

Gauge Unification of Integrable Nonlinear Systems

A. Kundu

Theoretical Nuclear Physics Division, Saha Institute of Nuclear Physics,
92 A.P.C. Road, Calcutta 700009, India

Gauge equivalence of generalized NLS type equations is established with its possible application to find out soliton solutions and with the generation of new integrable systems. Through certain gauge choices some realistic models are exactly solved. Explicit auto BT for different classes of equations are also obtained by gauge transformation.

1. INTRODUCTION

Some twenty years back discovery of nonlinear integrable field models created immense excitements. In fact when in 1967 Kruskal et al. [1] showed the Korteweg-de Vries (KdV) equation to be completely integrable with soliton solutions having beautiful properties, it was not considered by all to have any universal appeal. Subsequently, however, mainly with the works of Zakharov and Shabat [2] and AKNS [3] the existence of a whole class of such systems was revealed. Recent years have now witnessed a rapid growth of such members in the family of integrable systems. In the present day, however, the situation is somewhat reversed. There is already a large collection of members in the integrable circle, seemingly all with their own originality and independence, demanding individual care and analysis. Besides these bonafide members there is also a vast number of candidates from the real world with their nonlinearity, awaiting their recognition at least 'near' to some integrable members. In this complicated affair of today, there is then a natural urgency to work out some sort of unification scheme to group together integrable members of the same class by finding out their inter-relationship and pinpointing some genuine representatives, from which others originate. Thus individual treatment of each one of them, as is the standard practice of today, would become unnecessary and full information about only a few basic equations would be sufficient to solve the rest, generated from them. The aim of our article is to present such a unification scheme through gauge equivalence , to trace out genuinely independent equations, find out gauge relations between different systems and recognize real world candidates for their membership of the integrable family. Though our ambitious programme is not yet completed and we confine here only to nonlinear Schrödinger (NLS) type of equations, we are able to achieve a breakthrough in this line recently [4-8]. This scheme also enabled us to use the same machinary to generate a number of new integrable equations, some of them coinciding with the known nonlinear realistic systems or coming near to them, thus solving exactly few long standing hydrodynamic problems. Along with the above idea of gauge transformation (GT) between different systems we also use it for another aspect, i.e., for finding Bäcklund transformation (BT) between different solutions of the same equation.

Interest in gauge equivalence between different nonlinear dynamical systems and the corresponding models was boosted up after the remarkable works of Lakshmanan [9], Pohlmeyer [10] and Zakharov and Takhtajan [11]. On the one hand, sine-Gordon (SG) and nonlinear σ model was linked [10] and extended [12] and on the other hand Landau-Lifshitz equation (LLE) was gauge related to nonlinear Schrödinger equation (NLSE) and generalized [13-14]. However, besides the nonlinear systems mentioned above, there are various other NLS like equations in the integrable family, e.g., NLSE of repulsive type [15], attractive-repulsive NLSE [16] and also systems like derivative NLSE [17], mixed NLSE [18], Chen-Lee-Liu equation [19], modified derivative NLS [20], Gerdjikov-Ivanov equations [21], etc. It is natural to ask whether they are interrelated and whether there exist any Landau-Lifshitz type equation, gauge equivalent to them. Besides, these integrable members, there are also some well-known hydrodynamic equations with high nonlinearity, such as Johnson equation [22], Beney's first [23] and second [24] kind of long-short wave interaction equations. Whether integrability of these real models, which mostly received approximate treatment, may be recognized? Our aim is to answer to the above questions (Fig. 1) by generalizing the gauge equivalence scheme and proposing a LLE with noncompact Grassmannian manifold SU(p,q)/S(v(r,s) x v(u,v)). Through H-gauge transformation we are also able to connect realistic systems with integrable models and thus solve them exactly.

The organization of the paper is as follows. In Sec. 2 the general scheme of gauge equivalence (GE) is outlined and demonstrated for NLS and LLE. In Sec. 3, GE is applied to noncompact manifolds recovering old and yielding new results. In Sec. 4 we demonstrate the applicability of GE for extracting soliton solutions and other informations. Section 5 finds generalized LLE equivalent to derivative and mixed NLSE's. In Sec. 6 we generate through H-gauge transformation a hierarchy of higher-order equations, connecting a number of integrable systems. In

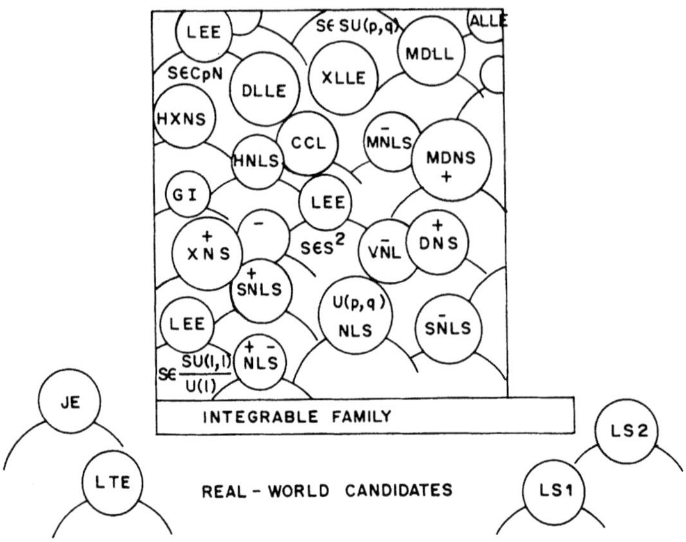

Fig. 1 Are there any interrelation between these members? Are there some genuinely fundamental members?

Sec. 7 we connect realistic equations with integrable systems and solve them exactly. Section 8 demonstrates the applicability of GE for finding BT even where earlier attempt [25] failed. Section 9 is the concluding section.

2. GAUGE EQUIVALENCE OF NONLINEAR EVOLUTIONARY EQUATIONS

Let the linear problem or Lax pair associated with the given integrable equation $\hat{L}q = 0$ be expressed as

$$\Phi_x = U\Phi , \qquad \Phi_t = V\Phi, \tag{2.1}$$

where the Jost function Φ and U, V are complex matrix functions of the field $q(x,t)$, q_x, the independent variables x and t and the spectral parameter λ. The compatibility of system (2.1), i.e., $\Phi_{xt} = \Phi_{tx}$: $U_t - V_x + [U,V] = 0$ yields the original nonlinear system $\hat{L}q = 0$ by appropriate construction of U and V. For real λ, $\Phi \in G$, G being a compact or noncompact Lie group and under the local gauge transformation

$$g(x,t;\lambda_o) = \Phi(x,t;\lambda)\big|_{\lambda=\lambda_o} \in G \tag{2.2}$$

the Jost function changes as

$$\Phi \to \Phi'(x,t;\lambda,\lambda_o) = g^{-1}(x,t;\lambda_o)\,\Phi(x,t;\lambda) \tag{2.3}$$

and the corresponding new linear system is

$$\Phi'_x = U'\Phi', \quad \Phi'_t = V'\Phi' \tag{2.4}$$

with

$$U' = g^{-1}Ug - g^{-1}g_x = g^{-1}(U-U_o)g, \quad g_xg^{-1} = U_o = U|_{\lambda=\lambda_o}$$

$$V' = g^{-1}Vg - g^{-1}g_t = g^{-1}(V - V_o)g, \quad g_tg^{-1} = V_o = V|_{\lambda=\lambda_o} \tag{2.5}$$

The compatibility of (2.4) gives now the GE equation

$$U'_t-V'_x + [U',V'] = g^{-1}\{(U_t-V_x+[U,V]) - (U_{ot}-V_{ox}+[U_o,V_o])\}g = 0 \tag{2.6}$$

relative to a new field S. There exists also another type of spectral parameter dependent GT: $\tilde{\Phi}(x,\lambda) = G(q,q',\lambda)\,\Phi(x,\lambda)$, which generates an auto Bäcklund transformation (BT) mapping a solution q to a different solution q' of the same system.

Let us now look into the well-established [11] GE between NLSE and LLE to make the picture clear. Scalar attractive NLSE

$$i\psi_t + \psi_{xx} + 2|\psi|^2\psi = 0 \tag{2.7}$$

is given by the Lax pair

$$U = A_0 + \lambda A_1, \quad V = B_o + \lambda B_1 + \lambda^2 B_2 , \tag{2.8}$$

where $A_1 = i\sigma_3$, $A_0 = \bar{\psi}\sigma^+ - \psi\sigma^-$, $B_1 = 2A_0$, $B_2 = 2A_1$ and $B_0 = -i(|\psi|^2\sigma_3 + \bar{\psi}_x\,\sigma^+ + \psi_x\sigma^-)$, where $\sigma^\pm = \sigma_1 \pm i\sigma_2$ and σ_i, $i = 1,2,3$ are Pauli matrices. If $\tilde{\Phi} = g^{-1}\Phi$ with $g = \Phi_+|_{\lambda=\lambda_o}$, where $\Phi_\pm$ is the Jost functions with $\Phi_\pm \xrightarrow[x\to\pm\infty]{} \exp(i\sigma_3 x)$ and scattering matrix $T(\lambda) = \Phi_+^{-1}\Phi_-$, then defining $S_+ = g^{-1}\sigma_3 g$ we get the GT Lax operators to be

$$U' = g^{-1}[(A_o + \lambda A_1) - (A_o + \lambda_o A_1)]g = i(\lambda - \lambda_o)S$$

and

$$V' = g^{-1}[(\lambda-\lambda_o)B_1 + (\lambda^2-\lambda_o^2)B_2]g = 2(\lambda^2-\lambda_o^2)S + (\lambda-\lambda_o)SS_x, \tag{2.9}$$

where we have used

$$SS_x = -S_xS = g^{-1}g_x - g^{-1}\sigma_3 g_x g^{-1}\sigma_3 g = g^{-1}g_x - g^{-1}\sigma_3 U_o \sigma_3 g = g^{-1}B_1 g. \tag{2.10}$$

(2.9) yields the GE LLE: $S_t = \frac{1}{2i}[S,S_{xx}] - 8\lambda_o S_x$. Note that for the choice $g = \Phi_-(x,\lambda_o)$ one similarly gets (2.10) $S_- = g_-^{-1}\sigma_3 g_-$ which is related to S_+ as $S_- = T_o^{-1} S_+ T_o$. But if we try to set $\Phi'_\pm = g_\pm^{-1}\Phi_\pm$, $\Phi'_\pm$ would satisfy different equations like $\Phi'_{\pm x} - U'_\pm \Phi'_\pm$ thus violating the Jost function condition. On the other hand with the choice $\Phi'_\pm = g_+^{-1}\Phi_\pm$ we immediately get the relation between the scattering matrices:

$$T'(\lambda) = \Phi'^{-1}_+ \Phi'_- = \Phi_+^{-1} g_+ g_+^{-1} \Phi_- = T(\lambda), \qquad (2.11)$$

which helps to bypass the individual IST investigation of a system, if one knows the corresponding information of its gauge equivalent counterparts. The integrability property is also obviously preserved. Another way of looking at GT for 2×2 matrices is through the trihedral E_a, as used by Pohlmeyer, Eichenherr and others, where one introduces $E_a = \mathbf{e}_a\boldsymbol{\sigma} = g^{-1}\sigma_a g$, a = 1,2,3 with the relation $E_a E_b = g_{ab} + i f_{abc} E_c$, $E_a^+ = \sigma_3 E_a \sigma_3$, where g_{ab} is a symmetric and f_{abc} is an antisymmetric tensor and with the equation $E_{ax} = A_{ab} E_b$, $E_{at} = B_{ab} E_b$. Hence in our case $S = E_3 = \mathbf{S}\boldsymbol{\sigma} = g^{-1}\sigma_3 g$, $E_\pm = g^{-1}\sigma_\pm g$ with the equation

$$S_x = 2(\bar{\psi}E_+ + \psi E), \; E_{+x} = -2i\lambda E_+ - \psi S, \; E_x = 2i\lambda E - \bar{\psi} S \qquad (2.12a)$$

and

$$S_t = 2(-i\bar{\psi}_x + 2\lambda\bar{\psi})E_+ + 2(i\psi_x + 2\lambda\psi)E \, ,$$

$$E_{+t} = 2i(|\psi|^2 - 2\lambda^2)E_+ - (i\psi_x + 2\lambda\psi)S. \qquad (2.12b)$$

The 0(3)-invariants are $tr(S_x^2)$, $tr(S_t,S_x)$ and $tr(S_t^2)$. Eliminating $E_\pm$ the final equation for S is again given by LEE (2.10').

3. GENERALIZED LLE WITH NONCOMPACT GRASSMANNIAN MANIFOLD AND GAUGE EQUIVALENT NLSE

We propose a generalized Landau-Lifshitz equation (LLE) with non-compact manifold:

$$S_t = \frac{1}{2i}[S,S_{xx}], \qquad S \in M = G/H,$$

$$G = SU(p,q), \quad H = S(U(u,v) \otimes U(r,s)), \qquad (3.1)$$

where u+v = m, s+r = n, u+r = p, v+s = q, p+q = m+n = N and find its gauge equivalent system using the above scheme. The corresponding linear system is

$$U = i\lambda S, \quad V = 2i\delta^2\lambda^2 S + \tfrac{1}{2}\lambda[S,S_x], \quad \delta = (m+n)/2mn \tag{3.2}$$

with S satisfying $S^2 = aI + bS$, $\bar{S} = \Gamma S^+\Gamma = S$, where $\Gamma = \mathrm{diag}(\Gamma_1,\Gamma_2)$, $\Gamma_1 = \mathrm{diag}(I_u,-I_v)$ and $\Gamma_2 = \mathrm{diag}(I_r,-I_s)$, a, b are constants and I_i is a i×i matrix. G/H being a symmetric space, it is always possible to express $S = g\Sigma g^{-1}$, $g \in G$ and $\Sigma = \mathrm{diag}(I_m/m, -I_n/n)$, which gives on using $S_\mu = g[\Sigma,L_\mu]g^{-1}$, $L_\mu = g^{-1}g_\mu$, $\tfrac{1}{2}[S,S_x] = 2g\Sigma[\Sigma,L_1]g^{-1}$ the transformed operators

$$U' = g^{-1}Ug - g^{-1}g_x = i\lambda\Sigma - A,$$

$$V' = g^{-1}Vg - g^{-1}g_t = 2i\delta^2\lambda^2\Sigma + \lambda\Sigma[\Sigma,A] - B \tag{3.3}$$

with $L_1 = A$, $L_0 = B$. Taking A and B in the explicit form

$$A = \begin{pmatrix} A_1 & \bar{\psi} \\ -\psi & A_2 \end{pmatrix}, \quad B = i\delta\begin{pmatrix} b_1 & \bar{\psi}_x+\bar{\psi}A_2-A_1\bar{\psi} \\ \psi_x+\psi A_1-A_2\psi & b_2 \end{pmatrix}$$

with $\bar{\psi} = \Gamma_1\psi^+\Gamma_2$, $\bar{A}_i = -A_i$, $\mathrm{tr}(A_1+A_2) = 0$, $A_i = H^i_x$ and $b_1 = (\psi\bar{\psi}-\mu I_m)+H^1_t$, $b_2 = -(\psi\bar{\psi}-\mu(m/n)I_n) + H^2_t$. The compatibility of (3.3) yields the gauge equivalent matrix NLSE

$$i\psi_t + \psi_{xx} + 2(\psi\bar{\psi}\psi - \mu\psi) + 4i\delta\lambda_o\psi = 0. \tag{3.4}$$

It is worth mentioning that in deriving (3.4) we have set simply $A_1 = i\lambda_o I_m/m$, $A_2 = -i\lambda_o I_n/n$. In general, contrary to the usual belief, LLE is GE not only to NLSE but also to all its H-gauge equivalent partners (see Fig. 2), which may also be obtained by different choices of function H^i. We, therefore, conclude that LLE (3.1) with noncompact manifold is gauge equivalent to matrix NLSE (3.4) with internal symmetry group $U(u,v) \otimes U(r,s)$ along with all its H-GT systems. Note that for trivial boundary condition on fields ψ, λ_o is arbitrary real and may be trivial [11], but for a nontrivial boundary condition, which is important in noncompact cases λ_o is nontrivial, since it should always be from the real spectrum [6]. The following particular cases recover old results, yield new relations and answer some of the questions raised in Sec. 1.

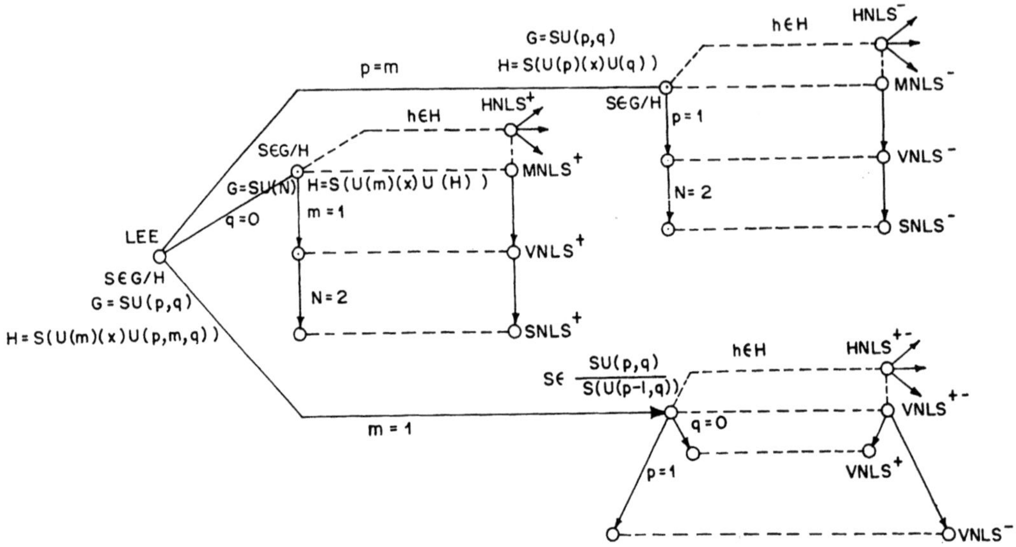

Fig. 2 Gauge equivalence between generalized LLE with noncompact Grassmannian manifold and NLSE along with its H-gauge equivalent systems and their different reductions. (±) represents 'attractive' or 'repulsive' while (+-) represents 'attractive-repulsive' type cases

3.1 Examples

Let $u = m$, $v = 0$, $r = p-m > 0$, that is,

$$S \in SU(p,q)/S(U(m)\times U(p-m,q)) \tag{3.5}$$

We further suppose that (i) $p = N$, $q = 0$, $\Gamma_1 = \Gamma_2 \to 1$, hence $\bar{S} = S^+$ recovering equivalence of LLE with compact manifold $S \in SU(N)/S(U(m)\otimes U(n))$ [12] and consequently for $m = 1$ with $S \in CP^N$ [13] and for $N = 1$ with standard $S \in S^2$ [11]. For extracting the relevance of noncompactness we suppose that (ii) $p = m$, $q = n$ giving $\Gamma_1 = I$, $\Gamma_2 = -I$, $\bar{\psi}=-\psi^+$, that is LLE with $SU(p,q)/S(U(p)\otimes U(q))$ is GE to matrix NLS of repulsive type. $p = 1$ connects LLE with $SU(1,N-1)/U(N-1)$ and vector NLS (VNLS) of repulsive type

$$i\psi_t + \psi_{xx} - 2\Big(\sum_{a=1}^{N-1} |\psi^a|^2 - \mu\Big)\psi = 0. \tag{3.6}$$

N=2 gives a $SU(1,1)/U(1)$ version of LLE and related repulsive scalar NLS [15]. (iii) $m = 1$ reduces (3.5) to LLE with noncompact manifold $SU(p,q)/U(p-1,q)$ and GE attractive-repulsive VNLS [16]

$$i\psi_t + \psi_{xx} + 2\Big(\sum_{a=1}^{p-1} |\psi^a|^2 - \sum_{b=1}^{q} |\psi^b|^2 - \mu\Big)\psi = 0, \tag{3.7}$$

which in limiting cases leads again to VNLS of attractive or repulsive type. The above equivalence is schematically depicted in Fig. 2.

It may also be shown that [4] SU(2) LLE with easy axis ($\Delta>0$) anisotropy (ALLE) is gauge equivalent to attractive SNLS, while the anisotropic SU(I,I) LLE to repulsive SNLS and easy plane ($\Delta<0$) case to AKNS system.

4. APPLICATION OF GE FOR SOLITON SOLUTION

Most of the early works, besides establishing the GE, largely ignored the possible applications of such a beautiful relationship. We aim to utilize the GE for extracting information of LEE without the associated tedius IST calculation but only using well-investigated properties of its GE NLSE. As shown in §2, the Jost function of LLE $\Phi'_{\pm}$ may be expressed through that of NLSE Φ_a as $\Phi'_{\pm} = \Phi_0^{-1}\Phi_{\pm}$, where

$$\Phi_{\pm} = \begin{pmatrix} \Phi^1_{\pm} & \varepsilon\Phi^{2*}_{\pm} \\ \Phi^2_{\pm} & \Phi^{1*}_{\pm} \end{pmatrix}$$

is the Jost solution of spectral problems connected with NLSE and $\varepsilon = \mp 1$ correspond to NLSE of attractive and repulsive type, respectively relating to LLE with $S \in SU(2)/U(1)$ and $SU(1,1)/U(1)$ manifolds. As shown before the scattering matrices of GE system are identical. The field solution of LLE may be expressed through NLSE Jost solutions as

$$S = \Phi^{-1}_{o+}\sigma_3\Phi_{o+} = \frac{1}{\Delta_o}(S^3\sigma^3 + S^-\sigma^- + S^+\sigma^+),\ \Delta_o = \det\Phi_o,$$

$$S_3 = |\Phi^1_o|^2 + \varepsilon|\Phi^2_o|^2,\quad S^- - -2\Phi^1_o\Phi^2_o,\quad S^- = -\varepsilon S^{+*}. \tag{4.1}$$

4.1 SU(2)/U(1) LLE Solution from NLSE of Attractive Type

The solution to LLE is normally extracted through tedious IST calculations [26]. We, however, recover the soliton solution through GE using the known results of NLSE [2] given by the Jost function for N-soliton solution:

$$\tilde{\Phi}(\lambda,x) = \begin{pmatrix} \Phi^{2*}_{+} \\ \Phi^{1*}_{+} \end{pmatrix} = \exp(-i\lambda x)\begin{pmatrix} 0 \\ 1 \end{pmatrix} + \sum_{n=1}^{N} \frac{C_n\Phi_n}{\lambda-\lambda_n}\exp(i\lambda_n x), \tag{4.2}$$

where $\Phi_n = \Phi(x,\lambda_n)$, λ_n being the discrete spectrum. For simplicity we consider $N = 1$ giving 1-soliton solution of NLSE as

$$\psi(x) = -2 \sum_{n=1}^{N} C_n \Phi_n'(x) \exp(i\lambda_n x)\Big|_{N=1} = -2i\eta \exp(i\gamma) \operatorname{sech} y, \qquad (4.3)$$

where $\lambda_1 = \xi + i\eta$, $\gamma = 2\xi x + 4(\xi^2-\eta^2)t + \Phi_o$ and $y = 2\eta(x-4\xi t-x_o)$. On the other hand from (4.2) one obtains

$$\Phi_o^1 = -\rho(\lambda_o) \exp[i(\lambda-\lambda_o)x] \operatorname{sech} y,$$

$$\Phi_o^2 = [1 - \rho(\lambda_o)\exp(-y)\operatorname{sech} y] \exp(-i\lambda_o x), \qquad (4.4)$$

with $\rho(\lambda_o) = i\eta/(\lambda-\lambda_o)$, which gives directly from (4.2) the LLE soliton solution

$$S_3 = 1 - 2|\rho|^2 \operatorname{sech}^2 y,$$

$$\arg(S^+) = 2\lambda_o x - \gamma - \tan^{-1}\{[(\lambda_o - \xi)/\eta]\coth y\} \qquad (4.5)$$

coinciding for $\lambda_o = 0$ with the result of ref. 26 found through direct IST.

4.2 Noncompact SU(1,1)/U(1) LLE Solution Through Repulsive Type NLSE

The noncompact LLE model may be given by the Hamiltonian

$$H = \int_{-\infty}^{\infty} \operatorname{tr}(S_x^2)dx = \frac{1}{2}\int_{-\infty}^{\infty}({S_x^3}^2 - {S_x^2}^2 - {S_x^1}^2)dx \qquad (4.6)$$

with ${S^3}^2 - {S^2}^2 - {S^1}^2 = 1$ and $S = S^a\tau_a \in SU(1,1)/U(1)$. This is a new model, which may have physical applications. For finding its soliton solution we may apply again the GE with NLSE of repulsive type established here and use the known IST information of the latter system [15]. The IST programme of repulsive NLSE is rather complicated due to nontrivial boundary condition $\lim_{|x|\to\infty}|\psi|^2 \to \mu^2$. Here $\xi = \pm(\lambda^2 - \mu^2)^{\frac{1}{2}}$ actually serves the role of spectral parameter and $\xi(\lambda)$ is defined on a 2-sheeted Riemann surface with cuts at $(-\infty,-\mu)$ and (μ,∞). The bound states are given at $\xi_n = (\lambda_n^2 - \mu^2)^{\frac{1}{2}} = i\nu_n$ with $|\lambda_n|^2 < \mu^2$. The Jost solution corresponding to 1-soliton of NLSE is given by [15]

$$\Phi = \exp(-i\xi x)\begin{pmatrix} 1 \\ \lambda-\xi \end{pmatrix}\left[\begin{pmatrix} 1 \\ 1 \end{pmatrix} - \frac{1}{\nu + i\xi}\begin{pmatrix} K_1 & K_2 \\ K_2^* & K_1^* \end{pmatrix}\right], \qquad (4.7)$$

where

$$K_1(y) = \nu f(y), \quad K_2(y) = \nu(\gamma - i\nu)f(y), \quad f(y) = [1+\exp(2y)]^{-1},$$

$$\xi^2 = \lambda^2 - 1, \quad \lambda_1 = \gamma, \quad \nu^2 = 1 - \gamma^2 \text{ and } y = \nu(x-x_o-2\gamma t),$$

$$\psi(x,t) = [(\gamma + i\nu)^2 + \exp(2y)]f(y). \tag{4.8}$$

Using $\varepsilon = 1$ and (4.7) one gets from (4.1) the soliton solution for (4.6) LLE in the form

$$S_3 = 1 + (2\gamma^2/\nu^2)\tanh^2 y ,$$

$$\arg(S^+) = 2\xi x + \tan^{-1}[(\gamma/\nu)\tanh y], \tag{4.9}$$

where we have chosen $\lambda_o - \xi_o = \gamma$ for simplicity, which, however puts restriction on the soliton velocity $u = 2\gamma = 2(\lambda_o - \xi_o)$, $2 > u \geq 0$. We also have the restriction $|\lambda_o| > \mu = 1$ since λ_o must be from the continuous spectrum.

5. NLS TYPE EQUATIONS AND GAUGE EQUIVALENT EXTENDED LLE

Using the technique similar to that applied above for standard NLSE, we may also find new extended LLE, gauge equivalent to various known NLS type equations [5].

5.1 DLL Gauge Generated from DNS

The scalar derivative NLS (DNS)

$$\text{(DNS)}: \quad q_t + q_{xx} \pm i\alpha(|q|^2 q)_x = 0 \qquad \alpha > 0 \tag{5.1}$$

may be given by the linear system

$$U = -i\alpha\lambda^2\sigma_3 + \alpha\lambda A$$

$$V = (-2i\alpha^2\lambda^4 \pm i\alpha^2|q|^2\lambda^2)\sigma_3 + 2\alpha^2\lambda^3 A + \alpha\lambda B, \tag{5.2}$$

where $A = q\sigma^+ \pm q^*\sigma^-$, $B = (iq_x \mp \alpha|q|^2 q)\sigma^+ + (\pm iq_x^* + \alpha|q|^2 q^*)\sigma^-$. Defining as before $S = g^{-1}\sigma_3 g$ one gets $SS_x = 2\lambda_o\alpha\, g^{-1}Ag$, $SS_x^2 = -4\lambda_o^2\alpha^2 g^{-1}\sigma_3 A^2 g$ and $S(S_t - 2\alpha\lambda_o^2 S_x) = 2\alpha\lambda_o g^{-1}Bg$. Repeating now the above procedure for GT (2.5) one obtains the new system

$$U' = g^{-1}[-i\alpha(\lambda^2-\lambda_o^2)\sigma_3 + \alpha(\lambda-\lambda_o)A]g$$

$$= -i\alpha(\lambda^2-\lambda_o^2)S + (\frac{1}{2\lambda_o})(\lambda-\lambda_o)SS_x$$

$$V' = g^{-1}[-2i\lambda^2\alpha^2(\lambda^4-\lambda_o^4)\sigma_3 + 2\alpha^2(\lambda^3-\lambda_o^3)A - i\alpha^2(\lambda^2-\lambda_o^2)\sigma_3 A^2+\alpha(\lambda-\lambda_o)B]g$$

$$= 2i\alpha^2(\lambda^4-\lambda_o^4)S + (\alpha/\lambda_o)(\lambda^3-\lambda_o^3)SS_x + i\,\frac{\lambda^2-\lambda_o^2}{4\lambda_o^2}\,SS_x^2$$

$$+ \frac{\lambda-\lambda_o}{2\lambda_o}(SS_t - 2\alpha\lambda_o^2 SS_x) \qquad (5.3)$$

yielding the extended LLE (DLL) [5]:

$$(\text{DLL}):\quad S_t + \frac{1}{2i}[S,S_{xx}] - 4\alpha\lambda_o^2 S_x + \frac{1}{4\alpha\lambda_o^2}\,S_x^3 = 0,\ \alpha > 0 \qquad (5.4)$$

with $S \in SU(2)/U(1)[SU(1,1)/U(1)]$ corresponding to +(-) signs. Note that the integrable system (5.4) has also been found recently through an altogether different method [27].

5.2 XLL Gauge Generated from XNS

The known mixed NLS (XNS), a hybrid of DNS and NSE,

$$(\text{XNS}):\quad iq_t + q_{xx} \pm\beta|q|^2q \pm i\alpha(|q|^2q)_x = 0,\ \alpha > 0,\ \beta > 0 \qquad (5.5)$$

given by somewhat complicated U, V operators [18] is gauge transformed similarly to [5]

$$U' = g^{-1}\{\delta\sigma_3 + aA\}g$$

$$= \delta S + \frac{a}{c}\,SS_x$$

$$V' = g^{-1}\{ k\sigma_3 + bA + aB +\gamma\sigma_3A^2 \}g$$

$$= kS + \frac{b}{c}SS_x - \frac{\gamma}{c^2}SS_x^2 + \frac{a}{c}(SS_t - 2\,\frac{\bar{b}}{c}\,SS_x), \qquad (5.6)$$

where $c, \delta, a, k, b, \gamma, \bar{b}$ are different expressions depending on λ, λ_o, α and B [5]. The flatness condition of (5.6) yields gauge equivalent LLE type equation in the form (XLL)†

†XLL given in refs. 5,6 may be simplified to this form.

$$(\text{XLL}): \quad S_t + \frac{1}{2i}[S, S_{xx}] + \gamma S_x + \rho S_x^3 = 0, \tag{5.7}$$

where $\gamma = 4\lambda_o\{\sqrt{2B} - \alpha\lambda_o\}$ and $\rho = \alpha/(2\alpha\lambda_o - \sqrt{2\beta})^2$. The ± sign as before corresponds to $S \in SU(2)/U(1)$ ($SU(1,1)/U(1)$) cases. One gets also the invariant relations

$$\text{tr}(S_x^2) = \pm 2c^2 |q|^2, \quad \text{tr}(S_x S_t) = c^2\{ \mp i(qq_x^* - q^*q_x) \pm 4\frac{\bar{b}}{c}|q|^2 - 2\alpha|q|^4\}. \tag{5.8}$$

Note that XLL (5.7) has exactly the same form as DLL (5.4) but only with different coefficients, which clearly reduces to the corresponding coefficients of LLE and DLL in particular cases $\alpha = 0$, $\beta \neq 0$ and $\alpha \neq 0$, $\beta = 0$, respectively. The coincidence of XLL and DLL reflects the fact that their gauge equivalent Schrödinger type equations, e.g., XNS and DNS are U(1)-gauge related (see§6.1), which does not change the corresponding LLE system as shown in §6.

5.3 MDLL Equivalent to DMNS

A modified DNS (MDNS) given by

$$(\text{MDNS}): \quad iq_t + (q/\rho)_{xx} = 0, \quad \rho = (1 \pm |q|^2)^{\frac{1}{2}} \tag{5.9}$$

corresponds to the linear system

$$U = -i\lambda\sigma_3 + \lambda A, \quad V = 2\lambda^2 D + \lambda B \tag{5.10}$$

with A as in (5.2) and

$$B = i\begin{pmatrix} 0 & (q/\rho)_x \\ \pm(q^*/\rho)_x & 0 \end{pmatrix}, \quad D = \rho^{-1}\begin{pmatrix} -i & q \\ \pm q^* & i \end{pmatrix}.$$

Repeating the above procedure we now get the gauge transformed operators as

$$\begin{aligned} U' &= (\lambda-\lambda_o)g^{-1}\{-i\sigma_3 + A\}g \\ &= (\lambda-\lambda_o)(-iS + \frac{1}{2\lambda_o}SS_x), \\ V' &= g^{-1}\{2(\lambda^2-\lambda_o^2)D + (\lambda-\lambda_o)B\}g \\ &= 2(\lambda^2-\lambda_o^2)\chi S(\frac{1}{2\lambda_o}S_x - i) + (\lambda-\lambda_o)S(S_t - 2\lambda_o\chi S_x)/2\lambda_o \end{aligned} \tag{5.11}$$

yielding a new LLE type integrable equation (MDLL) [5]

$$(\text{MDLL}): \quad S_t = i(\chi S S_x)_x + 4\lambda_o \chi S_x \tag{5.12}$$

with $\chi = (1+\mathrm{tr}(S_x^2)/8\lambda_o^2)^{-\frac{1}{2}}$ with $S \in SU(2)/U(1)[SU(1,1)/U(1)]$ corresponding to +(-) signs in (5.9).

6. UNIFICATION THROUGH H-GAUGE TRANSFORMATION

It is interesting to note that from the matrix NLSE (3.4) one can generate through the local H-gauge transformation $\psi \to h_2^{-1}\psi\, h_1$ relative to group element $h = \mathrm{diag}(h_1,h_2) \in H = S(U(u,v)\otimes U(s,t))$ a class of new higher-order equations

$$i\psi_t+\psi_{xx}+2(\psi\bar{\psi}\psi-\mu\psi)+4i\lambda_o\delta\psi_x -[\tfrac{i}{\delta}(\psi H_t^1-H_t^2\psi)+2(\psi_x H_x^1-H_t^2\psi_x)$$
$$+ (\psi H_{xx}^1-H_{xx}^2\psi)+2(H_x^2\psi H_x^1)] + (H_x^2)^2\psi + \psi(H_x^1)^2 = 0, \tag{6.1}$$

where $H_\mu^i = h_i^{-1}\partial_\mu h_i$. Since under $h(x,t) \in$ H-gauge transformation, the model field is invariant: $S' = g'\Sigma\, g'^{-1} = gh\Sigma h^{-1}g^{-1} = g\Sigma g^{-1} = S$, the LLE system remains unchanged under such transformation, whereas the equivalent matrix NLSE changes as $U_i' = h^{-1}U_i h+(i\lambda\Sigma-h^{-1}h_i)$. We consider below the simplest case of $h = \exp(i\theta\sigma_3) \in U(1)$, $G = SU(2)(SU(1,1))$ and show how certain new hierarchy of integrable equations can be generated unifying different systems for various particular choices.

6.1 U(1)-Gauge Generated Higher Order Integrable Systems

The XNS (5.5) is transformed to the higher-order equation

$$iQ_t+Q_{xx}+\beta|Q|^2Q + i\alpha(|Q|^2Q)_x + 2\{(\theta_t-2\theta_x^2-i\theta_{xx}+\alpha\theta_x|Q|^2)Q -2i\theta_x Q_x\} = 0,$$

$$Q = e^{2i\theta}q. \tag{6.2}$$

It is evident, that a choice of $\theta = \frac{V}{4}(x + \frac{V}{2}t)$ and a Galilian transformation with $V = -2\beta/\alpha$ reduces (6.2) to DNS (5.1) and thus establishes gauge equivalence between XNS and DNS while for $\alpha = 0$ we directly get NLS. We may, therefore, consider gauge generation from later two systems only.

(i) Gauge generation from NLS: One interesting choice for Θ_μ, e.g., $\Theta_x^{(j)} = \rho_j$, $\Theta_t^{(j)} = \mathfrak{I}_j$ where $(\rho_k)_t = (\mathfrak{I}_k)_x$ yielding conservation laws from the Ricatti-equation generates a hierarchy of integrable systems [8]

$$iQ_t + Q_{xx} + \beta|Q|^2Q + 2\{(\mathfrak{I}_j - 2\rho_j^2 - i\rho_{jx})Q - 2i\rho_j Q_x\} = 0, \quad j = 1,2,\ldots.$$

One may easily construct for these systems exact solutions, Lax pairs, infinite conservation laws, etc. [8]. In the simplest case of j = 1 we have $\Theta_x^{(1)} = -\delta|q|^2$, $\Theta_t^{(1)} = i\delta\,(qq_x^* - q^*q_x)$ yielding the new integrable system with fifth-order nonlinearity

$$iQ_t + Q_{xx} + \beta|Q|^2Q + 4\delta^2|Q|^4Q + 4i\delta(|Q|^2)_xQ = 0$$

reducible to a linear equation $iq_t + q_{xx} = 0$ for $\beta = 0$. For choice $\Theta = \Theta(x)$ one gets a NLS with variable dependent coefficients. Other choices of Θ will be discussed in the next section.

(ii) Gauge generation from DNS: Similar choice as the above gives again another hierarchy of integrable systems generated from DNS [8]. The simplest case gives Johnson type equation

$$iQ_t + Q_{xx} + i\alpha(|Q|^2Q)_x + \delta(4\delta+\alpha)|Q|^4Q + 4i\delta(|Q|^2)_xQ = 0,$$

which for the choice $\delta = -\frac{\alpha}{4}$ reduces to Chen-Lee-Liu [19] equation (CLL).

$$\text{(CLL):}\quad iQ_t + Q_{xx} + i\alpha|Q|^2Q_x = 0$$

and for $\delta = -\frac{\alpha}{2}$ to Gerdjikov-Ivanov (GI) equation [21]

$$\text{(GI):}\quad iQ_t + Q_{xx} + \frac{\alpha}{2}|Q|^4Q - i\alpha Q^2Q_x^* = 0.$$

Thus we have established the gauge relation between a large number of systems. Consequently, the soliton solutions and Lax pairs for these systems may also be trivially constructed from those of DNS thus avoiding lengthy sophisticated approaches [28,31].

7. EXACT SOLUTION OF REALISTIC MODELS THROUGH GE

Johnson derived [22] the following fifth order nonlinear equation (JE) for the amplitude of the fundamental wave.

$$\text{(JE):}\quad iA_\tau - a_1A_{\zeta\zeta} - a_2|A|^2A + a_3|A|^4A + ia_4|A|^2A - ia_5A(|A|^2)_\zeta - a_6A\Theta_T = 0, \quad (7.1)$$

where $\Theta_\zeta=\delta|A|^2$ and a_i are real numbers. Previous attempt [22] to solve this equation was both tedious and approximate. We, however, notice that (7.1) may be gauge transformed to XNS

$$iq_\tau-a_1q_{\zeta\zeta}-a_2|q|^2q - i\alpha(|q|^2q)_\zeta = 0 \tag{7.2}$$

with $\alpha = \delta a_1(a_6+4) + a_5$ and a constraint $\delta a_1+a_4+a_5+3a_6+4 = 0$. Thus (7.1) can be solved exactly extracting the Lax pair and different soliton solutions [8].

In describing the mechanism of interaction between long and short waves Benney has derived the following famous equations. The first of this kind may be given by [23]

$$(LS1):\quad Q_t = -iQ_{xx} + k_3A_xQ + \bar{k}_3AQ_x-ik_4A^2Q+2is|Q|^2Q = 0, \tag{7.3}$$

$$A_t = 2s(|Q|^2)_x$$

where A(x,t) describes the long wave, while Q(x,t) is associated with the short wave. The second equation due to Benney is [24]

$$(LS2):\quad L_t + c_1L_x = \varepsilon(\alpha(|S|^2)_x + \varphi LL_x),$$

$$S_t + c_gS_x - i\delta LS = \varepsilon(i\beta S_{xx} + i\gamma|S|^2S + mLS_x + nSL_x), \tag{7.4}$$

with L and S describing the long and short waves, respectively. We observe again through gauge equivalence, that for certain parameter choice both the above equations become exactly solvable and connected with the NLS equation. In particular, for the choice $\bar{k}_3 = 2k_3 = -2\alpha$, $k_4 = -\alpha^2$ by U(1) - GT described above (7.3) is reduced for $q = Qe^{-2i\Theta}$ with $\Theta_x = \frac{\alpha}{2}A$, $\Theta_t = \alpha s(|Q|^2)$ to NLSE

$$iq_t + q_{xx} + 2\bar{s}|q|^2q = 0, \quad A_t = 2s(|q|^2)_x \tag{7.3'}$$

where $\bar{s} = -s(1+\alpha)$, yielding easily the explicit Lax pair through $U_i' = h^{-1}U_ih - h^{-1}h_i$, $h = \exp(i\Theta\sigma_3)$ and soliton solutions. For example, when $\bar{s} > 0$ the exact 1-soliton solution of (7.3) may be given by

$$A = \frac{4\nu}{v|1+\alpha|}\operatorname{sech}^2 2\nu z,\quad Q = \frac{2\nu}{\sqrt{\bar{s}}}\operatorname{sech}2\nu z\ \exp i\left(\omega t+kz - \frac{2\tanh 2\nu z}{2\nu|1+\alpha|}\right)$$

where $z = x - vt$. Note that for $\alpha = -1$ one gets $\bar{s} = 0$ reducing (7.3) to a linear equation. It is worth mentioning that for a more restricted set of parameters the solvability of (7.3) has also been established[23] through much involved prolongation theory. A similar GT with $q = Se^{2i\Theta}$, where $\Theta_x = L - c_g/4$ and $\Theta_t = -c_1 L + \varepsilon(\alpha|S|^2 + \frac{\varphi}{2} L^2) + c_g^2/8$ transforms (7.4) directly into the NLS equation of the form $iq_t + q_{xx} + \bar{\gamma}|q|^2 q = 0$ with $\bar{\gamma} = \varepsilon(\gamma - 2\alpha)$ and the choice $\delta = 2(c_g - c_1)$, $4\beta = 2n = m = 4\varepsilon^{-1}$ and $\varepsilon\varphi = 4$. However, if $\varepsilon\varphi - 4 = \rho \neq 0$ one should include in (7.4) a higher order nonlinear term like $\rho L^2 S$ to restore its gauge equivalence with the NLSE.

Following the same line of argument one can show that the Langmuir type equation (LTE) with some additional nonlinearity

$$iQ_t + Q_{xx} + nQ - [(n^2 + in_x)Q + 2inQ_x] = 0$$

$$n_t - n_x = -\frac{\beta}{2}(|Q|^2)_x$$

is also reducible to NLS for the choice

$$\Theta_x = \frac{1}{2}n, \qquad \Theta_t = \frac{1}{2}(n - \beta|Q|^2).$$

Now we may depict the result of gauge unification scheme in Fig. 3.

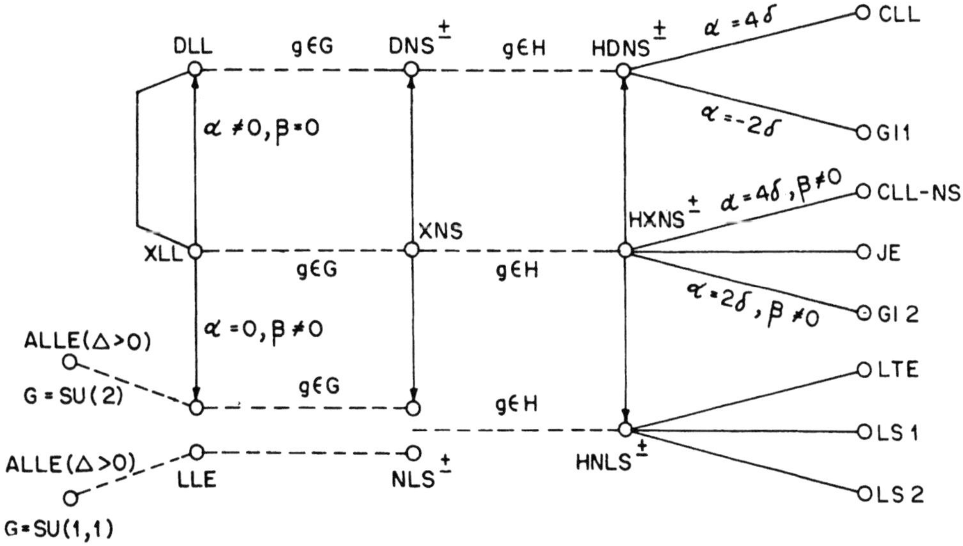

Fig. 3 Extended gauge equivalence scheme of different LLE and NLS type equations. This establishes the relationship between different members of the integrable family as well as real world candidates thus answering the questions raised in Fig. 1. Here H = U(1) and G = SU(2) (or SU(1,1) corresponds to attractive (+) (or repulsive (-)) case

8. EXPLICIT AUTO BT THROUGH GT

As mentioned in Sec. 2, another interesting application of GT is the auto BT for integrable systems, which is otherwise a tricky problem [29] demanding 'guess' work. Here the parameter dependent gauge element $G(q,q',\lambda)$ satisfies the equation

$$G_{x_i}(q,q',\lambda) = U_i(q',\lambda)G(q,q',\lambda)-G(q,q',\lambda)U_i(q,\lambda), \tag{8.1}$$

where $x_i = (x,t)$, $i = 0,1$. For finding out explicit Bäcklund Transform (BT) for some concrete systems like KdV, sG, NLS, DNLS, LLE, modified DNLS, we may express the elements $G^{ab} = \sum_{n=0}^{N} (\alpha_n^{ab})/(i\lambda)^n$ and from (8.1) choose N by matching coefficients of equal powers in $(i\lambda)^n$ to get consistent closed set of equations for α^{ab}. The value of N and explicit solutions would depend on particular form of U_1[7].

8.1 BT for AKNS System

For AKNS system we have $U_1 = \begin{pmatrix} i\lambda & q \\ r & -i\lambda \end{pmatrix}$, which results in the choice $\alpha^{11} = \alpha_o+2i\lambda$, $\alpha^{22} = \delta_o - 2i\lambda$, $\alpha^{12} = \beta_o = \alpha^{21} = \gamma_o$, where $\alpha_o,\beta_o,\gamma_o,\delta_o$ are independent of λ. This leads to the relation $\alpha_o v = u_x - \frac{c}{2}(u+v)$, $\alpha_{o_x} = \frac{1}{2}(r'(u+v)-r(u-v))$ with $u = q'+q$, $v = q'-q$. For deducing now the BT for particular systems we have to imply different reductions of AKNS, e.g., for KdV $r = -1$, $q = -w_x$, for sG $r = -q = \varphi_x$, for NLS $r = \pm q^*$, etc. leading to explicit BT's for these systems [7].

8.2 BT for Kaup Newell Problem

We have U_1 given by (5.2) dictating the choice $\alpha^{11} = 1 + \alpha_2\lambda^2$, $\alpha^{22} = -1 +\delta_2\lambda^2$, $\alpha^{12} = \beta_1\lambda$, $\alpha^{21} = \gamma_1\lambda$ leading to the relation

$$i(q'_x\delta_2-q_x\delta_2) = -2(q'+q) + (q'r'-qr)(q'\delta_2+q\alpha_2),$$

which yields easily for $r = \pm q^*$ the BT for DNS (5.2). Since the mixed DNLS (5.5) is shown in Sec. 6 to be GE to DNS, the BT for XNS is also easily obtained.

8.3 BT for WKI Problem: Modified DNS

Earlier attempt [25] for finding BT of modified DNS was not successful. We, however, notice that since LLE and NLS are gauge equivalent, one may construct the BT of the former system through the latter, which takes the form [7] $S' = B_1SB_1^{-1}$ where S', S are different solutions of LLE and $B_1 = \psi_o^{-1}G_o^{-1}\sigma_3\psi_o$, where G_o is the BT-gauge of NLS and ψ_o is its Jost function at $\lambda = \lambda_o$. Using now the fact [30] that a change of variable $iq = S^-/S^3$, $X = -\int S^3\,dx$ transforms LLE to (5.9) we may derive BT for the latter system [7].

9. CONCLUSION

We have demonstrated that through gauge equivalence method it is possible to 'unify' a large number of integrable systems, pinpointing a few basic ones. Using this machinery one may also obtain full information of a system, for example, know its Lax pair, soliton solutions, and Jost functions without doing any tedious IST investigation, but from its gauge equivalent, well investigated counterpart. Application of GE also allows us to generate new integrable systems along with their Lax pair and sometimes solve exactly existing realistic systems. It also opens up an elegant way of finding explicit BT of a large class of systems. One of our interesting observations is that some realistic nonlinear equations may turn into exactly integrable systems and sometimes they do so when some still higher nonlinear terms are added, whereas the usual practice is to throw away such terms in real models to give the equations some 'elegancy'. Our investigation shows however that in certain systems one may be able to gauge transform them to some known integrable models without neglecting any of their nonlinear terms (e.g., HNLS, LS1, LS2, Johnson equation), while in some other cases higher order nonlinear terms are necessary to add for restoring the integrability. Some models may even be linearized through such procedure. Therefore, in handling realistic equations the possibility of application of GE should be carefully explored.

In spite of some encouraging achievements, our ambitious programme of gauge unification is far from being complete. The extension of such equivalence to the corresponding quantum models is indeed a challenging problem, which is being taken up by us at present.

ACKNOWLEDGEMENT

I would like to extend my sincere thanks to Prof. M. Lakshmanan for giving me the opportunity to participate in this excellent Winter School and also for suggesting me the possible simplification of the XLL equation.

REFERENCES

[1] C. S. Gardner, J. M. Greene, M. D. Kruskal and R. M. Miura, Phys. Rev. Lett. **19** (1967) 1095.

[2] V. I. Zakharov and A. B. Shabat, Zh Eksp. Theor. Fiz. **61** (1971)118.

[3] M. J. Ablowitz, D. J. Kaup, A. C. Newell and H. Segur, Stud. Appl. Math. **53** (1974) 249.

[4a] A. Kundu, Lett. Math. Phys. **6** (1982) 479.
[4b] A. Kundu and O. Pashaev, J. Phys. **C16** (1983) L585.
[5] A. Kundu, J. Math. Phys. **25** (1984) 3493.
[6] A. Kundu, J. Phys. **A19** (1986) 1303.
[7] A. Kundu, J. Phys. A (1986) (in press).
[8] A. Kundu, Physica D (1986) (in press).
[9] M. Lakshmanan, Phys. Lett. **61A** (1977) 53.
[10] K. Pohlmayer, Comm. Math. Phys. **46** (1976) 207.
[11] V. I. Zakharov and L. A. Takhtajan, Teor. Math. Fiz. **38** (1979)26.
[12] H. Eichenherr and J. Honnerkamp, J. Math. Phys. **22** (1981) 374; J. Honnerkamp, J. Math. Phys. **22** (1981) 277.
[13] S. J. Orfanidis, Phys. Rev. **D21** (1981) 1513.
[14] R. Sasaki and Th. W. Ruijgrok, Physica **113A** (1982) 388.
[15] V. I. Zakharov and A. B. Shabat, Zh. Eksp. Teor. Fiz. **61** (1973)118.
[16] V. Makhankov, Phys. Lett. **81A** (1981) 156.
[17] D. J. Kaup and A. C. Newell, J. Math. Phys. **19** (1978) 798.
[18] M. Wadati, K. Konno and Y. H. Ichikawa, J. Phys. Soc. Japan **46** (1979) 1965.
[19] H. H. Chen, Y. C. Lee and C. S. Liu, Phys. Scr. **20** (1979) 490.
[20] M. Wadati, K. Konno and Y. H. Ichikawa, J. Phys. Soc. Japan **47** (1979) 1698.
[21] V. S. Gerdzikov and M. I. Ivanov, Preprint E2-82-959 JINR, Dubna (1982).
[22] R. S. Johnson, Proc. Roy. Soc. Lond. **A357** (1977) 131.
[23] A. C. Newell, SIAM J. Appl. Math. **35** (1978) 650.
[24] D. J. Benney, Stud. Appl. Math. **56** (1977) 81.
[25] M. Boiti, F. Pempinelli and G. Z. Tu, Prog. Theor. Phys. **69** (1983) 48.
[26] L. A. Takhtajan, Phys. Lett. **64A** (1977) 235.
[27] A. V. Mikhailov and A. B. Shabat, Phys. Lett. **116A** (1986) 191.
[28] R. Dodd and A. Fordy, Dublin Inst. preprint D/AS-STP 82-08(1982).
[29] R. M. Miura, ed., **Bäcklund Transformations**, Lecture Notes in Mathemathics, No. 515 (Springer, Berlin, 1976).
[30] M. Wadati and K. Sogo, J. Phys. Soc. Japan **52** (1983) 394.
[31] N. N. Bogolyubov, A. K. Prikarpatskii, A. M. Kurbatov and V. G. Samolienko, TMF **65** (1985) 271.

Prolongation Structure in One and Two Dimensions

A. Roy Chowdhury

High Energy Physics Division, Department of Physics,
Jadavpur University, Calcutta 700032, India

A computational basis for the prolongation theory applicable to soliton equations is given with several examples included as illustrations.

1. INTRODUCTION

One of the most important developments in theoretical physics during the past twenty years is the understanding of the complete integrability property of a class of nonlinear partial differential equations(nlpde)[1]. The initial attempts for the systematic analysis of such equations started with the famous papers of Gardner, Greene, Kruskal and Miura [2], and also with that of Lax [3]. The observation of Lax, that these integrable equations are representable as the consistency condition of two linear equations was of utmost importance for the fast development of the theory of solitons. However for a long time there was no concrete way to arrive at a Lax pair for a given nlpde, though given the Lax pair the problem can be analysed exhaustively. Then the now famous work of AKNS (Ablowitz, Kaup Newell, Segur) [4] indicated a reverse way of getting a class of integrable evolution equations starting from a given Lax pair. The first detailed study of inverse scattering was also taken up in this paper. One should of course mention the pioneering work of Zakharov and Shabat [5] also. But the problem of getting a Lax pair from a given nlpde was still not solved.

The basic problem was attacked and solved in the ingenious article of Whalquist and Estabrook [6]. Later a simplified version was also suggested by Corones [7], who showed how a pseudopotential can be associated with a given nlpde. Later, different applications of the method have been given by Gibbon [8], Fordy [9], Dodd [10], Roy Chowdhury [11] and others [12].

2. RUDIMENTS OF THE THEORY OF DIFFERENTIAL FORMS

The whole formalism of Whalquist and Estabrook is based on the Cartan's calculus of differential forms. Here we describe tersely the basic rules for the operations on forms. Since we will be working mainly in two spatial dimensions, we will use (x,y,t) as the set of independent variables and $u(x,y,t)$ as the dependent nonlinear field variable.

We start with the definition of differential forms:

(i) A differential dx, dt, du or an expression fdx, gdt, etc., is called one form.

(ii) An expression of the form $dx\wedge dt$, $du\wedge dx$, $dy\wedge du$ is called two forms, where $\wedge$ is the antisymmetric wedge product.

(iii) In fact one can also define the 1-form, 2-form, 3-form as the integrands in respectively of a single, double or triple integral. That is fdx of $\int fdx$, $gdxdy$ of $\int\int gdxdy$ and $hdxdydz$ of $\iiint hdxdydz$.

(iv) The symbol "d" stands for exterior derivative and acts in the following manner: $d(fdx) = df\wedge dx$, $d(gdxdt) = dg\wedge dx\wedge dt$.

(v) It possesses the basic antisymmetric property: $dx\wedge dt = -dt\wedge dx$, $dx\wedge dx = 0$.

(vi) In general for any coordinate representation, $df = f,_i\, dx^i$, $f,_i = \frac{\partial f}{\partial x_i}$.

(vii) $d(dw) = 0$, Poincáre Lemma.

(viii) For any vector v, and forms w and σ : $v.(w\wedge\sigma) = (v.w)\wedge\sigma + (-1)^p w\wedge(v.\sigma)$.

(ix) For forms w and σ , $w\wedge\sigma = (-1)^{pq}\ \sigma\wedge w$, where p = rank of w, q = rank of σ.

3. nlpde AS THE SECTION OF A SET OF FORMS

The starting point of the WE formulation is to write the nonlinear differential equation as a collection of differential forms. The process of obtaining a particular equation from a set of forms is called sectioning. In short it means that if one chooses a particular coordinate representation then the "coordinate free" writing of the differential form will reproduce the particular equation. In order to use a minimum amount of abstraction we illustrate this with examples. Let us consider the KdV equation

$$q_t + qq_x + q_{xxx} = 0. \tag{1}$$

The first step is to define a set of primary sets of variables by considering the derivatives of q, upto n-th degree (n being one less than the highest degree of derivative of q occurring in the equation), as independent set. In this case we consider,

$$u = q_x; \quad p = u_x = q_{xx}. \tag{2}$$

So (u,p,q) form the primary set. To recast these in the differential form language we set,

$$\alpha_1 = u dx\wedge dt - dq\wedge dt, \quad \alpha_2 = p dx\wedge dt - du\wedge dt \tag{3}$$

whence the equation (1) is

$$q_t + q_u + p_x = 0. \tag{4}$$

So we have

$$\alpha_3 = dq\wedge dx - qu dx\wedge dt - dp\wedge dt \tag{5}$$

and (1) is equivalent to $\{\alpha_1, \alpha_2, \alpha_3\}$ if we choose a coordinate such that

$$du = u_x dx + u_t dt, \quad dq = q_x dx + q_t dt. \tag{6}$$

The above statement is quite easy to prove. From (6) and (3), we get,

$\alpha_1 = u dx\wedge dt - (q_x dx + q_t dt)\wedge dt = (u - q_x) dx\wedge dt$. So $\alpha_1 = 0$ implies $u = q_x$.

Consider a more complicated situation for an equation of the form [12]

$$u_{xxx} - \frac{3}{2}\alpha^2 u^2 u_x + 3\,\partial_x^{-1} u_{tt} - 3u_x \partial_x^{-1} u_t = 0. \tag{7}$$

In this case we set

$$u_x = p, \quad r = p_x = u_{xx}, \quad u_t = -\frac{4}{3}w_x; \quad \frac{1}{4}r_x - w_t - \frac{3}{2}u^2 p + 2pw = 0. \tag{8}$$

These forms can be written as

$$\beta_1 = du\wedge dt - p dx\wedge dt, \quad \beta_2 = dp\wedge dt - r dx\wedge dt, \quad \beta_3 = du\wedge dx - \frac{4}{3} dw\wedge dt,$$

$$\beta_4 = dw\wedge dx + \frac{1}{4} dr\wedge dt + (2pw - \frac{3}{2}u^2 p) dx\wedge dt. \tag{9}$$

The verification can be done following the same procedure as above.

4. CLOSED IDEAL

The next stage of the prolongation analysis makes use of an important aspect of the differential forms, which is the closure of the set of forms under exterior differentiation. In mathematical language we write

$$d\alpha_i = \sum_{j=1}^{n} f_j^i \wedge \alpha_j \qquad (10)$$

As an example, let us consider equations (3), (4) and compute $d\alpha_i$. Using $d(fdx) = df\wedge dx$, $d(dg) = 0$, we get $d\alpha_1 = du\wedge dx\wedge dt = dx\wedge \alpha_2$, $d\alpha_2 = dp\wedge dx\wedge dt = dx\wedge\alpha_3$ and $d\alpha_3 = u\alpha_1\wedge dx + q\ \alpha_2\wedge dx$. So the f_j^i's are either dx or udx or qdx. On the other hand for equations (9) we obtain

$$d\beta_1 = -dp\wedge dx\wedge dt = -\beta_2\wedge dx, \quad d\beta_2 = -dr\wedge dx\wedge dt = 4\beta_4\wedge dx, \; d\beta_3 = 0,$$

$$d\beta_4 = (2w - \tfrac{3}{2}u^2)\beta_2\wedge dx + 2p\beta_4\wedge dt + 3pu\beta_1\wedge dx.$$

5. THE PROLONGATION ANALYSIS

After the closure property of the set of basis forms has been established the final stage is set for the computation of the prolongation structure. Here we actually search for a set of one forms ω_k,

$$\omega_k = dy_k + F_k dx + G_k dt, \qquad (13)$$

where F_k and G_k depend on the premitive set (u,p,q,...), independent variables (x,t) and also on some new dependent variables y_k (which are called the prolongation variables), in such a way that

$$d\omega_k = \sum_{i}^{m} g_i^k \alpha_i + \sum_{i=1}^{n} \eta_i^k \wedge \omega_i, \qquad (14)$$

where n is the number of prolongation variables, m is the number of basic defining forms α's and η_i^k is some set of one forms. Equation(14)is actually an extension of the closure condition elaborated above. Actually they lead to an overdetermined set of partial differential equations for F and G which are not always linear. Fortunately the nonlinear part always has certain commutator-like structure and almost always solvable. We will be using these features in the following paragraphs by some explicit examples. Incidentally when F^k, G^k do not depend on y's themselves, ω^k is called a potential. However if F and G depend on the y's, they are called pseodopotentials. To illustrate the above ideas we consider again the case of KdV equation.

An important observation at this stage is that η_i can be chosen in various ways, on which depends the generality of the prolongation forms ω_k. In the usual case we take

$$\eta_i = (a_i dx + b_i dt). \qquad (15)$$

On the other hand, one can also set [13] $\eta_i = a_i dx + b_i dt + c_i dz + d_i dp + e_i dq$. Equating to zero the coefficients of basic two forms $dx\wedge du$, $du\wedge dp$, $dp\wedge dt$, etc., in equation (14), with α_1, α_2, α_3 as in (3), (4), (5) we arrive at

$$F^k_{,u} = 0; \quad F^k_{,p} = 0; \quad F^k_{,q} + G^k_{,p} = 0 \ , \tag{16a}$$

$$uG^k_{,q} + pG^k_{,u} - 12uG^k_{,p} + G^i \frac{\partial F^k}{\partial y_i} - F^i \frac{\partial G^k}{\partial y_i} = 0 . \tag{16b}$$

In (16) we have used

$$d\omega_k = d(dy_k + F_k dx + G_k dt) = dF_k \wedge dx + dG_k \wedge dt$$

$$= \frac{\partial F_k}{\partial \sigma_i} d\sigma_i \wedge dx + \frac{\partial G^k}{\partial \sigma_i} d\sigma_i \wedge dt, \quad \sigma_i = \{q,u,p,x,t,y_k\}. \tag{17}$$

Equation (16) will determine the dependence of F and G on (u,p,q). Differentiating repeatedly (16b) and utilising (16a), we get $F_{qqq} = 0$, $F_p = F_u = 0$; $G_{qqqq} = 0$; $G_{uuu} = 0$; $G_{pp} = 0$. Then it is not difficult to envisage the following structure of F^k and G^k,

$$F^k = X^k_1(y) + qX^k_2(y) + q^2X^k_3(y); \quad G^k = -2(p+6q^2)X^k_2(y) +$$
$$3(u^2-9q^3-2qp)X^k_3(y) + 8X^k_4(y) + 8qX^k_5(y) + 4q\ X^k_6(y)+4uX^k_7(y) , \tag{18}$$

where X^k_i,s are dependent only on the prolongation variables. If we plugg in these F^k and G^k's in equation (16b) and equate coefficients of q^2, q^3, uq, pq, etc., we then get the following incomplete Lie algebra

$$[X_1,X_3] = [X_2,X_3] = [X_1,X_4] = [X_2,X_6] = 0$$

$$[X_1,X_2] = -X_7; \quad [X_1,X_7] = X_5; \quad [X_2,X_7] = X_6$$

$$[X_1,X_5] + [X_2,X_4] = 0; \quad [X_3,X_4] + [X_1,X_6] = -X_7. \tag{19}$$

As a next example, we consider the equation of a relativistic string in a curved space-time. The governing equations are [14]

$$\varphi_{xx} - \varphi_{tt} = e^{2\varphi}\cos 2\chi + Ke^{-2\varphi}, \quad \chi_{xx} - \chi_{tt} = e^{2\varphi}\sin 2\chi, \tag{20}$$

where K is the curvature of the embedding space. We can also write (2) as

$$\varphi_{1tt} - \varphi_{1xx} = \frac{1}{2}e^{-2i(\varphi_1+\varphi_2)} - \frac{1}{2}e^{2i(\varphi_1-\varphi_2)} = A,$$

$$\varphi_{2tt} - \varphi_{2xx} = -\frac{1}{2}e^{2\varphi_2} + \frac{1}{2}(e^{2(\varphi_1+\varphi_2)} + e^{2(\varphi_1-\varphi_2)}) = B \qquad (21)$$

when a special value of K is chosen.

The basis variables are now defined to be $\varphi_{2t} = z$, $\varphi_{2x} = p$, $\varphi_{1t} = q$, $\varphi_{1x} = r$. Then it is easy to observe that the basic two forms are

$$r_1 = d\varphi_2\wedge dx - zdt\wedge dx, \quad r_2 = d\varphi_1\wedge dx - qdt\wedge dx,$$

$$r_3 = d\varphi_2\wedge dt - pdx\wedge dt, \quad r_4 = d\varphi_1\wedge dt - rdx\wedge dt,$$

$$r_5 = dz\wedge dx + dp\wedge dt - Adt\wedge dx, \quad r_6 = dq\wedge dx + dp\wedge dt - Bdt\wedge dx. \qquad (22)$$

The closure of the ideal generated by r_i's can again be tested and we search for ω in the form

$$\omega_k = dy_k + F_k(z,p,q,r,\varphi_1,\varphi_2,y_i)dx + G_k(z,p,q,r,\varphi_1,\varphi_2,y_i)dt.$$

Following the procedure laid down previously we get

$$G_z = 0 = G_q \ ; \quad F_p = 0 = F_r \ ;$$

$$[F,G] = -F_{\varphi_2}z - F_{\varphi_1}q + G_{\varphi_2}p + G_{\varphi_1}r - AF_2 - BF_q. \qquad (23)$$

We choose F ang G as

$$F^k = zX_1^k(y) + qX_2^k(y) + F_3^k(\varphi_1,\varphi_2,y_k),$$

$$G^k = rX_2^k(y) + pX_1^k(y) + G_3^k(\varphi_1,\varphi_2, y_k). \qquad (24)$$

Equation (23) then implies

$$[X_1,G_3] = -F_{3\varphi_2} \ ; \quad [F_3,X_2] = G_{3\varphi_1},$$

$$[X_2,G_3] = -F_{3\varphi_1}; \quad [F_3,X_1] = G_{3\varphi_2} \qquad (25)$$

along with

$$[F_3,G_3] = -AX_1(y) - BX_2(y), \qquad (26)$$

$$G_3 = e^{\varphi_1-\varphi_2}Y_3(y) + e^{-(\varphi_1+\varphi_2)}Y_4(y) + e^{\varphi_2}Y_5(y) \qquad (27)$$

which immediately leads to the incomplete Lie algebra

$$[X_1,Y_3] = X_3, \quad [X_2,Y_3] = -X_3, \quad [X_1,Y_1] = X_4;$$

$$[X_2,Y_4] = X_4, \quad [X_2,Y_5] = 0, \quad [X_1,X_3] = Y_3;$$

$$[X_1,X_4] = Y_4, \quad [X_2,X_5] = 0, \quad [X_1,X_5] = -Y_5;$$

$$[X_3,Y_3] = \tfrac{1}{2}(X_1-X_2), \quad [X_4,Y_4] = \tfrac{1}{2}(X_1+X_2);$$

$$[X_5,Y_5] = -\tfrac{1}{2}X_1, \quad [X_1,X_2] = 0, \quad [X_3,Y_4] + [X_4,Y_3] = 0;$$

$$[X_3,Y_5] + [X_5,Y_3] = 0, \quad [X_4,Y_5] + [X_5,Y_4] = 0. \tag{28}$$

Lastly let us consider a new integrable system, which is a generalisation of Liouville equation [15] written as

$$\varphi_{xt} = -e^{\alpha+\beta-2\varphi},$$

$$\alpha_{tt} = \alpha_t\varphi_t - (\alpha_t)^2 - e^{\beta-\varphi}\beta_x,$$

$$\beta_{xx} = \beta_x\varphi_x - (\beta_x)^2 - e^{\alpha-\varphi}\alpha_t. \tag{29}$$

If we set $\alpha_t = q$, $\beta_x = r$, $\varphi_t = s$, $\varphi_x = p$, the basic two forms become

$$\alpha_1 = d\varphi\wedge dt - p\,dx\wedge dt, \quad \alpha_2 = d\alpha\wedge dx - q\,dt\wedge dx;$$

$$\alpha_3 = d\beta\wedge dt - r\,dx\wedge dt, \quad \alpha_4 = d\varphi\wedge dx - s\,dt\wedge dx;$$

$$\alpha_5 = dp\wedge dx + e^{\alpha+\beta-2\varphi}dt\wedge dx, \quad \alpha_6 = dq\wedge dx - q\,d\varphi\wedge dx + q^2dt\wedge dx - e^{\beta-\varphi}d\beta\wedge dt;$$

$$\alpha_7 = dr\wedge dt - r\,d\varphi\wedge dt + r^2dx\wedge dt + e^{\alpha-\varphi}d\alpha\wedge dx, \quad \alpha_8 = ds\wedge dt + e^{\alpha+\beta-2\varphi}dt\wedge dx. \tag{30}$$

In this case F^k, G^k are seen to be

$$F^k = pX_1^k(y) + e^{\alpha-\varphi}X_2^k(y) + e^{2\varphi}X_3^k(y) + qe^{\alpha-3\varphi}X_4^k(y),$$

$$G^k = sX_5^k(y) + e^{\beta-\varphi}X_6^k(y) + e^{2\varphi}X_7^k(y) + re^{\beta-3\varphi}X_8^k(y) \tag{31}$$

along with the Lie algebra

$$[X_1,X_5] = 0\ ,\quad [X_3,X_5] = -2X_3,\quad [X_1,X_6] = -X_6;$$

$$[X_3,X_6] = 0,\quad [X_1,X_7] = 2X_7,\quad [X_3,X_7] = 0;$$

$$[X_1,X_8] = -2X_8,\quad [X_3,X_8] = X_6,\quad [X_2,X_5] = X_2;$$

$$[X_4,X_5] = 2X_4,\quad [X_2,X_6] = X_1 - X_5,\quad [X_1,X_6] = X_8;$$

$$[X_2,X_7] = 0,\quad [X_4,X_7] = -X_2,\quad [X_2,X_8] - X_4,\quad [X_4,X_8] = 0. \qquad (32)$$

6. CLOSURE OF THE ALGEBRA

In all the examples cited above it is pertinent to observe that the Lie algebra generated is never closed by itself. Of course there are some examples where an exception can be seen [16]. In the former case there is no rule to attain this closure. However one may follow either one of the following two broad strategies:

(1) First, to find a scaling, Galiean, Lorentz or conformal type of symmetry of the original nlpde. Then impose these on ω_k, to obtain on automorphism of F and G. This will have some nontrivial implications regarding the unknown part of the Lie algebra. As an example, let us consider the case of the extended Liouville equation (29). It remains invariant under the transformation

$$x' \to \zeta x\ ,\quad t' \to \zeta^{-1} t,\quad \varphi' \to \varphi;$$

$$\alpha' \to \alpha - 3\log\zeta,\quad \beta' \to \beta + 3\log\zeta. \qquad (33)$$

Then we impose these on ω_k and get $F^{k'} \to \zeta^{-1}F^k$; $G^{k'} \to \zeta G^k$, which lead to the automorphism

$$X_1' \to X_1,\quad X_2' \to \zeta^2 X_2,\quad X_3' \to \zeta^{-1}X_3,\quad X_4' \to \zeta X_4,$$

$$X_5' \to X_5,\quad X_6' \to \zeta^{-2}X_6,\quad X_7' \to \zeta X_7,\quad X_8' \to \zeta^{-1}X_8. \qquad (34)$$

Now equation (32) does not give any information about $[X_2,X_4]$. If we study the transformation of $[X_2,X_4]$ under (34) we get $[X_2',X_4'] \to \zeta^3[X_2,X_4]$. But there is no generator with such a scaling property in (34). So its immediate implication is $[X_2,X_4] = 0$. On the other hand, if we consider $[X_3,X_4]$, under (34) we get $[X_3',X_4'] \to [X_3,X_4]$ and generators with such scaling behaviour (with no factor of ζ), are X_1 and X_5. So we infer

$$[X_3, X_4] = \alpha X_1 + \beta X_5 \tag{35}$$

and α and β may be fixed by Jacobi identity.

(2) The other route is the adhoc procedure adopted by WE who closed the algebra by hand. The main algebraic reason for the effectiveness of WE procedure was analysed by Shadwick [18]. Basically the procedure of closing off is to obtain the various forms of embedding of SL(2,R), SL(3,R) or SU(3), etc., in the incomplete Lie algebra. Such a procedure is described in detail in references [9] and [10].

Once the Lie algebra is closed, then one can successfully obtain the linearization, Lax pairs and Bäcklund transformations. For details see [6-19].

7. EQUATIONS IN TWO SPACE VARIABLES

After the success of prolongation theory in (1+1) dimensions, extension was made for (2+1) dimensional systems such as the K-P equation, Davey-Stewartson equation by Morris [19], though the IST equations for these were known *a priori* from the research of Dryuma [20] and Zakharov and Shabat [21]. Here we indicate the procedure by considering a novel application. We can summarize the basic rules for higher dimension as follows:

(1) In n-dimension the nlpde's are equivalent to n-forms α^i,

$$\alpha^A = \frac{1}{n}\,\alpha^A_{\mu_1\mu_2\ldots\mu_n}\, dx^{\mu_1}\wedge dx^{\mu_2}\wedge \ldots dx^{\mu_n} .$$

(2) They satisfy closure as before $d\alpha^A \subset I$; $I = \{\alpha^A\}$.

(3) Then we introduce the prolongation variables y's and (n-1) forms

$\Omega^i = \beta^i_j \wedge \omega^j$, $i = 1,2,3,\ldots,\dim y^j$, β^i_j are (n-2) forms to be determined, $\beta^i_j = \frac{1}{(n-2)}\, b^i_{j\,\mu_1\mu_2\ldots\mu_{n-2}}\, dx^{\mu_1}\wedge dx^{\mu_2}\wedge \ldots dx^{\mu_{n-2}}$

and ω^i is the connection one form defined as $\omega^i = dy^i + \Sigma\Lambda^{ik}\, dx^k$.

(4) We again impose the closure for the extended set $I' = d\Omega^i \subset I'$.

An important and novel equation is the Benjamin-Ono equation [21],

$$u_t + uu_x + \frac{P}{\pi} \int \frac{u_{x'x'}}{x'-x} dx' = 0 \tag{36}$$

which can be written as a 3-d system following reference [22] as

$$u_{xx} + u_{yy} = 0,$$

$$u_t + uu_x + u_{xy} = 0. \tag{37}$$

The basic three forms are [24]

$$\alpha_1 = du\wedge dy\wedge dt - pdx\wedge dy\wedge dt, \quad \alpha_2 = -du\wedge dx\wedge dt - rdx\wedge dy\wedge dt,$$

$$\alpha_3 = dp\wedge dy\wedge dt - dr\wedge dx\wedge dt, \quad \alpha_4 = dr\wedge dy\wedge dt + updx\wedge dy\wedge dt + du\wedge dx\wedge dy. \tag{38}$$

We then set

$$\omega^i = dy^i + F^i dx + G^i dy + H^i dt, \tag{39}$$

F, G, H depend on (u,p,r,x,y,t,y_i) and

$$\Omega^j = (a_k^j\, dx + b_k^j\, dy + c_k^j\, dt)\wedge w^k \tag{40}$$

with $[a_\alpha^j, b_\alpha^j] = 0$; $[b_k^i, c_k^i] = 0$; $[c_\alpha^j, a_\beta^j] = 0$.

Then Ω^j are written as

$$\Omega^j = a_k^j\, dx\wedge dy^k + b_k^j\, dy\wedge dy^k + c_k^j\, dt\wedge dy^k$$
$$+ L_k^j\, dx\wedge dy + M_k^j\, dt\wedge dx + N^j\, dt\wedge dy.$$

We then demand $d\Omega^j = \Sigma\alpha_i f_i^j + (\lambda dx + \mu dy + \nu dt)\wedge \Omega^j$ which will then lead to equations for L, M and N. Details of such a computation can be found in refs. [25]-[27].

8. THE CONCEPT OF CONSTANT COEFFICIENTS IDEAL

A few years back an excellent method was suggested by K. Harrison [25] for circumventing the difficulty of fixing the arbitrariness in the form of F and G. He suggested that if it is possible to choose a special class of two forms so that the nlpdes can be converted into an ideal of differential forms with constant coefficients (C.C. Ideal) then the process of obtaining the form of F and G can be streamlined and one can arrive at the incomplete Lie algebra quickly. Let us refer back to equation (22) and choose a new set of forms equivalent to the nlpde's,

$$\varepsilon_1 = dt - dx, \quad \varepsilon_2 = zdx + pdt, \quad \varepsilon_3 = qdx + rdt,$$

$$\varepsilon_4 = e^{2\varphi_2}(dx + dt), \quad \varepsilon_5 = e^{-2(\varphi_1+\varphi_2)}(dx + dt), \quad \varepsilon_6 = e^{2(\varphi_1+\varphi_2)}(dx+dt). \tag{40a}$$

Then we can observe that

$$d\varepsilon_1 = 0, \quad d\varepsilon_2 = \frac{1}{4}[\varepsilon_4\wedge\varepsilon_1 - \varepsilon_5\wedge\varepsilon_1 - \varepsilon_6\wedge\varepsilon_1],$$

$$d\varepsilon_3 = \frac{1}{4}[\varepsilon_6\wedge\varepsilon_1 - \varepsilon_5\wedge\varepsilon_1], \quad d\varepsilon_4 = 2\varepsilon_4\wedge\varepsilon_2,$$

$$d\varepsilon_5 = -2[\varepsilon_5 \wedge \varepsilon_2 + \varepsilon_5 \wedge \varepsilon_3], \quad \varepsilon_4 \wedge \varepsilon_5 = \varepsilon_4 \wedge \varepsilon_6 = \varepsilon_5 \wedge \varepsilon_6 = 0. \tag{41}$$

It is interesting to note that these closure conditions do not involve any variable coefficients so that (20) is equivalent to the C.C. ideal generated by (41). The one form ω is now written as

$$\omega^k = -y_k + (\Sigma B^i \varepsilon_i) y_k, \tag{42}$$

B^i being numerical matrices so that $d\omega = 0$ leads to

$$B^i d\varepsilon^i - \frac{1}{2}[B^i, B^k]\varepsilon_i\wedge\varepsilon_k = 0 \tag{43}$$

which immediately implies

$$[B_1, B_2] = 0; \quad [B_2, B_3] = 0; \quad [B_1, B_3] = 0;$$

$$[B_2, B_4] = -B_4; \quad [B_1, B_4] = -\frac{1}{2}B_2; \quad [B_3, B_5] = iB_5;$$

$$[B_1, B_5] = \frac{1}{2}(B_2+B_3); \quad [B_2, B_6] = 4B_6;$$

$$[B_1, B_6] = \frac{1}{2}(B_2-B_3); \quad [B_3, B_4] = 0; \quad [B_3, B_5] = 4B_5; \quad [B_1, B_6] = 4B_6. \tag{44}$$

The same procedure can also be illustrated with the basis of sine-Gordon equation. It is written as, $\varphi_{xt} = \sin\varphi$, Set $r = \varphi_x$, $r_t = \sin\varphi$.

The usual two forms are $\alpha = d\varphi \wedge dt - r dx \wedge dt$ and $\beta = dr \wedge dx - \sin\varphi \, dx \wedge dt$. But a set of C.C. ideals can be constructed by choosing the basis given by

$$\xi_1 = dx, \quad \xi_2 = r dx, \quad \xi_3 = \sin\varphi \, dt, \quad \xi_4 = \cos\varphi \, dt;$$

$$\xi_1 \wedge \xi_2 = \xi_3 \wedge \xi_4 = 0, \quad d\xi_1 = 0, \quad d\xi_4 = \xi_3 \wedge \xi_2,$$

$$d\xi_2 = \xi_3 \wedge \xi_1; \quad d\xi_3 = \xi_2 \wedge \xi_4 \tag{45}$$

is closed but with constant coefficients. We again write $\omega^k = -dy^k + (B^i \xi_i) y^k$. Then $d\omega^k = 0$ implies

$$[B^2, B^3] = -B^4, \quad [B^1, B^3] = -B^2, \quad [B^2, B^4] = B^3, \quad [B^1, B^4] = 0 \tag{46}$$

leading to a Lax pair for sine-Gordon equation.

9. USE OF PROLONGATION STRUCTURE FOR OBTAINING BÄCKLUND TRANSFORMATION

Lastly we only mention another important use of prolongation theory. One can use the forms ω^k to deduce a B.T. [26] of the particular equation. The basic principle is to assume that the new field variable depends on the old one and also on the primitive variables along with the prolongation variables y_k. That is, $u' = u'(\varphi, u, p, \ldots, x, t, y_k)$. For KdV, we write for the new set of forms as

$$\alpha_1' = du' \wedge dt - z' dx \wedge dt, \quad \alpha_2' = dz' \wedge dt - p' dx \wedge dt, \quad \alpha_3' = -du' \wedge dx$$

and demanding that these be in the ring of the prolonged ideal, we can obtain the BT, $u' = -u - y^2 + \lambda$. Detailed discussions can be found in refs. [26]-[29] for other systems.

10. CONCLUSION

In our above exposition we have tried to give a computational basis for the prolongation theory that may be useful for a beginner. There are many references for the geometric or differential geometric background for the prolongation structure, which we have not touched at all. Finally, it is pleasure to thank the organisers, Prof. M. Lakshmanan and Prof. P. K. Kaw for giving me this opportunity to deliver this talk at the Winter School organized by SERC (DST, Government of India).

REFERENCES

[1] See for example, C. Rebbi and G. Soliani, **Solitons and Particles** (World Scientific Press, Singapore, 1983).

[2] C. S. Gardner, J. M. Greene, M. D. Kruskal, R. M. Miura, Phys. Rev. Lett. **19** (1967) 1095.

[3] P. D. Lax, Commn. Pure Appl. Math. **21** (1968) 467.

[4] M. J. Ablowitz, D. J. Kaup, A. C. Newell, and H. Seger, Stud. App. Math. **53** (1974) 249.

[5] V. E. Zhakharov and A. B. Shabat, Sov. Phys. JETP, **34** (1972) 62.

[6] H. Whalquist and F. B. Estabrook, J. Math. Phys. **16** (1975) 1; Phys. Rev. Lett. **37** (1973) 1386.

[7] J. Coronoes, Some Heuristic Comments on Soliton, Lecture Notes in Physics, Vol. 810 (Springer, Berlin, 1979), p. 23.

[8] J. D. Gibbon, Lecture Notes in Maths. 775 (Springer, Berlin).

[9] R. K. Dodd and A. Fordy, Proc. Roy. Soc. **A385** (1983) 389.

[10] R. K. Dodd and J. D. Gibbon, Proc. Roy. Soc. **A359** (1978) 411.

[11] A. Roy Chowdhury and T. Roy, J. Math. Phys. **21** (1980) 189; J. Phys. **A12** (1979) 189; J. Math. Phys. **20** (1979) 1559.

[12] D. J. Kaup, Physica **1D** (1980) 391; M. Lakshmanan, Physics Lett. **64A** (1978) 354; J. Math. Phys. **20** (1979) 1667 ; App. Sci. Research **37** (1981) 127.

[13] A. Roy Chowdhury and S. Ahmad, Prog. Theo. Phys. **75** (1986) 1250.

[14] P. Molino, J. Math. Phys. **25** (1984) 2222.

[15] A. Roy Chowdhury and Shibani Sen, Prog. Theo. Phys. **76** (1986) 1.

[16] A. Roy Chowdhury and K. De Archan, J. Math. Phys.

[17] A. Roy Chowdhury and S. Ahmad, Phys. Rev. **D32** (1985) 2780.

[18] W. F. Shadwick, J. Math. Phys. **21** (1980) 454.

[19] H. C. Morris, J. Math. Phys. **17** (1976) 1870.

[20] V. Dryuma, Sov. Phys. JETP **19** (1974) 387.

[21] V. E. Zakharov and A. B. Shabat, Func. Analysis App. **13** (1979)166.

[22] T. B. Benjamin and H. Ono, J. Fluid Mech. **25** (1966) 241; J. Phys. Soc. Jpn. **39** (1975) 1082.

[23] B. Grammaticos, B. Dorizzi and A. Ramani, Phys. Rev. Lett. **53** (1984) 1.

[24] A. Roy Chowdhury and S. Ahmad, ICTP (Trieste, Italy), Preprint No. IC/86/61.

[25] B. K. Harrison, J. Math. Phys. **24** (1983) 2178.

[26] R. K. Dodd and H. C. Morris, in Lecture Notes in Maths., Vol. 810 (Springer, Berlin), pp. 63, 95.

[27] A. Roy Chowdhury, Phys. Scr. **29** (1986) 289.

[28] C. Rogers and W. F. Shadwick, **Bäcklund Transformations and Their Applications** (Academic Press, New York, 1982).

[29] H. C. Morris, J. Math. Phys. **18** (1977) 533; Int. J. Theor. Phys. **16** (1977) 227; Chris Radford, J. Math. Phys. **27** (1986) 1266.

Integrable Equations in Multi-Dimensions (2+1) are Bi-Hamiltonian Systems

A.S. Fokas[1] *and P.M. Santini*[1;2]

[1]Department of Mathematics and Computer Science and Institute for Nonlinear Studies, Clarkson University, Potsdam, NY 13676, USA
[2]Permanent Address: Universitá Degli Studi, Roma, Istituto di Fisica "Guglielmo Marconi", Piazzale delle Scienze, 5, I-1-00185 Roma, Italy

Recent developments by the authors in finding the recursion operators and the bi-Hamiltonian formulation of a large class of nonlinear evolution equations in (2+1)-dimensions is reviewed. The general theory associated with factorizable recursion operators in multidimensions is discussed. Both gradient and non-gradient master-symmetries are simply derived and their general theory is developed, using the Kadomtsev-Petviashvili equation as an example.

1. INTRODUCTION

Ablowitz, Kaup, Newell and Segur [1], following ideas of Lax [2] were the first to solve in the concrete case of the Dirac problem the following question: Given a linear eigenvalue problem find all nonlinear equations that are related to it. They found that associated with a given eigenvalue problem there exists a hierarchy of infinitely many equations. This hierarchy is generated by a certain linear operator. This operator is the squared eigenfunction operator of the underlying linear eigenvalue problem. The operator generating the KdV hierarchy (i.e., the squared eigenfunction operator of the Schrödinger eigenvalue problem) was found by Lenard. For other eigenvalue problems see [3]-[10].

Olver [11] established the group theoretical origin of the above hierarchy: Finding the hierarchy associated with a given equation is equivalent to finding the non-Lie point symmetries of the given equation. He thus interpreted the squared eigenfunction operator as an operator mapping symmetries onto symmetries; this lead to a simple mathematical characterization of the recursion operator Φ. Olver was thus the first to establish that certain integrable nonlinear equations possess infinitely many symmetries. This motivates the following question: Is there an algorithmic way for generating equations possessing infinitely many symmetries? Fuchssteiner [12] discovered such a way: If an operator Φ has a certain mathematical property called hereditary then the equations $u_t = \Phi^n u_x$, n integer, possess infinitely many symmetries. From the above discussion it follows that both linear eigenvalue problems and hereditary operators yield hierarchies of equations possessing infinitely many symmetries. Actually Anderson and the author [13], following ideas of Fuchssteiner, have shown that eigenvalue problems algorithmically imply hereditary operators.

Equations solvable by the Inverse Scattering Transform are Hamiltonian systems. Magri, in a pioneering paper [14], realized that integrable Hamiltonian systems have additional structure: They are bi-Hamiltonian systems. Actually the underlying hereditary operator can be factorized in terms of the two associated Hamiltonian operators. The theory of factorizable hereditary operators has been further developed by Fuchssteiner and the author [15] and by Gel'fand and Dorfman [16].

The understanding of the central role played by factorizable hereditary operators for equations in 1+1, motivated a search for hereditary operators for equations in 2+1. However, in this direction several negative results have appeared in the literature. For example, Zakharov and Konopelchenko [17], in an interesting paper proved that recursion operators (of a certain type naturally motivated from the results in 1+1) did not exist in multidimensions. A similar result has been proved for the Benjamin-Ono (BO) equation [18]. It should be noted that the BO equation has more similarities [19] with the Kadomtsev-Petviashvili (KP) equation than with the KdV equation. Fuchssteiner and the author [18] after failing to find a recursion operator for the BO introduced the concept of the master-symmetries τ. Subsequently Oevel and Fuchssteiner [20] found a master-symmetry for the KP equation. The τ theory for equations in 2+1 has been developed by Dorfman [21] and Fuchssteiner [22]. However, the τ is not related to the underlying isospectral problem and also cannot be used to construct a second Hamiltonian operator. This is a serious drawback: several prominent investigators, for example Gel'fand [23] have considered the existence of a bi-Hamiltonian formulation as fundamental to integrability. Without finding a recursion operator Φ, one cannot find the second Hamiltonian operator. Several investigators have noticed that master-symmetries also exist for equations in 1+1. The theory for the master-symmetries T in 1+1 was developed by Oevel [24] (see also [25]) and is more satisfactory than the theory in 2+1: If one assumes that an equation is invariant under scaling then there exists a one-to-one constructive relationship between T and the recursion operator Φ.

Recently P. M. Santini and the author [26]-[28] have found the recursion operator and the bi-Hamiltonian formulation of a large class of equations in 2+1. They have also established the general theory associated with factorizable recursion operators in multidimensions. Furthermore, both gradient and non-gradient (the 2+1 analogue of T) master-symmetries are simply derived and their general theory is developed.

2. MASTER SYMMETRIES

In this section we review certain aspects of non-gradient master-symmetries in 1+1 and gradient master-symmetries in 2+1.

Definition 2.1
A function τ is a master-symmetry of the equation $q_t = K$ iff the map

$$[\cdot,\tau]_L \quad \text{where } [a,b]_L \doteq a'[b] - b'[a] \tag{2.1}$$

maps symmetries onto symmetries (prime denotes Fréchet derivative).

The first example of a master-symmetry was given for the Benjamin-Ono equation

$$q_t = Hq_{xx} + 2qq_x, \quad (Hf)(x) \doteq \frac{1}{\pi}\int_R \frac{d\xi f(\xi)}{\xi - x}. \tag{2.2}$$

It was shown in [18] that if $\tau \doteq x(Hq_{xx} + 2qq_x) + q^2 + \frac{3}{2} Hq_x$ and σ_n is a symmetry then $\sigma_{n+1} \doteq [\sigma_n,\tau]$ is also a symmetry. It was further shown in [18] that $D^{-1}\tau$ is a gradient function ($\tau'D + D\tau'^* = 0$).

Master-symmetries are intimately related to time-dependent non-Lie-point symmetries [25]. Indeed, the first non-Lie-point time-dependent symmetry is a natural candidate for a master-symmetry: Consider the evolution equation $q_t = K^{(1)}$ and let $K^{(2)}, K^{(3)}, \ldots$ denote its time-independent nón-Lie-point symmetries. Let

$$\Sigma^{(2)} = tK^{(2)} + \tau \tag{2.3}$$

be a time-dependent non-Lie-point symmetry. Then

$$K^{(2)} + [tK^{(2)} + \tau,\, K^{(1)}]_L = 0, \text{ or } K^{(2)} = [K^{(1)},\tau]_L$$

and τ is a candidate for a master-symmetry.

2.1 Master-symmetries for equations in 1+1

Lemma 2.1

Let

$$S_i \doteq \Phi'[K_i] + [\Phi, K_i'], \quad i = 1,2. \tag{2.4}$$

If Φ is hereditary, i.e., if $\Phi'[\Phi v]w - \Phi\Phi'[v]w$ is symmetric with respect to v, w, then

$$\Phi^{n+m}[K_1,K_2]_L = [\Phi^n K_1, \Phi^m K_2]_L + \Phi^n\Big(\sum_{r=1}^{m} \Phi^{m-r} S_1 \Phi^{r-1}\Big)K_2 - \Phi^m\Big(\sum_{r=1}^{n} \Phi^{n-r} S_2 \Phi^{r-1}\Big)K_1, \tag{2.5}$$

m,n are non-negative integers.

Proof

See Theorem 2.1 of [28].

Corollary 2.1

Assume that τ_0 is a scaling of both K and of the hereditary operator Φ, i.e.,

$$[K,\tau_0] = \alpha K, \quad \Phi'[\tau_0] + [\Phi, \tau_0'] = \beta\Phi. \tag{2.6}$$

Then

(i) $(\alpha + n\beta)\Phi^{n+1}K = [\Phi^n K, \Phi\tau_0]_L$, (2.7)

i.e., $\Phi\tau_0$ is a master-symmetry for $q_t = K$.

(ii) $(\alpha + n\beta)\Phi^{n+m}K = [\Phi^n K, \Phi^m \tau_0]_L$, (2.8)

i.e., $\Phi^m\tau_0$ is a master-symmetry of order m for $q_t = K$.

(iii) $\Sigma \doteq (\alpha + n\beta)t\Phi^{n+1}K + \Phi\tau_0$ is a symmetry of $q_t = \Phi^n K$.

Proof

(i) Apply Theorem 2.1 with $K_1 = K,\ K_2 = \tau_0,\ [K_1,K_2]_L = \alpha K,\ L_1 = 0,\ L_2 = \beta\Phi$.

(ii) Similar to (i).

(iii) Use the definition of a symmetry.

In the above we derive T from Φ. Now we obtain Φ from T.

Lemma 2.2

Let Φ be a hereditary operator such that $\Phi\Theta = \Theta\Phi^+$, where Θ is a constant, invertible, skew-symmetric operator. Then

$$(\Theta T)' + \Theta(\Phi T)'^{+}\Theta^{-1} = \Phi(T' + \Theta(T')^{+}\Theta^{-1}) + \Theta S^{+}\Theta^{-1}, \qquad (2.9)$$

where

$$S^{+} \doteq \Phi'^{+}[T] + [T'^{+}, \Phi].$$

Proof

See [28].

Theorem 2.1

(i) If the hereditary operator Φ admits the scaling τ_0 then $\Phi\tau_0$ is a master-symmetry for the hierarchy generated by Φ.

(ii) Assume that the hereditary operator Φ admits the scaling τ_0 and that it also satisfies $\Phi\Theta = \Theta\Phi^+$, where Θ is a constant, invertable, skew-symmetric operator which also admits the scaling τ_0. Then

$$\Phi = (\Phi\tau_0)' + \Theta(\Phi\tau_0)'^{+}\Theta^{-1}. \qquad (2.10)$$

Proof

(i) If Φ admits a scaling and K is generated from Φ then K also admits a scaling. Hence Corollary 2.1 implies (i) above.

(ii) Since Φ admits a scaling, Φ^+ also admits a scaling, hence S^+ is proportional to Φ^+, thus $\Theta S^+\Theta^{-1}$ is proportional to Φ. Furthermore, since Θ admits the scaling τ_0, $\tau_0'\Theta + \Theta\tau_0'^{+} = \alpha\Theta$, thus $\tau_0' + \Theta(\tau_0')^{+}\Theta^{-1}$ equals a constant. Hence (2.9) implies (2.10).

EXAMPLES

1. $\Phi = D + q + q_x D^{-1}$ is the hereditary operator associated with Burgers equation. It admits the scaling $q \to \alpha q$, $x \to \alpha^{-1}x$, i.e., $\tau_0 = q + xq_x$. Thus $x(q_{xx} + 2qq_x) + q^2$ is a master-symmetry of Burgers equation.

2. $\Phi = D^2 + 4q + 2q_x D^{-1}$ admits the scaling $q \to \alpha q$, $x \to \alpha^{-2}x$, i.e., $\tau_0 = q + 2xq_x$. Thus $T = \Phi\tau_0$ is a master-symmetry of the KdV.

3. If $\tau_0 = q + 2xq_x$, then $\tau_0' + D(\tau_0')^{+}D^{-1} = -3$. Hence if T is the master-symmetry of KdV,

$$\Phi = T' + D(T')^{+}D^{-1}$$

is the recursion operator of the KdV.

2.2 Gradient Master-Symmetries for Equations in 2+1

A straightforward generalization of Theorem 2.1 to equations 2+1 fails: (i) Φ could not be found; (ii) the known master-symmetries τ were gradient functions, hence $\tau' + \Theta(\tau')^{+}\Theta^{-1} = 0$. It will be shown in §3 that for equations in 2+1: (i) suitable generalizations of Φ, denoted by Φ_{12} can be found; (ii) there exist non-gradient master-symmetries T_{12} (for example for the KP $T_{12} = \Phi_{12}^{2}\delta(y_1-y_2)$, where δ denotes the Dirac delta function). Hence a generalization of Theorem 2.1 to equations in 2+1 is given in §3.

One can still develop a theory for master-symmetries without using the connection with the recursion operator Φ: see [21],[22].

3. SYMMETRIES FOR EQUATIONS IN 2+1

In this section we review the theory recently developed by Paolo Santini and the author. We use the KP as an illustrative example and quote the basic theorems when needed. We hope that this form of presentation will aid the non-expert reader to become familiar with the notions and methods developed in [26]-[28]. We advise the non-expert reader to read [15] before reading this paper since many of the results presented here are two dimensional generalizations of results given in [15].

3.1 Derivation of Recursion Operators

Given an isospectral eigenvalue problem there exists a simple algorithmic way of obtaining a recursion operator. This approach involves three steps: compatibility, an integral representation of a certain differential operator, and an expansion in terms of delta functions. Let us consider the eigenvalue equation

$$w_{xx} + q(x,y)w + \alpha w_y = 0, \quad \alpha \text{ is a constant} \tag{3.1}$$

and for convenience of notation we suppress the t-dependence. Using vector notation, (3.1) yields

$$W \doteq \begin{pmatrix} w \\ w_x \end{pmatrix}, \quad W_x = \begin{pmatrix} 0 & 1 \\ -\hat{q} & 0 \end{pmatrix} W, \quad \hat{q} \doteq q + D_y;\ D_y = \frac{\partial}{\partial y}. \tag{3.2}$$

3.1a Compatibility

Associated with $W_x = UW$ we look for compatible flows $W_t = VW$ where

$$V = \begin{pmatrix} A & 2C \\ B & E \end{pmatrix}, \quad A,\ B,\ C,\ E \text{ polynomials in } D_y.$$

Compatibility implies the **operator** equation

$$U_t = V_x - [U,V],$$

or

$$\begin{pmatrix} 0 & 0 \\ -\hat{q}_t & 0 \end{pmatrix} = \begin{pmatrix} A_x & 2C_x \\ B_x & E_x \end{pmatrix} - \left[\begin{pmatrix} 0 & 1 \\ -\hat{q} & 0 \end{pmatrix}, \begin{pmatrix} A & 2C \\ B & E \end{pmatrix}\right].$$

Solving in the above equation for A, B, E in terms of C we obtain the following operator equation:

$$q_t = C_{xxx} + [\hat{q},C]_x^+ + [\hat{q},C_x]^+ + [\hat{q},D^{-1}[\hat{q},C]] + A_0\hat{q} - \hat{q}A_0, \tag{3.3}$$

where
$[\ ,\ ]$ is a commutator, $[\ ,\ ]^+$ is an anticommutator, A_0
is an operator such that

$$A_{0_x} = 0 \text{ and } (D^{-1}f)(x,y) = \int_{-\infty}^{x} f(\xi,y)d\xi .$$

In what follows we take $A_0 = 0$ (the general case is considered in [27]).

3.1b An Integral Representation

The crucial step is to use an integral representation for the differential operator C:

$$(Cf)(x,y_1) = \int_R dy_2 T(x,y_1,y_2) f(x,y_2). \tag{3.4}$$

Let

$$q_i \doteqdot q(x,y_i), \quad D_i \doteqdot D_{y_i}, \quad i = 1,2, \quad T_{12} \doteqdot T(x,y_1,y_2). \tag{3.5}$$

Equation (3.4) implies similar integral representations for all quantities appearing on the RHS of (3.3). For example

$$(\hat{q}_1 C)f = \int_R dy_2 \{q_1 T_{12} + \alpha(D_1 + D_2)T_{12}\} f_2 .$$

For

$$(q_1C)f = \int_R dy_2(q_1T_{12})f,\quad D_1(Cf) = (D_1C)f + Cf_{y_1} = \int_R dy_2 T_{12_{y_1}} f_2,$$

$$Cf_{y_1} = \int_R dy_2 T_{12} f_{2_{y_2}} = -\int_R dy_2 T_{12_{y_2}} f_2.$$

Thus

$$(D_1C)f = \int_R dy_2(T_{12_{y_1}} + T_{12_{y_2}})f_2.$$

Similarly

$$(\hat{q}_1C \pm C\hat{q}_1)f = \int_R dy_2(q_{12}^{\pm}T_{12})f_2,$$

where the operators $q_{12}^{\pm}$ are defined by

$$q_{12}^{\pm} \doteq q_1 \pm q_2 + \alpha(D_1 \mp D_2). \tag{3.6}$$

Using the above integral representations in (3.3) we obtain

$$\delta_{12}q_{1_t} = T_{12_{xxx}} + (q_{12}^{+}T_{12})_x + q_{12}^{+}T_{12_x} + q_{12}^{-}D^{-1}q_{12}^{-}T_{12},\quad \delta_{12} \doteq \delta(y_1-y_2)$$

or

$$\delta_{12}q_{1_t} = D\Psi_{12}T_{12},\quad \Psi_{12} \doteq D^2 + q_{12}^{+} + D^{-1}q_{12}^{+}D + D^{-1}q_{12}^{-}D^{-1}q_{12}^{-}. \tag{3.7}$$

Let us introduce the operator Φ_{12} via

$$D\Psi_{12} = \Phi_{12}D,\quad \Phi_{12} \doteq D^2 + q_{12}^{+} + Dq_{12}^{+}D^{-1} + q_{12}^{-}D^{-1}q_{12}^{-}D^{-1}. \tag{3.8}$$

Thus

$$\delta_{12}q_{1_t} = D\Psi_{12}T_{12} = \Phi_{12}DT_{12}. \tag{3.9}$$

3.1c Expansions in terms of delta functions

We expand T_{12} in the form

$$T_{12} = \sum_{j=0}^{n} \delta_{12}^{j}T_{12}^{(j)};\qquad \delta_{12}^{j} \doteq \frac{d^j}{dy_1^j}\,\delta(y_1-y_2). \tag{3.10}$$

It turns out that Ψ_{12} admits a simple commutator relationship with respect to $h_{12} = h(y_1-y_2)$. Actually the following **operator** equation is valid

$$[\Psi, h_{12}] = 4\alpha h_{12}';\qquad h_{12}' \doteq \frac{d}{dy_1}\,h_{12}. \tag{3.11}$$

Hence equation (3.7) yields

$$\delta_{12}q_{1_t} = \sum_{j=0}^{n} D\Psi_{12}\delta_{12}^{j}T_{12}^{(j)} = \sum_{j=0}^{n} \delta_{12}^{j}D\Psi_{12}T_{12}^{(j)} + 4\alpha \sum_{j=1}^{n+1} \delta_{12}^{j}DT_{12}^{(j-1)}.$$

Thus

$$T_{12_x}^{(n)} = 0, \quad T_{12}^{(j-1)} = -\frac{1}{4}\Psi_{12}T_{12}^{(j)}, \quad \delta_{12}q_{1_t} = \delta_{12}D\Psi_{12}T_{12}^{(0)}. \tag{3.12}$$

Letting $T_{12}^{(n)} = 1$ we have the following proposition:

Proposition 3.1

The isospectral equation

$$w_{xx} + \hat{q}w = 0, \quad \hat{q} \doteq q + \alpha D_y; \quad \alpha \text{ constant} \tag{3.13}$$

is associated with the equations

$$q_{1_t} = \beta_n \int_R dy_2 \delta_{12} D\Psi_{12}^{n+1}.1 = \beta_n \int_R dy_2 \delta_{12}\Phi_{12}^{n}(\Phi_{12}D).1, \ \beta_n \text{ constant} \tag{3.14}$$

where

$$\Psi_{12} \doteq D^2 + q_{12}^{+} + D^{-1}q_{12}^{+}D + D^{-1}q_{12}^{-}D^{-1}q_{12}^{-}, \quad \Phi_{12} \doteq D^2 + q_{12}^{+} + Dq_{12}^{+}D^{-1} + q_{12}^{-}D^{-1}q_{12}D^{-1} \tag{3.15}$$

and Ψ_{12}, Φ_{12} are related via $D\Psi_{12} = \Phi_{12}D$. The operators $q_{12}^{\pm}$ are defined by

$$q_{12}^{\pm} \doteq \hat{q}_1 \pm \hat{q}_1^{*}, \quad \hat{q}_1 \doteq q_1 + \alpha D_{y_1}, \quad \hat{q}_1^{*} \doteq q_2 - \alpha D_{y_2}. \tag{3.16}$$

(The notation $\hat{q}_1^{*}$ is justified, since $\hat{q}_1^{*}$ is indeed the adjoint of $\hat{q}_1$, see §3.2)

EXAMPLE

1. Equation (3.14) with $n = 0$ and $\beta_0 = 1/2$ implies $q_{1_t} = q_{1_x}$.

2. Equation (3.14) with $n = 1$ and $\beta_1 = 1/2$ implies the KP equation

$$q_{1_t} = q_{1_{xxx}} + 6q_1q_{1_x} + 3\alpha^2D^{-1}q_{1_{y_1y_1}}. \tag{3.17}$$

Remark 3.1

(i) The operators Φ_{12} and Ψ_{12} with $y_2 = y_1$ and $\alpha = 0$ reduce to Φ and Φ^{+} respectively, where Φ is the recursion operator of the KdV.

(ii) The starting symmetry $(\Phi_{12}D).1$ is given by $q_{1_x} + q_{2_x} +$

$(q_1-q_2)D^{-1}(q_1-q_2) + \alpha D^{-1}(q_{1y_1} - q_{2y_2})$. Thus it reduces to q_{1_x}, the starting symmetry of the KdV, when $y_2 = y_1$.

3.2 A New Directional Derivative and a New Bilinear Form

Recall that Φ generates symmetries and Φ^+ generates conserved covariants. Similarly, it will turn out, that Φ_{12} and Φ_{12}^* generate extended symmetries and extended conserved covariants respectively. To define these extended notions we need to introduce a new bilinear form and a new directional derivative:

(i) <u>A new bilinear form</u>

$$\langle g_{12}, f_{12}\rangle \doteq \int_{R^3} dx\, dy_1 dy_2 \text{ trace } g_{21}f_{12}, \tag{3.18}$$

where f_{12} and g_{12} are matrix valued functions of x, y_1, y_2 and obviously the trace is dropped if f_{12}, g_{12} are scalars. In association with the above form we define L_{12}^* to be the adjoint of L_{12} iff

$$\langle L_{12}^* g_{12}, f_{12}\rangle = \langle g_{12}, L_{12} f_{12}\rangle. \tag{3.19}$$

We recall that the usual bilinear form and the usual adjoint are defined by

$$(g,f) \doteq \int_{R^2} dx\, dy \text{ trace } gf, \quad (L^+g,f) = (g,Lf), \tag{3.20}$$

where f, g are matrix valued functions of x, y.

EXAMPLE

1. The adjoint of $\hat{q}_1$ is given by $\hat{q}_1^* = q_2 - \alpha D_2$

2. $(q_{12}^+)^* = q_{12}^+, \quad (q_{12}^-)^* = -q_{12}^-$ (3.21)

3. $\Phi_{12}^* = \Psi_{12}$.

Note that the fastest way to compute the adjoint of an operator L_{12} is to evaluate the adjoint as usually and then interchange $1 \leftrightarrow 2$.

Let I be a functional given by

$$I = \int_{R^2} dx\, dy_1 \text{ trace } \rho_{11} = \int_{R^3} dx\, dy_1 dy_2 \delta_{12} \text{ trace } \rho_{12}. \tag{3.22}$$

The extended gradient of this functional is defined by

$$\langle \mathrm{grad}_{12} I, .\rangle \doteq I_d[.] = \int_{R^3} dx\, dy_1 dy_2 \delta_{12} \rho_{12_d}[.], \tag{3.23}$$

where subscript d denotes a suitable directional derivative. It is easily seen that a function γ_{12} is an extended gradient function, i.e., it has a potential I, iff

$$\gamma_{12_d} = \gamma^{*}_{12_d}. \tag{3.24}$$

Also

$$(\mathrm{grad}\ I, .) \doteq I_f[.] = \int_{R^2} dx\, dy\, \rho_f[.] \tag{3.25}$$

and γ is a gradient function iff $\gamma_f = \gamma^{+}_f$.

(ii) A new directional derivative

Recall the crucial integral representation

$$(\hat{q}_1 f)(x,y_1) = \int_R dy_3 q(x,y_1,y_3) f(x,y_3).$$

Allowing f also to depend on y_2 we obtain

$$\hat{q}_1 f_{12} = \int_R dy_3 q_{13} f_{32}.$$

The above mapping between an operator and its kernel induces a mapping between derivatives: Let subscript d denote the new directional derivative. Then

$$\hat{q}_{1_d}[\sigma_{12}] f_{12} = \int_R dy_3 \sigma_{13} f_{32}.$$

The integral representation for $\hat{q}_1$ also induces, via (3.18) an integral representation for the adjoint of $\hat{q}_1$:

$$\langle g_{21}, \hat{q}_1 f_{12}\rangle = \int_{R^3} dy_1 dy_2 dx\, g_{21} \int_R dy_3' q_{13'} f_{3'2} = \int_{R^4} dy_3' dy_2 dy_1 dx\ .\ g_{23'} g_{3'1} f_{12}$$

$$= \int_{R^3} dy_1 dy_2 dx\, G_{21} f_{12},$$ where we have used $3' \leftrightarrow 1$, and

$$G_{21} = \int_R dy_3' g_{23'} q_{3'1}, \quad \text{thus } G_{12} = \int_R dy_3' g_{13'} q_{3'2}.$$

Thus

$$\hat{q}_1^* f_{12} = \int_R dy_3 q_{32} f_{13}.$$

Furthermore, the $\hat{q}_1^*$ mapping induces a mapping between derivatives. Thus

$$\hat{q}_1 f_{12} \doteq (q_1 + \alpha D_1) f_{12} = \int_R dy_3 q_{13} f_{32}, \quad \hat{q}_1^* f_{12} \doteq (q_2 - \alpha D_2) f_{12} = \int_R dy_3 q_{32} f_{13} \qquad (3.26)$$

$$\hat{q}_{1_d}[\sigma_{12}] f_{12} = \int_R dy_3 \sigma_{13} f_{32}, \quad \hat{q}^*_{1_d}[\sigma_{12}] f_{12} = \int_R dy_3 \sigma_{32} f_{13}. \qquad (3.27)$$

The above derivatives with respect to $\hat{q}_1$ and $\hat{q}_1^*$ imply the following derivatives with respect to q_{12}^+, q_{12}^-:

$$q^{\pm}_{12_d}[\sigma_{12}] f_{12} \doteq \int_R dy_3 (\sigma_{13} f_{32} \pm \sigma_{32} f_{13}). \qquad (3.28)$$

Furthermore, using the chain rule and (3.28), if an operator K_{12} depends only on q_{12}^+, q_{12}^- its directional derivative $L_{12_d}[\sigma_{12}]$ is well defined. This derivative is linear, and satisfies the Leibnitz rule. Also, using (3.28) it follows that the directional derivative in the direction of δ_{12} reduces to the usual total Fréchet derivative:

$$K_{12_d}[\delta_{12} F_{12}] = K_{12_f}[F] \doteq K_{12_{q_1}}[F_{11}] + K_{12_{q_2}}[F_{22}], \qquad (3.29)$$

where the subscript f stands for a Fréchet derivative and

$$K_{12_{q_i}}[F_{ii}] = \frac{\partial}{\partial \varepsilon} K_{12}(q_i + F_{ii}, q_j)\Big|_{\varepsilon=0}, \quad i,j = 1,2, \; i \neq j. \qquad (3.30)$$

Operators which depend only on $q_{12}^{\pm}$ are called admissible. Similarly, a function K_{12} is called admissible if it can be written in the form $K_{12} = \hat{K}_{12} H_{12}$, where $\hat{K}_{12}$ is an admissible operator and H_{12} is an appropriate function (for the KP, $H_{12} = H(y_1, y_2)$).

EXAMPLE

The function $\hat{M}_{12}\delta_{12} \doteq Dq_{12}^+\delta_{12} + q_{12}^- D^{-1} q_{12}^- \delta_{12}$ is an admissible function since the operator $\hat{M}_{12}$ depends only on $q_{12}^{\pm}$, and $\delta_{12} = \delta(y_1 - y_2)$. It is easy to compute its directional derivative:

$$(\hat{M}_{12}\delta_{12})_d[\sigma_{12}] = D\sigma^+_{12}\delta_{12} + \sigma^-_{12}D^{-1}q^-_{12}\delta_{12} + q^-_{12}D^{-1}\sigma^-_{12}\delta_{12},$$

where $\sigma^{\pm}_{12}f_{12} = \int_R dy_3(\sigma_{13}f_{32} \pm \sigma_{32}f_{13})$. Hence $(\hat{M}_{12}\delta_{12})_d[\sigma_{12}] = 2D\sigma_{12}$.

3.3 Isospectral Problems Yield Hereditary Operators

Using the same methods as in 1+1, it can be shown that if the extended gradient $(G_\lambda)_{12}$ of the eigenvalue λ of an isospectral problem satisfies

$$\Psi_{12}(G_\lambda)_{12} = \mu(\lambda)(G_\lambda)_{12}, \tag{3.31}$$

then $\Phi_{12} \doteq \Psi^*_{12}$ is a hereditary operator. (One must again assume completeness, a proof of which should follow a two dimensional version of the method developed in [6].)

EXAMPLE

Consider the isospectral problem

$$V_{1_{xx}} + (\hat{q}_1 - \lambda)V_1 = 0. \tag{3.32}$$

Taking the directional derivative of the above it follows that

$$(D^2 + \hat{q}_1 - \lambda)V_{1_d}[f_{12}] + (\hat{q}_{1_d}[f_{12}] - \lambda_d[f_{12}])V_1 = 0.$$

Multiplying the above by V^+_1, where V^+_1 satisfies the adjoint of (3.32), integrating with respect to dx dy_1, and assuming $\int_{R^2} dx\, dy_1\, V_1V^+_1 = 1$, we obtain

$$\lambda_d[f_{12}] = \langle \mathrm{grad}_{12}\lambda, f_{12}\rangle = \int_{R^2} dx\, dy_1\, V^+_1\hat{q}_{1_d}[f_{12}]V_1.$$

Using (3.26) to evaluate $\hat{q}_{1_d}[f_{12}]$ it follows that

$$(G_\lambda)_{12} \doteq \mathrm{grad}_{12}\lambda = V_1V^+_2. \tag{3.33}$$

It is easy to show that Φ^*_{12} as defined by (3.7) satisfies

$$\Phi^*_{12}V_1V^+_2 = 4\lambda V_1V^+_2. \tag{3.34}$$

Hence Φ_{12} is a hereditary operator.

Remark 3.2

Konopelchenko and Dubrovsky [29] were the first to establish the importance of working with $V(x,y_1)V^+(x,y_2)$, as opposed to $V(x,y)V^+(x,y)$. They also found a linear equation satisfied by $V_1V_2^+$. However, they failed to recognize that this equationcould actually yield the recursion operator of the entire associated hierarchy of nonlinear equations. Indeed, they used the above equation to obtain "local" recursion operators. Thus the question of studying the remarkably rich structure of these recursion operators in particular its connection to symmetries, conservation laws, and bi-Hamiltonian operators were not even posed.

3.4 Starting Symmetries

The theory of symmetries for equations in 1+1 is based on the existence of "starting" symmetries K^0, which via Φ generate infinitely many symmetries. For example, for the KdV $K^0 = q_x$. For equations in 2+1 we find that the starting symmetries K_{12}^0 have the following important properties: (i) Can be written in the form $\hat{K}_{12}^0 H_{12}$, where $\hat{K}_{12}^0$ is an admissible operator and H_{12} is an appropriate function. (ii) The starting operators $\hat{K}_{12}^0$ have simple commutator properties with respect to $h_{12} = h(y_1-y_2)$. (iii) The Lie algebra of the starting operator $\hat{K}_{12}^0$ acting on functions H_{12} is closed. (iv) Using (ii) and the fact that Φ_{12} also admits a simple commutator relationship with h_{12}, it can be shown that $\delta_{12}\Phi_{12}^n\hat{K}_{12}^0 . 1 = \sum_{\ell=0}^{n} b_{n,\ell}\Phi_{12}^{n-\ell}\, \hat{K}_{12}^0 .\, \delta_{12}^{\ell}$, where $b_{n,\ell}$ are appropriate constants; hence $\delta_{12}\, \Phi_{12}^n\, \hat{K}_{12}^0 .\, 1$ are admissible functions. It is thus clear that in 1+1 one considers the Lie algebra of functions K^0, while in 2+1 one considers the Lie algebra of operators $\hat{K}_{12}^0$. This richer algebraic structure of equations in 2+1 can be exploited in a variety of ways. For example, different choices of H_{12} yield both time-independent and time-dependent symmetries. Furthermore, all these symmetries correspond to gradient functions.

We now discuss (i)-(iv) above for the concrete case of the KP: It should be first noted that given an operator Φ_{12} there exists an algorithmic way of finding its starting symmetries: One looks for operators $\hat{S}_{12}$ such that $\hat{S}_{12}H_{12} = 0$ but $\Phi_{12}\hat{S}_{12}H_{12} = \hat{K}_{12}^0 H_{12} \neq 0$. It can be shown that if a starting symmetry is constructed in the above way and Φ_{12} is hereditary then Φ_{12} is a strong symmetry for this starting symmetry.

(i) For the KP there exist two starting symmetries:

$$\hat{M} \doteq Dq_{12}^+ + q_{12}^- D^{-1} q_{12}^-,\quad \hat{N} \doteq q_{12}^-,\quad H_{12} \doteq H(y_1,y_2) \tag{3.35}$$

corresponding to $\hat{S}_{12} = D$ and $\hat{S}_{12} = D(q_{12}^-)^{-1}D$ respectively.

(ii) The following operator equations are valid:

$$[\hat{M}_{12},h_{12}] = 2\alpha Dh'_{12}, \quad [\hat{N}_{12},h_{12}] = 0. \tag{3.36}$$

(iii) The Lie algebra of $\hat{M}_{12}$, $\hat{N}_{12}$ is given by

$$[\hat{N}_{12}H_{12}^{(1)},\hat{N}_{12}H_{12}^{(2)}]_d = -\hat{N}_{12}H_{12}^{(3)}, \quad [\hat{N}_{12}H_{12}^{(1)},\hat{M}_{12}H_{12}^{(2)}]_d = -\hat{M}_{12}H_{12}^{(3)},$$

$$[\hat{M}_{12}H_{12}^{(1)},\hat{M}_{12}H_{12}^{(2)}]_d = -\Phi_{12}\hat{N}_{12}H_{12}^{(3)}, \tag{3.37}$$

where

$$[K_{12}^{(1)},K_{12}^{(2)}]_d \doteq K_{12_d}^{(1)}[K_{12}^{(2)}] - K_{12_d}^{(2)}[K_{12}^{(1)}], \tag{3.38}$$

$$H_{12}^{(3)} \doteq [H_{12}^{(1)},H_{12}^{(2)}]_I \doteq \int_R dy_3(H_{13}^{(1)}H_{32}^{(2)} - H_{13}^{(2)}H_{32}^{(1)}). \tag{3.39}$$

Let us derive (3.37a):

$$q_{12_d}^{-}[q_{12}^{-}H_{12}^{(2)}]H_{12}^{(1)} = \int_{R^2} dy_3dy_3' \Big\{(q_{13'}H_{3'3}^{(2)} - q_{3'3}H_{13'}^{(2)})H_{32}^{(1)}$$

$$- H_{13}^{(1)}(q_{33'}H_{3'2}^{(2)} - q_{3'2}H_{33'}^{(2)})\Big\}$$

$$= \int_{R^2} dy_3dy_{3'}\Big\{H_{32}^{(1)}[q_1-q_3-\alpha(D_1+D_3)]H_{13}^{(2)} - H_{13}^{(1)}[q_3-q_2-\alpha(D_3+D_2)]H_{32}^{(2)}\Big\}.$$

Hence

$$[q_{12}^{-}H_{12}^{(1)},q_{12}^{-}H_{12}^{(2)}]_d = -\Big\{q_1-q_2 + \alpha(D_1+D_2)\Big\}[H_{12}^{(1)},H_{12}^{(2)}]_I.$$

Remark 3.3

The bracket (3.39) can also be traced back to the integral representation of $\hat{q}_1$ (see [27]).

(iv) Equations (3.36) and the operator equation (see (3.11))

$$[\Phi_{12},h_{12}] = 4\alpha h'_{12} \tag{3.40}$$

imply

$$\delta_{12}\Phi_{12}^n\hat{N}_{12}\cdot 1 = \sum_{\ell=1}^{n}(-4\alpha)^\ell\binom{n}{\ell}\Phi_{12}^{n-\ell}\hat{N}_{12}\delta_{12}^\ell, \tag{3.41}$$

$$\delta_{12}\Phi_{12}^n\hat{M}_{12}\cdot 1 = \sum_{\ell=1}^{n} b_{n,\ell}\,\Phi_{12}^{n-\ell}M_{12}\delta_{12}^\ell, \quad b_{n,\ell} \doteq (-4\alpha)^\ell\sum_{j=0}^{\ell}2^{-j}\binom{n-j}{\ell-j}. \tag{3.42}$$

Let us indicate how the above equations can be derived: Introducing an operator $\mathbb{D}$, which commutes with all admissible operators $\hat{K}_{12}$ and which has the property that

$$\mathbb{D}.h_{12} = h'_{12},$$

it follows that

$$\delta_{12}\Phi_{12}^{n}\hat{N}_{12}.1 = (\Phi_{12} - 4\alpha\mathbb{D})^{n}\,\delta_{12}\hat{N}_{12}.1 = (\Phi_{12} - 4\alpha\mathbb{D})^{n}N_{12}.\delta_{12}$$

$$= \sum_{\ell=1}^{n}(-4\alpha)^{\ell}\binom{n}{\ell}\Phi_{12}^{n-\ell}\,\hat{N}_{12}.\,\delta_{12}^{\ell}.$$

To derive equation (3.42) note that

$$\delta_{12}\Phi_{12}^{n}\hat{M}_{12}1 = (\Phi_{12}-4\alpha\mathbb{D})^{n}\,\delta_{12}\hat{M}_{12}.1 = (\Phi_{12}-4\alpha\mathbb{D})^{n}(\hat{M}.\delta_{12}-2\alpha\mathbb{D}.\delta'_{12})$$

$$= \sum_{\ell=0}^{n}(-4\alpha)^{\ell}\binom{n}{\ell}\Phi_{12}^{n-\ell}\,\hat{M}.\,\delta_{12}^{\ell} - 2\alpha\sum_{\ell=0}^{n}(-4\alpha)^{\ell}\binom{n}{\ell}\Phi_{12}^{n-\ell}D.\delta_{12}^{\ell+1}. \tag{3.43}$$

The next step is to express $\Phi_{12}^{j}D$ in terms of $\Phi_{12}^{j'}\,\hat{M}_{12}$, where j, j' are integers. This can be achieved as follows: It can be shown that $\Phi_{12}^{n+1}D.1 = \Phi_{12}^{n}\hat{M}.1$. This equation implies

$$\Phi_{12}^{n+1}D.h_{12} = \sum_{j=0}^{n}(2\alpha)^{j}\Phi_{12}^{n-j}\,\hat{M}_{12}.h_{12}^{j};\quad h_{12}^{j} \doteq \frac{d^{j}}{dy_{1}^{j}}h_{12}. \tag{3.44}$$

For example, multiplying $\Phi_{12}D.1 = \hat{M}_{12}.1$ by h_{12}, it follows that $(\Phi_{12}-4\alpha\,\mathbb{D})h_{12}D.1 = (\hat{M}_{12}-2\alpha\,\mathbb{D}).h$, or $\Phi_{12}D.h_{12} = \hat{M}_{12}.h_{12}$. Similarly, $\Phi_{12}^{2}D.1 = \Phi_{12}\hat{M}_{12}.1$ implies $\Phi_{12}^{2}D.h_{12} = \Phi_{12}\hat{M}_{12}.h_{12} + 2\alpha\hat{M}_{12}.h_{12}$, etc. Using (3.44) into (3.43) yields (3.42).

3.5 Basic Notions and Results

We consider exactly solvable evolution equations in the form $q_t = K(q)$, $q(x,y,t)$, on a normed space M of vector-valuedfunctions on $\mathbb{R}$; K is a suitable C^{∞} vector field on M. We assume that the space of smooth vector fields on M is some space S of C^{∞} functions on the plane vanishing rapidly as $x, y \to \pm\infty$. The above equation is a member of a hierarchy generated by Φ_{12}, hence more generally we shall study $q_t = K^{(n)}(q)$. Fundamental in our theory is to write these equations in the form

$$q_{1_t} = \int_R dy_2\, \delta_{12}\, \Phi_{12}^n \hat{K}_{12}^0 . 1 \doteq \int_R dy_2 \delta_{12} K_{12}^{(n)} = K_{11}^{(n)}, \tag{3.45}$$

(in the matrix case, 1 is replaced by the identity matrix I), where $K_{12}^{(n)}(q_1,q_2)$ belongs to a suitably extended space $\tilde{S}$, and $\tilde{S}^*$ denotes the dual of $\tilde{S}$. In the extended spaces $\tilde{S}$ and $\tilde{S}^*$ we define the new directional derivative (3.28) and the new bilinear form (3.18); the notions of the adjoint and of a gradient are well defined with respect to (3.18) (see (3.19), (3.23), (3.24)). In analogy with definition 2.1 we have:

Definition 3.1

(i) A function $\sigma_{12} \epsilon\, \tilde{S}$ is called an **extended symmetry** of

$$q_{1_t} = \int_R dy_2 \delta_{12} K_{12} = K_{11} \tag{3.46}$$

iff

$$\frac{\partial \sigma_{12}}{\partial t} + \sigma_{12_f}[K] - (\delta_{12}K_{12})_d[\sigma_{12}] = 0. \tag{3.47}$$

(ii) A function $\gamma_{12} \epsilon\, \tilde{S}^*$ is called an **extended conserved gradient** (i.e., it is the extended gradient of a conserved functional I) of (3.46) iff

$$\frac{\partial \gamma_{12}}{\partial t} + \gamma_{12_f}[K] + (\delta_{12}K_{12})_{d^*}[\gamma_{12}] = 0, \quad \gamma_{12_d} = \gamma_{12_{d^*}}. \tag{3.48}$$

Functions which satisfy (3.48a) are called **extended conserved covariants.**

(iii) An operator valued function $\Phi_{12}: \tilde{S} \rightarrow \tilde{S}$, is a **recursion operator** for (3.46) (or it is a **strong symmetry** for K_{12}) iff

$$\Phi_{12_f}[K] + [\Phi_{12}, (\delta_{12}K_{12})_d] = 0. \tag{3.49}$$

(iv) An operator valued function $\Phi_{12}: \tilde{S} \rightarrow \tilde{S}$, is a **hereditary operator** (or **Nijenhuis** or **regular**) iff

$\Phi_{12_d}[\Phi_{12}v_{12}]w_{12} - \Phi_{12}\Phi_{12_d}[v_{12}]w_{12}$ is symmetric with respect to v_{12}, w_{12}. (3.50)

(v) An operator valued function $\Theta_{12}: \tilde{S}^* \rightarrow \tilde{S}$ is a **Hamiltonian operator** iff it is skew symmetric, i.e., $\Theta_{12} = -\Theta_{12}^*$, and it satisfies $\langle a_{12}, \Theta_{12_d}[\Theta_{12}b_{12}]c_{12}\rangle$ + cyclic permutation = 0. (3.51)

(vi) Equation (3.46) is a **Hamiltonian system** iff it can be written in the form

$$q_{1_t} = \int_R dy_2 \delta_{12} \Theta_{12} f_{12}, \tag{3.52}$$

where Θ_{12} is a Hamiltonian operator and f_{12} is an extended gradient function, i.e., $f^*_{12_d} = f_{12_d}$. Associated with (3.52) we define the following Poisson bracket

$$\{I,H\} = \langle \mathrm{grad}_{12} I, \Theta_{12} \mathrm{grad}_{12} H \rangle . \tag{3.53}$$

In the above, subscripts f and d denote total Fréchet (see (3.29)) and directional (see (3.28)) derivatives respectively.

Remark 3.4

(i) Equation (3.47) can also be written as

$$\frac{\partial \sigma_{12}}{\partial t} + [\sigma_{12}, \delta_{12} K_{12}]_d = 0,$$

since $\sigma_{12_d}[\delta_{12} K_{12}] = \sigma_{12_f}[K]$. Similarly, $\Phi_{12_f}[K] = \Phi_{12_d}[\delta_{12} K_{12}]$.

(ii) Some of the above notions are well defined only if $(\delta_{12} K_{12})_d$ is well defined. However, for equations (3.45)

$$\delta_{12} K_{12}^{(n)} = \delta_{12} \Phi_{12}^n \hat{K}_{12}^0 . 1 = \sum_{\ell=0}^{n} b_{n,\ell} \Phi_{12}^{n-\ell} \hat{K}_{12}^0 . \delta_{12}^{\ell} .$$

Furthermore, by construction Φ_{12} and the starting operators $\hat{K}_{12}^0$ depend on the basic operators $q_{12}^{\pm}$. Hence $(\delta_{12} \hat{K}_{12}^{(n)})_d$ is well defined.

In analogy with the basic results in 1+1:

Theorem 3.1

(i) If Φ_{12} is a recursion operator for (3.46) then Φ_{12} maps extended symmetries onto extended symmetries and Φ^*_{12} maps extended conserved covariants onto extended conserved covariants.

(ii) If (3.46) is a Hamiltonian system then $\sigma_{12} = \Theta_{12} \gamma_{12}$.

(iii) If Φ_{12} is a hereditary operator and a recursion operator for $\hat{K}_{12}^0 . 1$ then Φ_{12} is a recursion operator for $q_{1_t} = \int_R dy_2 \delta_{12} \Phi_{12}^n \hat{K}_{12}^0 . 1$.

(iv) If $\Phi_{12} \doteqdot \Theta_{12}^{(2)}(\Theta_{12}^{(1)})^{-1}$, where $\Theta_{12}^{(1)} + \nu\Theta_{12}^{(2)}$ is a Hamiltonian operator for all values of the constant ν and $\Theta_{12}^{(1)}$ is invertible, then Φ_{12} is hereditary.

(v) If Φ_{12} as in (iv) and $\gamma_{12}^{0} \doteqdot (\Theta_{12}^{(1)})^{-1}\hat{K}_{12}^{0}.1$ is an extended gradient function then all $(\Phi_{12}^{*})^{m}\gamma_{12}^{0}$ are extended gradient functions.

EXAMPLE

The hereditary operator Φ_{12} of the KP equation is factorizable in terms of the Hamiltonian operators D and $\Phi_{12}D$. Hence each member of the KP hierarchy is a bi-Hamiltonian system, with respect to the following two Poisson brackets

$$\{I,H\} = \langle \mathrm{grad}_{12}I,\ \Theta_{12}^{(i)}\mathrm{grad}_{12}H\rangle, \quad i = 1,2$$

$$\Theta_{12}^{(1)} = D, \quad \Theta_{12}^{(2)} = \Phi_{12}D = D^3 + q_{12}^{+}D + Dq_{12}^{+} + q_{12}^{-}D^{-1}q_{12}^{-}.$$

3.6 Extended Symmetries

Lemma 3.1

(i) Let Φ_{12} be hereditary, then

$$[\Phi_{12}^{n}K_{12}^{(1)},\ \Phi_{12}^{m}K_{12}^{(2)}]_d = \Phi_{12}^{n+m}[K_{12}^{(1)},K_{12}^{(2)}]_d + \Phi_{12}^{m}(\sum_{r=1}^{n}\Phi_{12}^{n-r}S_{12}^{(2)}\Phi_{12}^{r-1})K_{12}^{(1)}$$
$$- \Phi_{12}^{n}(\sum_{r=1}^{m}\Phi_{12}^{m-r}S_{12}^{(1)}\Phi_{12}^{r-1})K_{12}^{(2)}, \tag{3.54}$$

where

$$S_{12}^{(i)} \doteqdot \Phi_{12_d}[K_{12}^{(i)}] + [\Phi_{12},K_{12_d}^{(i)}], \tag{3.55}$$

m, n are non-negative integers.

(ii) $\sigma_{12}^{(r)}$ is a time-dependent extended symmetry of order r of equation (3.46) iff

$$\sigma_{12}^{(r)} = \sum_{j=0}^{r} t^j\Sigma_{12}^{(j)}, \quad \Sigma_{12}^{(j)} \doteqdot -\frac{1}{j}[\Sigma_{12}^{(j-1)},\ \delta_{12}K_{12}]_d,\ j=1,\dots,r,$$

$$[\Sigma_{12}^{(r)},\ \delta_{12}K_{12}]_d = 0. \tag{3.56}$$

Proof

See [28].

We propose the following constructive approach to extended symmetries: Given an isospectral problem construct a recursion operator Φ_{12}. This operator must be hereditary (see §3.3). Then construct its starting symmetries operators, say $\hat{M}_{12}$, $\hat{N}_{12}$. The operator Φ_{12} is a strong symmetry of $\hat{M}_{12}$, $\hat{N}_{12}$ (see [27]). Compute the commutators of $\hat{M}_{12}$, $\hat{N}_{12}$, Φ_{12} with h_{12}. Use the commutator relationships to derive $\delta_{12}\Phi_{12}^{n}\hat{K}_{12}^{0}.1 = \sum_{\ell=0}^{n} b_{n,\ell}\Phi_{12}^{n-\ell}\hat{K}_{12}^{0}.\delta_{12}^{\ell}$, $\hat{K}_{12}$ is $\hat{M}_{12}$ or $\hat{N}_{12}$. Finally, compute the Lie algebra of $\hat{M}_{12}$, $\hat{N}_{12}$. This Lie algebra together with (3.54)-(3.56) yield infinitely many time-independent and time-dependent extended symmetries.

EXAMPLE

1. $\Phi_{12}^{m}\hat{M}_{12}.1$, $\Phi_{12}^{m}\hat{N}_{12}.1$ are extended symmetries of the KP hierarchy

$q_{1_t} = \int_R dy_2\delta_{12}\Phi_{12}^{n}\hat{M}_{12}.1$ (recall KP corresponds to n = 1). $\hat{M}_{12}$, $\hat{N}_{12}$ are defined in (3.35).

For

$$[\delta_{12}\Phi_{12}^{n}\hat{M}_{12}.1,\ \Phi_{12}^{m}\hat{M}_{12}.H_{12}]_d = [\sum_{\ell=0}^{n} b_{n,\ell}\Phi_{12}^{n-\ell}\hat{M}.\delta_{12}^{\ell},\Phi_{12}^{m}\hat{M}.H_{12}]_d$$

$$= \sum_{\ell=0}^{n} b_{n,\ell}\ \Phi_{12}^{m+n-\ell}[\hat{M}.\delta_{12}^{\ell},\hat{M}.H_{12}]_d = -\sum_{\ell=0}^{n} b_{n,\ell}\ \Phi_{12}^{m+n-\ell+1}N_{12}[\delta_{12}^{\ell},H_{12}]_d, \tag{3.57}$$

where we have used (3.54) (Φ_{12} is hereditary and it is also a strong symmetry for $\hat{M}_{12}H_{12}$, thus $S_{12}^{(1)} = 0$), and (3.37c). Taking $H_{12} = 1$ and using

$$[\delta_{12}^{\ell},1]_I = 0,$$

equation (3.57) implies $[\delta_{12}\Phi_{12}^{n}\hat{M}_{12}.1,\ \Phi_{12}^{m}\hat{M}_{12}.1]_d = 0$, i.e., $\Phi_{12}^{m}\hat{M}_{12}.1$ is an extended symmetry of the KP hierarchy. Similarly for $\Phi_{12}^{m}\hat{N}_{12}.1$, since

$$[\delta_{12}\Phi_{12}^{n}\hat{M}_{12}.1,\ \Phi_{12}^{m}\hat{N}_{12}.H_{12}]_d = \sum_{\ell=0}^{n} b_{n,\ell}\Phi_{12}^{m+n-\ell}M_{12}[\delta_{12}^{\ell},H_{12}]_I.$$

2. $\Phi_{12}^{m}\hat{M}_{12}.1$, $\Phi_{12}^{m}\hat{N}_{12}.1$ are extended symmetries of the hierarchy

$$q_{1_t} = \int_R dy_3\delta_{12}\Phi_{12}^{n}\hat{N}_{12}.1.$$

3. The KP hierarchy admits two hierarchies of t-dependent symmetries of order r given by (3.56) where

$$\Sigma_{12}^{(0)} = \hat{N}_{12}^{(m)}.H_{12}^{(r)}, \qquad H_{12}^{(r)} \doteq (y_1 + y_2)^r$$

$$\Sigma_{12}^{(2j)} = \Sigma \nu(r,2j,s)\hat{N}_{12}^{(m+2jn+j-\sum_{\ell=1}^{2j} 2s_\ell+1)}.H_{12}^{(r-\sum_{\ell=1}^{2j} 2s_\ell+1)},$$

$$\Sigma_{12}^{(2j-1)} = \Sigma \nu(r,2j-1,s)\hat{M}_{12}^{(m+(2j-1)n+j-1-\sum_{\ell=1}^{2j-1} 2s_\ell+1)}.H_{12}^{(r-\sum_{\ell=1}^{2j-1} 2s_\ell+1)},$$

the summation Σ is over $s_1, s_2, \ldots, s_j$, from zero to P_n, $P_n = (n-1)/2$ if n is odd, $(n-2)/2$ if n is even,
and

$$\Sigma_{12}^{(0)} = \hat{M}_{12}^{(m)}.H_{12}^{(r)},$$

$$\Sigma_{12}^{(2j)} = \Sigma \nu(r,2j,s)\hat{M}_{12}^{(m+2jn+j-\sum_{\ell=1}^{2j} 2s_\ell+1)}.H_{12}^{(r-\sum_{\ell=1}^{2j} 2s_\ell+1)},$$

$$\Sigma_{12}^{(2j-1)} = \Sigma \nu(r,2j-1,s)\hat{N}_{12}^{(m+(2j-1)n+j-\sum_{\ell=1}^{2j-1} 2s_\ell+1)}.H_{12}^{(r-\sum_{\ell=1}^{2j-1} 2s_\ell+1)}$$

with $j \geq 1$, $b_{n,\ell} = \sum_{s=0}^{\ell} \beta^{\ell-s}\tilde{\beta}^s \binom{n-s}{\ell-s} = (-4\alpha)^\ell \sum_{s=0}^{\ell} 2^{-s}\binom{n-s}{\ell-s}$ and

$$\nu(r,j,s) \doteq \frac{(-2)^j}{j!}\left(\prod_{\mu=1}^{j}\Theta\left(r - \sum_{\ell=1}^{j} 2s+1\right)\right)\left(\prod_{\ell=1}^{j} b_{n,2s_\ell+1}\right)\frac{r!}{\left(r - \sum_{\ell=1}^{j} 2s_\ell+1\right)!}.$$

For

Equation (3.56) implies that constructing a symmetry of order r is equivalent to finding a function $\Sigma_{12}^{(0)}$ with the property that its $(r+1)^{st}$ commutator with $\delta_{12}K_{12}$ is zero. This can be easily achieved by using suitable H_{12}'s. For example, let $H_{12} = y_1 + y_2$, then (3.57) implies:

$$[\delta_{12}\Phi_{12}^n\hat{M}_{12}.1,\ \Phi_{12}^m\hat{M}_{12}(y_1+y_2)]_d = -\sum_{\ell=0}^{n} b_{n,\ell}\ \Phi_{12}^{m+n-\ell+1}\hat{N}_{12}2\delta_{1,\ell},$$

since

$$[\delta_{12}^\ell, y_1+y_2]_I = 2\delta_{1,\ell} \quad \text{where } \delta_{1,\ell} = 0 \text{ if } \ell \neq 1 \text{ or } 1 \text{ if } \ell = 1.$$

Thus, using the fact that $[\delta^{\ell}_{12},\delta_{1,\ell}]_I = 0$ it follows that $\Phi^m_{12}\hat{M}_{12}(y_1+y_2)$ generates first order time-dependent symmetries

$$\Phi^m_{12}\hat{M}_{12}.(y_1+y_2) - 2b_{n,1}t\Phi^{m+n}_{12}\hat{N}_{12}.1.$$

Similarly, to generate r-order time-dependent symmetries use $\Phi^m_{12}\hat{M}_{12}.(y_1+y_2)^r$, since the commutator of $(y_1+y_2)^r$ with δ^{ℓ}_{12} produces $(y_1+y_2)^{r-\ell}$ and hence the r-th commutator of $(y_1+y_2)^r$ with δ^{ℓ}_{12} produces 1 which commutes with $\delta^{(\ell)}_{12}$:

$$[\delta^s_{12},(y_1+y_2)^r]_I = (1-(-1)^s)\Theta(r-s)\frac{r!}{(r-s)!}(y_1+y_2)^{r-s},$$

where $\Theta(r-s)$ denotes the Heaviside function with $\Theta(0) = 1$.

4. The hierarchy $q_{1_t} = \int_R dy_3\,\delta_{12}\Phi^n_{12}\hat{N}_{12}.1$ admits two hierarchies of t-dependent symmetries of order r given by (3.56) where

$$\Sigma^{(0)}_{12} = \hat{N}^{(m)}_{12}.H^{(r)}_{12}$$

$$\Sigma^{(j)}_{12} = \Sigma\nu(r,j,s)\hat{N}_{12}^{(m+jn-\sum_{\ell=1}^{j}2s_\ell+1)}.H_{12}^{(r-\sum_{\ell=1}^{j}2s_\ell+1)},$$

and by

$$\Sigma^{(0)}_{12} = \hat{M}^{(m)}_{12}.H^{(r)}_{12}$$

$$\Sigma^{(j)}_{12} = \Sigma\nu(r,j,s)\hat{M}^{(m+jn-\sum_{\ell=1}^{j}2s_\ell+1)}.H_{12}^{(r-\sum_{\ell=1}^{j}2s_\ell+1)},$$

where the summation Σ is over $s_1,s_2,\dots,s_j$ from zero to P_n, $j \geq 1$, $P_n = (n-1)/2$ if n is odd and $(n-2)/2$ if n is even. Also,

$$b_{n,\ell} = (-4\alpha)^{\ell}\binom{n}{\ell}.$$

The above extended symmetries, under the reduction $y_2 = y_1$ yield symmetries. This follows from the following theorem (see [27]).

Theorem 3.2

Assume that the admissible operators Φ_{12}, $\hat{K}^0_{12}$, satisfy

$$[\Phi_{12},\delta_{12}] = -\beta\delta'_{12},$$

$$[\hat{K}^0_{12},\delta_{12}] = -\tilde{\beta}\hat{S}_{12}\delta'_{12},$$

where β, $\tilde{\beta}$ are constants and $\hat{S}_{12}$ is such that $\hat{S}_{12_d}[.]H_{12} = 0$. Then

(i) If σ_{12} is an extended symmetry of (3.45), σ_{11} is a symmetry of (3.45).

(ii) If σ_{12} is an extended symmetry of (3.45) then $\sigma_{12} = \sigma(q_1,q_1)=0$ is an auto-Bäcklund transformation of (3.45), where q_1 and q_2 are viewed as two different solutions of (3.45).

(iii) If γ_{12} is an extended conserved covariant of (3.45), γ_{11} is a conserved covariant of (3.45).

(iv) If γ_{12} is an extended gradient function then γ_{11} is a gradient function.

EXAMPLE

Consider the extended symmetry of the KP

$$\hat{M}_{12}.1 = q_{1_x} + q_{2_x} + (q_1-q_2)D^{-1}(q_1-q_2) + \alpha D^{-1}(q_{1y_1} - q_{2y_2}).$$

Clearly $(\hat{M}_{12}.1)_{11} = 2q_{1x}$ which is a symmetry of the KP. Also, $\hat{M}_{12}.1 = 0$ is a well-known auto-Bäcklund transformation of the KP.

Remark 3.5

(i) It is quite interesting that both symmetries and Bäcklund transformations of an equation in 2+1 come from the same basic entity, the extended symmetry. Indeed, when $\alpha = 0$ the recursion operator Φ_{12} for the KP equation reduces to an operator that Calogero and Degasperis have introduced [30] and which generates the auto-Bäcklund transformations of the KdV equation.

(ii) Using the interpretation that $q^{\pm}b \doteqdot qb \pm bq$, q, b matrices, the recursion operator of the KP becomes the operator generating auto-Bäcklund transformations for the equations associated with the $N \times N$ matrix Schrödinger problem in one dimension (studied by Calogero and Degasperis [30]). This important connection is explained from the fact that certain 2+1 dimensional systems can be viewed as reductions of certain evolution equations non-local in y. These equations are directly connected to matrix evolution equations [28],[31].

3.7 Extended Conserved Gradients

Lemma 3.2

Assume that Θ_{12} is a Hamiltonian operator. Then,

$$[\Theta_{12}f_{12},\Theta_{12}g_{12}]_d = \Theta_{12}\mathrm{grad}_{12}\langle f_{12},\Theta_{12}g_{12}\rangle +$$

$$\Theta_{12}\left\{(f_{12_d} - f^*_{12_d})[\Theta_{12}g_{12}] - (g_{12_d} - g^*_{12_d})[\Theta_{12}f_{12}]\right\}. \qquad (3.58)$$

Proof

See [28].

One way of proving that Φ^*_{12} generates gradient functions is to use Theorem 3.1, (v). However, this requires that Φ^*_{12} is factorizable in terms of Hamiltonian operators. Alternatively, we propose the following constructive approach, which only uses one Hamiltonian operator: Construct the Lie algebra of the starting operators, say $\hat{M}_{12}$, $\hat{N}_{12}$. Then use this algebra and (3.58) to prove that all $\Phi^{*m}_{12}\,\Theta^{-1}_{12}\hat{K}^0_{12}\cdot H_{12}$ are gradients, provided that $\Theta^{-1}_{12}\hat{K}^0_{12}H_{12}$ is a gradient, where $\hat{K}^0_{12}$ is $\hat{M}_{12}$ or $\hat{N}_{12}$. Finally, use Theorem 3.2 to show that $(\Theta^{-1}_{12}\hat{K}^0_{12}H_{12})_{11}$ are gradients.

EXAMPLE

Consider the operators $\hat{M}_{12}$, $\hat{N}_{12}$ associated with the KP. Then

$$D^{-1}\Phi^{n+1}_{12}\hat{M}_{12}H^{(3)}_{12} = \mathrm{grad}\langle \Phi^n_{12}\hat{M}_{12}H^{(1)}_{12},\ D^{-1}\Phi_{12}\hat{N}_{12}H^{(2)}_{12}\rangle, \qquad (3.59)$$

$$D^{-1}\Phi^{n+1}_{12}\hat{N}_{12}H^{(3)}_{12} = \mathrm{grad}\langle \Phi^n_{12}\hat{M}_{12}H^{(1)}_{12},\ D^{-1}\hat{M}_{12}H^{(2)}_{12}\rangle . \qquad (3.60)$$

For

It is easy to verify that $D^{-1}\hat{M}_{12}H_{12}$ is an extended gradient. Then (3.58) and (3.37c) imply that $D^{-1}\Phi_{12}\hat{N}_{12}H_{12}$ is an extended gradient. Equation (3.37b) implies

$$[\Phi^n_{12}\hat{M}_{12}H^{(1)}_{12},\ \Phi_{12}\hat{N}_{12}H^{(2)}_{12}]_d = \Phi^{n+1}_{12}\hat{M}_{12}H^{(3)}_{12}$$

Since $D^{-1}\Phi_{12}\hat{N}_{12}H_{12}$, $D^{-1}\hat{M}_{12}H_{12}$ are extended gradients, the above equation with n = 0 and (3.58) imply that $D^{-1}\Phi_{12}\hat{M}_{12}H_{12}$ is an extended gradient. Similarly $D^{-1}\Phi^n_{12}\hat{M}_{12}$ is an extended gradient. Equation (3.60) follows in a similar manner using (3.37c).

Remark 3.6

It was shown in §3.6 that time-dependent symmetries of order r are generated via $\Phi_{12}^m \hat{M}_{12} H_{12}$, $\Phi_{12}^m \hat{N}_{12} H_{12}$ with $H_{12} = (y_1+y_2)^r$. The above results show that $D^{-1}\Phi_{12}^m \hat{M}_{12} H_{12}$, $D^{-1}\Phi_{12}^m \hat{N}_{12} H_{12}$ are gradient functions for arbitrary H_{12}. Hence the time-dependent symmetries correspond to gradient functions. However, the time-dependent symmetries are closely related to master-symmetries τ (see §2). Hence the master-symmetries τ correspond to gradient functions.

3.8 Non-gradient Master-symmetries

Lemma 3.3

Assume that the hereditary operator Φ_{12} satisfies $\Phi_{12}\Theta_{12} = \Theta_{12}\Phi_{12}^{*}$, where Θ_{12} is a Hamiltonian operator (if Φ_{12} is factorizable, then this equation follows). Assume for simplicity that $\Theta_{12_d} = 0$. Then

$$(\Phi_{12}^m T_{12})_d + \Theta_{12}(\Phi_{12}^m T_{12})_d^{*}\Theta_{12}^{-1} = \Phi_{12}^m(T_{12_d} + \Theta_{12}T_{12_d}^{*}\Theta_{12}^{-1}) +$$

$$\sum_{r=1}^{m} \Phi_{12}^{r-1}\ \Theta_{12}(\Phi_{12}^{*})^{m-r} S_{12}^{*}\Theta_{12}^{-1}, \tag{3.61}$$

where

$$S_{12}^{*} \doteq \Phi_{12_d}^{*}[T_{12}] + [T_{12_d}^{*}, \Phi_{12}]. \tag{3.62}$$

Proof

See [28].

The results of Lemmas 3.1-3.3 can be used to obtain non-gradient master-symmetries T_{12}. Such master-symmetries are explicitly related to recursion operators Φ_{12}. Indeed given Φ_{12} one computes T_{12} and given T_{12} one computes Φ_{12}. These formulae are the two dimensional analogues of the formulae given in §2.1. The basic idea is to find a T_{12} such that $T_{12_d} + \Theta_{12}T_{12_d}^{*}\Theta_{12}^{-1} = 0$ and $S_{12}^{*} = C.1$, C constant.

EXAMPLE

A master-symmetry of the KP hierarchy is given by $\Phi_{12}^2\delta_{12}$. Indeed

$$\Phi_{12} = \beta\left\{(\Phi_{12}^2\delta_{12})_d + D(\Phi_{12}^2\delta_{12})_d^{*}D^{-1}\right\}, \quad \beta \ \ \text{constant} \tag{3.63}$$

$$\Phi_{12}^{n+1}\hat{M}_{12}\cdot 1 = \beta_n[\Phi_{12}^n\hat{M}_{12}\cdot 1,\ \Phi_{12}^2\delta_{12}]_d,\ \beta_n \text{ constant} \tag{3.64}$$

$$D^{-1}\Phi_{12}^{m+n-1}\hat{M}_{12}\cdot 1 = \text{grad}_{12}\ \langle(\Phi_{12}^*)^n D^{-1}\hat{M}_{12}\cdot 1,\ \Phi_{12}^m\delta_{12}\rangle\ . \tag{3.65}$$

For

Φ_{12} and M_{12} are given by (3.8), (3.35a), $\Theta_{12} = D$. Let $T_{12} = \delta_{12}$. Then $T_{12_d} + \Theta_{12}T_{12_d}^*\Theta_{12}^{-1} = 0$ and $S_{12}^* = 4$ (see (3.62)). Thus, equation (3.61) with m = 2 implies (3.63).

To derive equation (3.64) use Lemma 3.1 with $K_{12}^{(1)} = \hat{M}_{12}\cdot 1$, $K_{12}^{(2)} = \delta_{12}$, $m = 2$, $S_{12}^{(1)} = 0$ (since Φ_{12} is a strong symmetry of $\hat{M}_{12}\cdot 1$) and $S_{12}^{(2)} = \Phi_{12_f}[\delta_{12}] = 4$. Thus

$$[\Phi_{12}^n\hat{M}_{12}\cdot 1,\ \Phi_{12}^2\delta_{12}]_d = \Phi_{12}^{n+2}\ [\hat{M}_{12}\cdot 1,\ \delta_{12}]_d + 4n\ \Phi_{12}^{n+1}\ \hat{M}_{12}\cdot 1.$$

However, $\Phi_{12}^{n+2}[\hat{M}_{12}\cdot 1,\ \delta_{12}]_d$ is proportional to $\Phi_{12}^{n+1}\hat{M}_{12}\cdot 1$, hence the above implies (3.64).

To derive equation (3.65), use Lemmas 3.1-3.3 to obtain the following general result: If Φ_{12} is a hereditary operator such that it is a stong symmetry for M_{12} and $\Phi_{12}\Theta_{12} = \Theta_{12}\Phi_{12}^*$, where Θ_{12} is a constant invertible Hamiltonian operator then

$$\Phi_{12}^m\left(\sum_{r=1}^{n}\Phi_{12}^{n-r}S_{12}\Phi_{12}^{r-1}\right)M_{12} + \left(\sum_{r=1}^{n}\Phi_{12}^{r-1}\Theta_{12}\Phi_{12}^{*m-r}S_{12}^*\Theta_{12}^{-1}\right)\Phi_{12}^n M_{12} + \Phi_{12}^{n+m}[M_{12},T_{12}]_d$$

$$= \Theta_{12}\ \text{grad}_{12}\langle\Theta_{12}^{-1}\Phi_{12}^n M_{12},\ \Phi_{12}^m T_{12}\rangle - \Phi_{12}^m(T_{12_d} + \Theta_{12}T_{12_d}^*\Theta_{12}^{-1})\Phi_{12}^n M_{12},$$

where

$$S_{12} \doteq \Phi_{12_d}[T_{12}] + [\Phi_{12},T_{12_d}],\quad S_{12}^* \doteq \Phi_{12_d}^*[T_{12}] + [T_{12_d}^*,\ \Phi_{12}^*].$$

Using $T_{12} = \delta_{12}$ in the above and noting that $S_{12} = S_{12}^* = 4$, $T_{12_d} + \Theta_{12}T_{12_d}^*\ \Theta_{12}^{-1} = 0$, $\Phi_{12}^{n+m}[M_{12},\ \delta_{12}]_d$ is proportional to $\Phi_{12}^{n+m-1}\hat{M}_{12}\cdot 1$ we obtain (3.65).

ACKNOWLEDGEMENTS

This work was partially supported by the Office of Naval Research under Grant Number N00014-76-C-0867 and the National Science Foundation under Grant Number MCS-8202117. One of the authors (A.S.F.) would like to thank Professor Lakshmanan and his colleagues for their kind hospitality.

REFERENCES

[1] M. J. Ablowitz, D. J. Kaup, A. C. Newell and H. Segur, Stud. Appl. Math. **53** (1974) 249; Phys. Rev. Lett. **30** (1983) 1262; Phys. Rev. Lett. **31** (1973) 125.
[2] P. D. Lax, Comm. Pure Appl. Math. **21** (1968) 467.
[3] W. Symes, J. Math. Phys. **20** (1979) 721.
[4] A. C. Newell, Proc. R. Soc. London Ser. **A365** (1979) 283.
[5] H. Flaschka and A. C. Newell, **Lecture Notes in Physics**, Vol. 38 (Springer, Berlin, 1975), p. 355.
[6] V. S. Gerdjikov, Lett. Math. Phys. **6** (1982) 315.
[7] M. Boiti, F. Pempinelli, G. Z. Tu, Il Nuovo Cimento **79B** (1984)231.
[8] M. Leo, R. Leo, G. Soliani, L. Solombrino, Lett. Al Nuovo Cimento **38** (1983) 45; M. Boiti, P. J. Caudrey, F. Pempinelli, Il Nuovo Cimento **83B** (1984) 71.
[9[F. W. Nijhoff, J. Van der Linden, G. R. Quispel, H. Capel, and J. Velthnizen, Physica **116A** (1983) 1.
[10] B. G. Konopelchenko, Phys. Lett. **75A** (1980) 447; **79A** (1980) 39; **95B** (1980) 83; **100B** (1981) 254; **108B** (1987) 26.
[11] P. J. Olver and P. Rosenau, On the 'Non-Classical' Method for Group-Invariant Solutions of Differential Equations, Univ. of Minnesota Math. Re. No. 85-125, 1985.
[12] B. Fuchssteiner, Nonlinear Anal. TMA **3** (1979) 849.
[13] A. S. Fokas and R. L. Anderson, J. Math. Phys. **23** (1982) 1066.
[14] F. Magri, J. Math. Phys. **19** (1979) 1156; **Nonlinear Evolution Equations and Dynamical Systems**, ed. M. Boiti, F. Pempinelli and G. Soliani, Lecture Notes in Physics, Vol. 120 (Springer, New York, 1980), p. 233.
[15] A. S. Fokas and B. Fuchssteiner, Lett. Nuovo Cimento **28** (1980) 299; B. Fuchssteiner and A. S. Fokas, Physica **4D** (1981) 47.
[16] I. M. Gel'fand and I. Ya. Dorfman, Funct. Anal. Appl. **13**(1979)13; **14** (1980) 71.
[17] V. E. Zakharov and B. G. Konopelchenko, Comm. Math. Phys. **94** (1984) 483.
[18] A. S. Fokas and B. Fuchssteiner, Phys. Lett. **A86** (1981) 341.
[19] A. S. Fokas and M. J. Ablowitz, Stud. Appl. Math. **68** (1983) 1.
[20] W. Oevel and B. Fuchssteiner, Phys. Lett. **A88** (1982) 323.
[21] I. Y. Dorfman, Deformation of Hamiltonian Structures and Integrable Systems (preprint).
[22] B. Fuchssteiner, Prog . Theoret. Phy. **70** (1983) 150; I. Ya Dorfman, Deformations of Hamiltonian Structures and Integrable Systems (preprints).
[23] I. Gel'fand (private communication).
[24] W. Oevel, A . Geometrical Approach to Integrable Systems Admitting Scaling Symmetries (preprint, 1986).
[25] H. H. Chen, Y. C. Lee, J. E. Lin, Physica **9D** (1983) 439; Phys. Lett. **91A** (1982) 381.
[26] A. S. Fokas and P. M. Santini, Stud. Appl. Math. **75** (1986) 179.
[27] P. M. Santini and A. S. Fokas, Recursion Operators and Bi-Hamiltonian Structures in Multidimensions I , INS #65, preprint 1986.
[28] A. S. Fokas and P. M. Santini, Recursion Operators and Bi-Hamiltonian Structures in Multidimensions II, INS #67, preprint 1986.
[29] B. G. Konopelchenko and V. G. Dubrovsky, Physica **16D** (1985) 79.
[30] F. Calogero and A. Degasperis, Nuovo Cimento **39B** (1977) 1.
[31] P. Caudrey, Discrete and Periodic Spectral Transforms Related to the Kadomtsev-Petviashvili Equation, preprint UMIST (1985).

Painlevé Analysis and Integrability Aspects of Nonlinear Evolution Equations

M. Lakshmanan[1] *and K.M. Tamizhmani*[2]

[1]Department of Physics, Bharathidasan University, Tiruchirapalli 620024, India
[2]Department of Mathematics, Pondicherry University, Pondicherry 605001, India

A brief review of the singularity structure aspects of the solutions of nonlinear ordinary differential equations and their generalization to partial differential equations leading to the Painlevé (P) property is given. It is pointed out that the Painlevé analysis leads naturally to Lax pairs, Bäcklund transformations, linearizations and Hirota's bilinearization of nonlinear evolution equations. Specifically we treat the Burgers', Liouville, Korteweg-de Vries, coupled nonlinear Schrödinger and Kadomtsev-Petviashvili equations as examples.

1. INTRODUCTION

During the past two decades or so, there has been considerable progress in the understanding of classes of nonlinear evolution equations leading to many fascinating new concepts such as solitons, Bäcklund transformations, generalized symmetries, etc. [1-3] (as exemplified by the various articles in this book). However, in this process an important question arises as to how to search, identify, characterize and classify the integrable nonlinear equations systematically and then understand the solution characteristics. In the last century, this question was analysed for ordinary differential equations by various authors [4-6] through the singularity structure analysis of the solutions in the complex plane. With the recent developments in soliton equations this analysis has again received much attention [3,7-19], and now many of the integrable dynamical systems are associated with the so-called Painlevé property, in that they are free from movable critical points/manifolds. The development of the singularity structure analysis can in fact be traced down to the following four main aspects:

1. The classification of first order and second order nonlinear ordinary differential equations (odes) which are free from movable critical

points achieved through the works of Fuchs, Painlevé and his co-workers [5] in the last century.

2. S. Kovalevskaya's (1886) investigation for finding the integrable cases of rigid body motion around a fixed point under the influence of gravity [6] through the singularity structure analysis.
3. Ablowitz, Ramani and Segur's (ARS) [3] conjecture in 1978 that every ode obtained by an exact reduction of a soliton system which is solvable by the inverse scattering method is of Painlevé (P) type (see below), a fact verified from the invariance point of view by Lakshmanan and Kaliappan [7].
4. Weiss, Tabor, and Carnevale's (WTC) [8] generalized version of the Painlevé property directly applicable to partial differential equations (pdes). Here the solutions of the pdes are required to be single-valued around movable singular manifolds in order that they be integrable.

In Sec. 2, we begin with a brief discussion of odes and their singularities, and the application of Painlevé analysis to them. We review the generalized WTC procedure for pdes and apply it to a class of nonlinear evolution equations in (1+1) dimensions in Sec. 3 and obtain the basic solitonic properties. Furthermore, in Sec. 4, we extend the application of P-analysis to the (2+1) dimensional Kadomtsev-Petviashvili (K-P) equation and derive the associated Bäcklund transformation and Lax-pair straight forwardly through the P-analysis.

2. PAINLEVÉ PROPERTY AND ORDINARY DIFFERENTIAL EQUATIONS: INTEGRABILITY

The singularities of an ode can be classified as (i) fixed and (ii) movable [4,5]. While the location of the former is fixed by the nature of the coefficients of the ode, the latter is a function of the integration constant or initial condition.

Consider for example the linear first order ode $\frac{dw}{dz} + z^{-1}w = 0$. It has the solution, $w = C\exp(1/z)$, C, arbitrary. So $z = 0$ is the fixed (essential) singular point. More generally, for an n-th order linear ode

$$\frac{d^n w}{dz^n} + P_1(z)\frac{d^{n-1}w}{dz^{n-1}} + \ldots + P_n(z)w = 0, \tag{1}$$

where $P_i(z)$, $i = 1,2,\ldots,n$, are all analytic at $z = z_o$, admits n linearly independent solutions in the neighbourhood of z_o, so that the general solution may be written as

$$w(z) = \sum_{i=1}^{n} c_i w_i(z), \tag{2}$$

where c_i's are integration constants. Here the singularities of the solution must be located at the singularities of the coefficients $P_i(z)$ which are all fixed and *apriori* known and that they do not depend on the constants of integration c_i, $i = 1,2,\ldots,n$, at all.

However, in the case of nonlinear odes the singularities have the additional property that they can depend on the integration constants and so, they can 'move'. Examples: $\frac{dw}{dz} + w^2 = 0$ has the solution $w = (z - z_o)^{-1}$, where z is a constant of integration. Thus, at $z = z_o$, w has a singularity, a pole of order one; it is movable because its location depends on z_o. Similarly, $\frac{dw}{dz} + w^3 = 0$, $w = \frac{i}{\sqrt{2}}(z - z_o)^{-1/2}$; $\frac{dw}{dz} + w \log^2 w = 0$, $w = \exp(\frac{1}{z-z_o})$; $\frac{dw}{dz} + w \exp(w) = 0$, $w = \log(\frac{1}{z-z_o})$, which admit movable algebraic branch point, essential singularity and logarithmic singularity respectively. In fact, Fuchs [4] had shown that the only first order equation which admits no movable critical points is the generalized Riccati equation

$$\frac{dw}{dz} = P_0(z) + P_1(z)w + P_2(z)w^2. \tag{3}$$

Following the work of Fuchs for first order odes, Painlevé, Gambier, Garnier and others classified 50 canonical nonlinear second order odes which do not exhibit movable critical singularities and which can be integrated in terms of elementary functions, including elliptic functions, and Painlevé transcendentals [5]. Also the nature of entire bounded functions and the constancy of integrals of motion, coupled with the works of Euler (1780), Lagrange (1788) and Fuchs (1884) prompted S. Kovalevskaya [6] to treat the dynamical problem of the motion of a spinning top fixed to a point in terms of meromorphic functions. She concluded that only for three parametric choices (including those found by Euler and Lagrange earlier) the system is free from movable critical points and hence integrable, a result which even today stands undisputed.

The classification of integrable dynamical systems through the singularity structure analysis was not utilized much for more than eight decades or so. The ARS conjecture [3] on the connection between the integrable soliton systems and the Painlevé equations (noted in Sec. 1) revived the interest again in the singularity structure analysis for dynamical systems recently. Such a revival has also led to the concept of strong and weak Painlevé properties [18] for dynamical systems, which are then associated with the integrability aspect. Here the strong P-property stands for solutions which are meromorphic around a movable singular point while the weak P-property [18] allows for a relaxation of the meromorphicity condition on the solution so that under necessary circumstances determined solely by the nonlinearity of the equation movable algebraic branch points are also allowed for algebraic integrals of motion [17] to exist. Some of the typical integrable dynamical systems which possess the P-property are (i) Henon-Heiles system [18], (ii) coupled polynomial nonlinear oscillators [19], and (iii) Toda-lattice, for specific parametric choices.

An n-th order ode of the form

$$\frac{d^n w}{dz^n} = F\left(z, w, \frac{dw}{dz}, \ldots, \frac{d^{n-1}w}{dz^{n-1}}\right), \tag{4}$$

where F is rational in $\left(\frac{d^{n-1}w}{dz^{n-1}}, \ldots, w\right)$ and locally analytic in z, can be analysed algorithmically for its singularity structure as follows. Let us look for a Laurent series solution of (4) in the neighbourhood of a movable singular point in the form

$$w = (z - z_o)^{q_j} \sum_{m=0}^{\infty} a_{j,m}(z - z_o)^m, \quad 0 < z - z_o < R, \quad j = 1,2,\ldots \tag{5}$$

and determine the allowed values of q_j, $a_{j,m}$ and the powers at which (n-1) arbitrary constants enter into the series (5). In order to avoid inconsistencies, it may be necessary to introduce further logarithmic terms in the above series. Finally, after verifying the existence of the above solution, one classifies the conditions under which the solution is free from movable critical singularities to within transformations, which are then the possible cases for integrability. The latter may be proved often by finding the appropriate integrals of motion by other methods.

3. THE PAINLEVÉ ANALYSIS FOR PARTIAL DIFFERENTIAL EQUATIONS: INTEGRABILITY

The natural extension of the P-property discussed in Sec. 2 to pdes was suggested by Weiss, Tabor and Carnevale (WTC) [8], who required that the solutions be single-valued around movable singularity **manifolds.**

The major difference between the P-analysis of odes and pdes is that now the singularities of the latter are in general not isolated, as the solutions are functions of several complex variables $(z_1, z_2, \ldots, z_n)$, but rather lie on manifolds determined by the condition

$$\varphi(z_1, z_2, \ldots, z_n) = 0. \tag{6}$$

Thus if $u = u(z_1, z_2, \ldots, z_n)$ is a solution of the pde

$$u_t + K(u) = 0, \tag{7}$$

then we require that in the neighbourhood of the manifold

$$u = \varphi^{\alpha} \sum_{j=0}^{\infty} u_j \varphi^j, \tag{8}$$

where $u_o \neq 0$, $u_j = u_j(z_1, z_2, \ldots, z_n)$ and $\varphi = \varphi(z_1, z_2, \ldots, z_n)$ are analytic functions of (z_j) in a neighbourhood of the manifold (6) and that α is a negative integer. By Cauchy-Kovalevskaya's theorem such an expansion of the general solution must have sufficient number of arbitrary functions equal to that of the order of the pde. Implimentation of this procedure is direct and follows algorithmically in a manner similar to that of the odes.

There are essentially four steps involved in the P-analysis of pdes: (i) Determination of the leading-order behaviours; (ii) Identification of the powers at which arbitrary functions can enter into the Laurent series called resonances; (iii) verifying that at the resonance values sufficient number of arbitrary functions exist without the introduction of movable critical manifolds; (iv) Establishing connections with the soliton and other integrability properties. The remarkable feature of the P-analysis, particularly for soliton equations, is that a natural connection exists between the P-property and the linearization property, Lax pairs, Bäcklund transformations, integrability, etc. [8-15]. In the following subsections we will explain each of the stages succinctly.

3.1 Leading order analysis

The analysis starts with the determination of the all possible value(s) of α and u_o in the expansion (8), and under what conditions α is a negative integer so that no movable critical manifolds enter at this stage. For each value of α, the homogeneous terms with the highest degree may balance each other. These terms are called **leading terms** (or dominant terms). The value of u_o can be determined by equating the coefficients of the dominant terms to zero and solving the resulting algebraic equation for u_o.

3.2 Resonance analysis

Next, one has to find the "resonance" values, j, that is the power(s) at which the coefficient(s) u_j of the term $\varphi^{j+\alpha}$ in the expansion (8) is arbitrary. To find these, we substitute (8) into Eq. (7), and obtain appropriate recursion relations for u_j and extract the coefficient $\tilde{Q}(j) = Q(j)u_j$ of the term $\varphi^{j+\alpha-N}$, where N is the order of the pde. Then $Q(j) = 0$ is called the **resonance equation**, for which -1 is always a root, which corresponds to the arbitrary nature of φ. In order to avaoid any movable critical singular manifolds, we require that these remaining roots are non-negative integers.

3.3 Arbitrary functions

Let j_s be the highest of the allowed resonance values. Then we substitue

$$u = \sum_{j=0}^{j_s} u_j \varphi^{j+\alpha} \tag{9}$$

into Eq. (7) and collect the coefficients of $\varphi^{j+\alpha-N}$ to obtain

$$Q(j)u_j + R_j = 0, \tag{10}$$

where R_j is a polynomial in the partial derivatives of φ and u_k's $(k = 0,1,\ldots,j-1)$. Since $Q(j) = 0$, for any resonance value j, R_j should identically vanish. In this case u_j is arbitrary. In case if it is not so, we have to introduce logarithmic terms of the form $a_j + b_j \log\varphi$ in the series. But due to this addition, logarithmic singularities will appear in the solution manifold. Thus, $R_j = 0$ is the condition to ensure that the solution is free from movable critical manifolds at a particular resonance value j. In this way, we can check that the general solution is free from movable critical manifolds.

3.4 Bäcklund transformation (BT), Lax pairs, etc.

The Bäcklund transformation of Eq. (7) can be obtained by truncating the expansion (8) at the 'constant' level term, by setting

$$u = u_o \varphi^{-n} + u_1 \varphi^{-n+1} + \ldots + u_n. \tag{11}$$

Then one can find an over-determined system of equations for φ and u_j, $j = 0,1,\ldots,n$, where u_n will satisfy Eq. (7). Upon solving the overdetermined system, the other soliton properties of Eq. (7) such as linearization, Lax pairs, etc., can also be obtained, in general [8-14].

4. EXAMPLES

In this section, we briefly illustrate the theory discussed above with some typical examples [8-14].

4.1 Burgers' equation

The Burgers' equation is of the form [2]

$$u_t + uu_x = \sigma u_{xx}. \tag{12}$$

Using Eq. (8) in Eq. (12), we can easily find from the leading order analysis that

$$\alpha = -1, \quad u_o = -2\sigma\varphi_x. \tag{13}$$

The recursion relations for u_j are found to be

$$u_{j-2,t} + (j-2)u_{j-1}\,\varphi_t + \sum_{m=0}^{j} u_{j-m}[u_{m-1,x} + (m-1)\varphi_x u_m]$$
$$= \sigma[u_{j-2,xx} + 2(j-2)u_{j-1,x}\,\varphi_x + (j-2)u_{j-1}\varphi_{xx} + (j-1)(j-2)u_j\varphi_x^2]. \tag{14}$$

The resonance analysis shows that the resonances occur at $j = -1,2$. In fact, $j = -1$ corresponds to the arbitrariness of the manifold $\varphi(x,t) = 0$.

From Eq. (14), we find that

$$j = 0\ ; \quad u_o = -2\sigma\varphi_x, \tag{15a}$$

$$j = 1\ ; \quad \varphi_t + u_1\varphi_x = \sigma\varphi_{xx}, \tag{15b}$$

$$j = 2\ ; \quad \frac{\partial}{\partial x}(\varphi_t + u_1\,\varphi_x - \sigma\,\varphi_{xx}) = 0. \tag{15c}$$

By (15b), (15c) is identically satisfied and hence u_2 is arbitrary. The coefficients u_j, $j > 2$ can then be obtained uniquely in terms of u_0, u_1, and u_2. Therefore, the general solution of (12) contains the required two arbitrary functions and that (12) possesses the Painlevé property [8] in the sense of WTC.

If now one sets the arbitrary function $u_2 = 0$, then all $u_j = 0$, $j \geq 2$, provided

$$u_{1t} + u_1 u_{1x} = \sigma u_{1xx}, \tag{16}$$

which is just the Burgers' equation. Thus, we obtain the Bäcklund transformation for Burgers' equation,

$$u = -2\sigma \frac{\varphi_x}{\varphi} + u_1, \tag{17}$$

where both u and u_1 satisfy the Burgers' equation and φ obeys (15b). When we consider the vacuum solution $u_1 = 0$ in Eq. (17), the well-known Cole-Hopf transformation results, which is the linearizing transformation for the Burgers' equation.

4.2 Liouville equation

We next consider the Liouville equation [11] in the form

$$u_{xt} = e^u . \tag{18}$$

Under the transformation

$$u = \log V, \tag{19}$$

(18) becomes

$$VV_{xt} - V_x V_t - V^3 = 0. \tag{20}$$

On expanding

$$V = \varphi^{-2} \sum_{j=0}^{\infty} V_j \varphi^j, \tag{21}$$

we find that

$$V_o = 2\varphi_x \varphi_t, \quad V_1 = -2\varphi_{xt} \tag{22a}$$

and

$$\begin{aligned} &V_0 V_{0xt} + V_0 V_{1x} \varphi_t + V_0 V_{1t} \varphi_x - 3V_0 V_1 \varphi_{xt} - V_1 V_{0x} \varphi_t - \\ &\quad V_{ot} V_1 \varphi_x + V_1^2 \varphi_x \varphi_t + 6\, V_0 V_2 \varphi_x \varphi_t - V_{0x} V_{0t} - \\ &\quad 3V_0^2 V_2 - 3V_0 V_1^2 = 0. \end{aligned} \tag{22b}$$

After making use of Eqs. (22a) in (22b), we observe that V_2 is arbitrary and thus Eq. (18) passes the Painlevé test [11]. Furthermore, a BT is obtained with

$$V = \frac{2\varphi_x\varphi_t}{\varphi^2} + V_2 \tag{23}$$

provided V_2 satisfies Eq. (20) and the linearized wave equation is true:

$$\varphi_{xt} = 0. \tag{24}$$

The solution of (24) is given by

$$\varphi = h(x) + g(t), \tag{25}$$

where h(x) and g(t) are arbitrary functions. Consequently, we obtain the general solution

$$u = \log\left(\frac{2g_t h_x}{(g+h)^2}\right) . \tag{26}$$

4.3 Korteweg-de Vries equation (K-dV)

The K-dV equation

$$u_t + uu_x + \sigma u_{xxx} = 0 \tag{27}$$

is easily shown to possess the Painlevé property. One can find the leading order and the resonance values as $\alpha = -2$ and $j = -1,4,6$ respectively. From the recursion relations it is easy to check that

$$j = 0 : u_0 = -12\sigma\varphi_x^2 \tag{28a}$$

$$j = 1 : u_1 = 12\,\sigma\varphi_{xx} \tag{28b}$$

$$j = 2 : \varphi_x\varphi_t + u_2\varphi_x^2 + 4\,\sigma\varphi_x\varphi_{xxx} - 3\sigma\varphi_{xx}^2 = 0 \tag{28c}$$

$$j = 3 : \varphi_{xt} + u_2\varphi_{xx} - u_3\varphi_x^2 + \sigma\varphi_{xxxx} = 0 \tag{28d}$$

$$j = 4 : \frac{\partial}{\partial x}(\varphi_{xt} + u_2\varphi_{xx} - u_3\varphi_x^2 + \sigma\varphi_{xxxx}) = 0 \tag{28e}$$

$$\begin{aligned} j = 5 : \; & u_{2t}+u_2u_{2x}+\sigma u_{2xxx}+u_3\varphi_t+u_1u_{3x}+u_3u_{1x}+u_0u_{4x}+u_{0x}u_4+u_0u_5\varphi_x+ \\ & u_1u_4\varphi_x+u_2u_3\varphi_x+\sigma(3u_{3xx}\varphi_x+6u_{4x}\varphi_x^2+3u_{3x}\varphi_{xx}+6u_5\varphi_x^2+6u_4\varphi_x\varphi_{xx}+ \\ & u_3\varphi_{xxx}) = 0. \end{aligned} \tag{28f}$$

Now it is clear that by the condition (28d), the determining Eq.(28e) is always satisfied. Similarly, we can verify that at $j = 6$ the compatibility condition is satisfied identically and so, u_6 is arbitrary. Thus, we conclude that the K-dV equation possesses the Painlevé property [8].

If the arbitrary functions u_4 and u_6 are chosen identically to be zero, and if we require that $u_3 = 0$, then we can verify that $u_j = 0$, $j \geq 3$, provided u_2 satisfies the K-dV equation. Thus, we get the BT for the K-dV equation in the form

$$u = \frac{\partial^2}{\partial x^2} \log\varphi + u_2 \tag{29}$$

with

$$\varphi_x\varphi_t + u_2\varphi_x^2 + 4\sigma\varphi_x\varphi_{xxx} - 3\sigma\varphi_{xx}^2 = 0, \tag{30a}$$

$$\varphi_{xt} + 12u_2\varphi_{xx} + \varphi_{xxxx} = 0 , \tag{30b}$$

and

$$u_{2t} + u_2u_{2x} + \sigma\, u_{2xxx} = 0. \tag{30c}$$

Eliminating u_2 from (30a) and (30b), we obtain after an integration

$$\frac{\varphi_t}{\varphi_x} + \sigma\{\varphi;\, x\} = \lambda , \tag{31}$$

where

$$\{\varphi;\, x\} = \left(\frac{\varphi_{xx}}{\varphi_x}\right)_x - \frac{1}{2}\left(\frac{\varphi_{xx}}{\varphi_x}\right)^2 \tag{32}$$

is the 'Schwarzian derivative' of φ and λ is a constant parameter. With the substitution $\varphi = v_1/v_2$, Eq. (31) can be rewritten in the form

$$\frac{\Psi_t}{\Theta_x} + \sigma\left(\{\varphi;\, x\} - \frac{2v_{2xx}}{v_2} + \frac{2\Theta_{xx}}{\Theta_x} - \frac{v_{2x}}{v_2}\right) = \lambda, \tag{33}$$

where $\Theta_x = v_2v_{1x} - v_1v_{2x}$, $\Psi_t = v_2v_{1t} - v_1v_{2t}$. If (v_1, v_2) satisfy a linear system (say)

$$v_{xx} = av \tag{34}$$

$$v_t = bv_x + cv \tag{35}$$

then we can check that $\Theta_{xx} = 0$ and $\Psi_t = b\Theta_x$. Using these values in Eq. (33) we get $a = -\frac{1}{6}(u_2 + \lambda)$, $b = -u_2/3 + (2/3)\lambda$ and $c = u_{2x}/6$. With these values, Eqs. (34) and (35) are just the Lax pair of the K-dV equation.

4.4 Coupled nonlinear Schrödinger equations

Our next example is a system of two coupled nonlinear Schrödinger (NLS) equations defined by

$$i\chi_{1t} = c_1\chi_{1xx} + 2\alpha|\chi_1|^2 \chi_1 + 2\beta|\chi_2|^2 \chi_1, \tag{36a}$$

$$i\chi_{2t} = c_2\chi_{2xx} + 2\gamma|\chi_2|^2 \chi_2 + 2\beta|\chi_1|^2 \chi_2. \tag{36b}$$

The system (36) is known to be integrable [16] for the following specific parametric restrictions

$$\alpha = \beta = \gamma \qquad c_1 = c_2 \tag{37}$$

$$\alpha = -\beta = \gamma \qquad c_1 = -c_2. \tag{38}$$

Now we will show that for exactly the same parametric choices, (37) and (38), the system (36) passes the Painlevé test [12].

We rewrite the Eqs. (36) as

$$-Q_t = c_1P_{xx} + 2\alpha(P^2 + Q^2)P + 2\beta(R^2 + S^2)P \tag{39a}$$

$$P_t = c_1Q_{xx} + 2\alpha(P^2 + Q^2)Q + 2\beta(R^2 + S^2)Q \tag{39b}$$

$$-S_t = c_2R_{xx} + 2\beta(P^2 + Q^2)R + 2\gamma(R^2 + S^2)R \tag{39c}$$

$$R_t = c_2S_{xx} + 2\beta(P^2 + Q^2)S + 2\gamma(R^2 + S^2)S, \tag{39d}$$

where $\chi_1 = P + iQ$ and $\chi_2 = R + iS$, P, Q, R, and S are reals. By expanding P, Q, R and S about the singularity manifold φ as

$$P = \varphi^{-1} \sum_{j=0}^{\infty} P_j\varphi^j, \quad Q = \varphi^{-1} \sum_{j=0}^{\infty} Q_j \varphi^j, \tag{40a}$$

$$R = \varphi^{-1} \sum_{j=0}^{\infty} R_j\varphi^j, \quad S = \varphi^{-1} \sum_{j=0}^{\infty} S_j\varphi^j \tag{40b}$$

one can easily find the following two possibilities:

Case 1 $c_1 c_2 \varphi_x^2 = 2(R_0^2 + S_0^2) \quad (c_1\gamma - c_2\beta)$ (41)

and the associated resonances are

$$j = -1,\ 0,\ 0,\ 1,\ 2,\ 3,\ 3,\ 4. \tag{42}$$

Case 2 $(R_0^2 + S_0^2)\ (c_1\gamma - c_2\beta) = 0$ (43)

and

$$j = -1,\ 0, 0, 0, 3,\ 3,\ 3,\ 4. \tag{44}$$

We can verify that for case 1 the associated series solution will have a lesser number of arbitrary functions only since none of the functions P_1, Q_1, R_1 and S_1 is arbitrary and so it does not correspond to the general solution. For case 2, the resonance values (44) require that three of the four functions P_0, Q_0, R_0 and S_0 are arbitrary. From the leading order analysis we have

$$\alpha(P_0^2 + Q_0^2) + \underset{\text{(twice)}}{\beta(R_0^2 + S_0^2)} = - c_1\varphi_x^2, \tag{45}$$

$$\beta(P_0^2 + Q_0^2) + \underset{\text{(twice)}}{\gamma(R_0^2 + S_0^2)} = - c_2\varphi_x^2 \tag{46}$$

and hence two, say P_0 and Q_0, of the four functions (P_0, Q_0, R_0, S_0) are arbitrary. Also, from Eq. (43) we can show that R_0 (or S_0) is arbitrary only for the conditions (37) and (38). Proceeding further we can establish the required arbitrariness at the other resonance values. Thus, Eqs. (36) possess the Painlevé property for the parametric choices of Zakharov and Schulman only.

We will discuss the connection between the corresponding BT and soliton solutions in Sec. 5. We remark that this analysis can be extended straightforwardly to the case of a set of N coupled nonlinear Schrödinger equations [12] as well.

4.5 Kadomtsev-Petviashvili equation

The Kadomtsev-Petviashvili (KP) equation is a two-dimensional generalization of the well-known Korteweg-de Vries equation and describes unidirectional weak (quadratic) nonlinear disturbances perturbed by weak transverse balancing fourth order dispersion. The system admits infinitely many non-trivial symmetries and constants of motion and a Lax pair exists. In this section, we will demonstrate the existence of the Painlevé property for the KP equation and obtain its Lax pair therefrom.

The KP equation is of the form [10]

$$\frac{\partial}{\partial x}(u_t + 6uu_x + u_{xxx}) + 3\sigma^2 u_{yy} = 0, \quad \sigma^2 = \pm 1. \tag{47}$$

Now, the Laurent series solution takes the form $u = \sum_{j=0}^{\infty} u_j \varphi^{j+\alpha}$ around the singularity manifold $\varphi(x,y,t) = 0$, where $\alpha = -2$ and $u_0 = -2\varphi_x^2$. Substituting the above series solution into Eq. (47), we find that the resonance values are $j = -1,4,5$ and 6. Then equating the lowest order coefficients to zero, we obtain,

$$j = 0 \quad : \quad u_0 = -2\varphi_x^2 \tag{48a}$$

$$j = 1 \quad : \quad u_1 = 12\varphi_{xx} \tag{48b}$$

$$j = 2 \quad : \varphi_x\varphi_t - 3\varphi_{xx}^2 + 4\varphi_x\varphi_{xxx} + 6u_2\varphi_x^2 + 3\sigma^2\varphi_y = 0 \tag{48c}$$

$$j = 3 \quad : \varphi_{xt} + 6u_2\varphi_{xx} - 6u_3\varphi_x^2 + 3\sigma^2\varphi_{yy} + \varphi_{xxxx} = 0 \tag{48d}$$

$$j = 4 \quad : \quad (\varphi_x\varphi_t - 3\varphi_{xx}^2 + 4\varphi_x\varphi_{xxx} + 6u_2\varphi_x^2 + 3\sigma^2\varphi_y^2)_{xx} = 0 \tag{48e}$$

$$j = 5 \quad : \quad (\varphi_{xt} + 6u_2\varphi_{xx} - 6u_3\varphi_x^2 + \varphi_{xxxx} + 3\sigma^2\varphi_{yy})_{xx} = 0 \tag{48f}$$

$$\begin{aligned} j = 6 \quad : \quad & u_3\varphi_{xt} + u_{2xt} + u_{3t}\varphi_x + u_{3x}\varphi_t + 6u_1u_{3xx} + 6u_2u_{2xx} + \\ & u_3u_2\varphi_{xx} + 6u_3u_{1xx} + 12u_2u_{3x}\varphi_x + 12u_{3x}u_{1x} + 6u_{2x}^2 + \\ & 12u_3u_{2x}\varphi_x + 12u_3^2\varphi_x^2 + u_{2xxxx} + u_3\varphi_{xxxx} + 4u_{3xxx}\varphi_x + \\ & 6u_{3xx}\varphi_{xx} + 4u_{3x}\varphi_{xxx} + 3\sigma^2 u_{2yy} + 6\sigma^2 u_{3y}\varphi_y + \\ & 3\sigma^2\varphi_{yy}u_3 = 0. \end{aligned} \tag{48g}$$

From the above sets of equations, one can verify that the functions u_4, u_5 and u_6 are arbitrary [9,10] and so the P-property holds.

If we choose the arbitrary functions u_4, u_5 and u_6 identically to be zero, it is easily seen that $u_j = 0$, $j \geq 7$ provided $u_3 = 0$, in which case u_2 satisfies the KP equation

$$\frac{\partial}{\partial x}(u_{2t} + 6u_2u_{2x} + u_{2xxx}) + 3\sigma^2 \frac{\partial^2 u_2}{\partial y^2} = 0. \tag{49}$$

Accordingly, we have the Bäcklund transformation

$$u = 2\frac{\partial^2}{\partial x^2} \log\varphi + u_2, \tag{50}$$

where φ satisfies Eqs. (48).

Eliminating u_2 from Eqs. (48c) and (48d), it is found that

$$\frac{\partial}{\partial x}\left(\frac{\varphi_t}{\varphi_x} + \{\varphi\ ;\ x\} + \frac{3}{2}\sigma^2 \frac{\varphi_y^2}{\varphi_x^2}\right) + 3\sigma^2 \frac{\partial}{\partial y}\left(\frac{\varphi_y}{\varphi_x}\right) = 0. \tag{51}$$

Integrating (51) with respect to x and inserting $\varphi = v_1/v_2$, we require that (v_1, v_2) satisfy

$$\sigma v_y = f_1 v_{xx} + f_2, \quad v_2 = f_3 v_{xxx} + f_4 v_x + f_5 v, \tag{52}$$

where f_1, f_2, f_3, f_4 and f_5 are functions of (x,t,y,u) to be determined. Finding the f_i's, we obtain finally the Lax-pair in the form

$$\sigma v_y + v_{xx} + uv = 0 \tag{53a}$$

$$v_t + 4v_{xxx} + 6uv_x + 3\left(u_x - \sigma \int_{-\infty}^{x} u_y \, dx'\right) = 0 \tag{53b}$$

the compatibility of which is the KP eq. (47).

5. INTERCONNECTION BETWEEN THE PAINLEVÉ PROPERTY AND HIROTA'S METHOD

Among the direct methods, Hirota's bilinearization technique is the most convenient one to construct soliton solutions of npdes. Essentially, the method advocates transforming the given equation into homogeneous bilinear forms and then by 'self-truncating' the series to obtain a set of bilinear equations. In this process, the reasons for both the choice of the initial transformation and the self-truncation of the series is rather obscure. From the P-analysis, it is clear that the truncation of the Laurent series at the constant level term results in the BT. For the vacuum solutions using the BT we can obtain in most cases the required Hirota's dependent variable transformation to transform the given equation into the bilinear operator form [13-15]. As an example we consider [12], the coupled nonlinear Schrödinger eq. (39). The BT for Eq. (39)

$$P = P_0\varphi^{-1} + P_1 , \quad Q = Q_0\varphi^{-1} + Q_1 \tag{54a}$$

$$R = R_0\varphi^{-1} + R_1 , \quad S = S_0\,\varphi^{-1} + S_1 \tag{54b}$$

in the vacuum case leads to the following set of equations:

$$[2c_1(\varphi_x^2 - \varphi\varphi_{xx}) + 2\alpha\Gamma]P_0 = [Q_{0t}\varphi - Q_0\varphi_t + c_1(P_{0xx}\varphi + P_0\varphi_{xx} - 2P_{0x}\varphi_x)]\varphi \tag{55a}$$

$$[2c_1(\varphi_x^2-\varphi\varphi_{xx}) + 2\alpha\Gamma]Q_0 = [P_0\varphi_t - P_{0t}\varphi + c_1(Q_{0xx}\varphi + Q_0\varphi_{xx} - 2Q_{0x}\varphi_x)]\varphi \tag{55b}$$

$$[2c_1(\varphi_x^2-\varphi\varphi_{xx}) + 2\alpha\Gamma]R_0 = [S_{0t}\varphi - S_0\varphi_t + c_1(R_{0xx}\varphi - 2R_0\varphi_{xt} - 2R_{0x}\varphi_x)]\varphi \tag{55c}$$

$$[2c_1(\varphi_x^2-\varphi\varphi_{xx}) + 2\alpha\Gamma]S_0 = [R_0\varphi_t - R_{0t}\varphi + c_1(S_{0xx}\varphi + S_0\varphi_{xx} - 2S_{0x}\varphi_x)]\varphi , \tag{55d}$$

where

$$\Gamma = P_0^2 + Q_0^2 + R_0^2 + S_0^2 .$$

Eqs. (55) can be reexpressed in terms of the Hirota's bilinear operators and after decoupling we obtain the form

$$(c_1D_x^2 - \mu)\varphi.\varphi = 2\alpha\Gamma \tag{56a}$$

$$D_tQ_0\varphi + c_1D_x^2P_0\varphi - \mu P_0\varphi = 0 \tag{56b}$$

$$-D_tP_0\varphi + c_1D_x^2Q_0\varphi - \mu Q_0\varphi = 0 \tag{56c}$$

$$D_tS_0\varphi + c_1D_x^2R_0\varphi - \mu R_0\varphi = 0 \tag{56d}$$

$$-D_tR_0\varphi + c_1D_x^2S_0\varphi - \mu S_0\varphi = 0 , \tag{56e}$$

where μ is a constant to be determined and

$$D_t^nD_x^m f.g = \left(\frac{\partial}{\partial t} - \frac{\partial}{\partial t'}\right)^n\left(\frac{\partial}{\partial x} - \frac{\partial}{\partial x'}\right)^m f(x,t)g(x',t')\Big|_{\substack{x'=x\\ t'=t}} .$$

In particular, from (56a), we get

$$P_0^2 + Q_0^2 + R_0^2 + S_0^2 = \frac{1}{\alpha}(c_1 \frac{\partial^2}{\partial x^2} \log\varphi - \frac{\mu}{2}), \qquad (57)$$

and for the choice (45) we obtain

$$P_0^2 + Q_0^2 - (R_0^2 + S_0^2) = \frac{1}{\alpha}(c_1 \frac{\partial^2}{\partial x^2} \log\varphi - \frac{\mu}{2}). \qquad (58)$$

Now expanding the functions φ, P_0, Q_0, R_0 and S_0 as power series [12] and using them in (57) we can construct the N soliton solutions. Similar analysis can be applied to the other systems as well.

6. CONCLUSIONS

In this review, we have briefly pointed out how the Painlevé analysis is a useful and systematic procedure for investigating the integrability properties of nonlinear partial differential equations. For a class of physically important equations we have demonstrated that this technique can in a straightforward manner lead to linearizations, Lax pairs and Hirota's bilinearization, showing the deep interconnections between the Painlevé analysis and integrability.

REFERENCES

[1] M. J. Ablowitz and H. Segur, **Solitons and the Inverse Scattering Transform** (SIAM, Philadelphia, 1981).

[2] P. J. Olver, **Applications of Lie Groups to Differential Equations** (Springer, New York, 1986).

[3] M. J. Ablowitz, A. Ramani and H. Segur, Lett. Nuovo Cim. **23**, (1978) 333; J. Math. Phys. **21** (1980) 715.

[4] E. Hille, **Ordinary Differential Equations in the Complex Domain** (Wiley, New York, 1976).

[5] E. L. Ince, **Ordinary Differential Equations** (Dover, New York, 1956).

[6] S. Kovalevskaya, Acta Math. **12** (1889) 177; R. Cooke, **The Mathematics of Sonya Kovalevskaya** (Springer, New York, 1984).

[7] M. Lakshmanan and P. Kaliappan, J. Math. Phys. **24** (1983) 795.

[8] J. Weiss, M. Tabor and G. Carnevale, J. Math. Phys. **24** (1983) 522.

[9] J. Weiss, J. Math. Phys. **26** (1985) 2174.

[10] W. Oevel and W-H. Steeb, Phys. Lett. **103A** (1984) 239.

[11] K. M. Tamizhmani and M. Lakshmanan, J. Math. Phys. **27** (1986) 2257.

[12] R. Sahadevan, K. M. Tamizhmani and M. Lakshmanan, J. Phys. **A19** (1986) 1783.

[13] K. M. Tamizhmani, "Geometrical, Group Theoretical and Singularity Structure Analysis Aspects of Certain Nonlinear Partial Differential Equations," Ph.D. Thesis, University of Madras (1986).

[14] J. D. Gibbon, P. Radmore, M. Tabor and D. Wood, Stud. In Appl. Math. **72** (1985) 39.
[15] M. Tabor and J. D. Gibbon, Physica **18D** (1986) 180.
[16] V. E. Zakharov and E. I. Schulman, Physica **4D** (1982) 270.
[17] H. Yoshida, Celest. Mech. **31** (1983) 363, 381.
[18] A. Ramani, B. Dorizzi and B. Grammaticos, Phys. Rev. Lett. **49** (1982) 1539; J. Math. Phys. **24** (1983) 2252.
[19] M. Lakshmanan and R. Sahadevan, Phys. Rev. **A31** (1984) 861; R. Sahadevan and M. Lakshmanan, Phys. Rev. **A33** (1985) 3563.

Generalised Burgers Equations and Connection Problems for Euler-Painlevé Transcendents

P.L. Sachdev

Department of Applied Mathematics, Indian Institute of Science, Bangalore 560012, India

Some recent results on a class of generalised Burgers' equations (GBE) are reviewed. Characteristically they reduce to nonlinear Euler-Painlevé equations, on similarity reduction, possessing single hump type solution. GBE's with variable viscosity coefficients are briefly discussed. It is pointed out that the Korteweg-de Vries type equations and GBE's behave somewhat analogously in their self-similar forms.

1. INTRODUCTION

There has been a pervading and persistent interest for a couple of decades in the study of the model nonlinear equation, called the Korteweg-de Vries equation,

$$u_t + uu_x + \frac{\delta}{2} u_{xxx} = 0. \qquad (1)$$

This describes a certain balance between the (simplest) dispersion and nonlinear convections. Equation (1) and its kindred class have three main unifying features: (1) they exhibit clean soliton interaction; (2) they can be exactly linearised through inverse scattering theory to linear integral equations of Gelfand-Levitan type; (3) their self-similar form, as it is or through some simple transformations, belongs to the class of nonlinear ODE's whose only movable singularities are poles; this characteristic of the ODE's (or PDE's) is called Painlevé property, after Painlevé who first studied second order nonlinear ODE's from their singularity structure point of view.

There is another distinguished model equation,

$$u_t + uu_x = \frac{\delta}{2} u_{xx}, \qquad (2)$$

called the Burgers equation, which has a certain kinship with (1). It had preceded the latter in its inception and investigation. Equation

(2) represents a balance between linear diffusion and nonlinear convection. Equations (1) and (2) have the same convective term but different higher order terms. Each of these equations has influenced the study of the other. Equation (2) is simpler and can, in fact, be exactly linearised to the heat equation by the Hopf-Cole transformation [1]. In contrast, Eq. (1) has no such simple transformation and attempts to linearise it by a Hopf-Cole type transformation have led to its further 'non-linearization' [2]. Equation (2) does not display soliton behaviour; indeed it has no solitary wave solution. Actually, Eq. (2) is a severe idealisation of reality. The equations which actually occur in applications are invariably more complicated. We quote two such equations:

$$u_t + u^\beta u_x + \lambda u^\alpha = \frac{\delta}{2} u_{xx}, \tag{3}$$

$$u_t + u^\alpha u_x + \frac{ju}{2t} = \frac{\delta}{2} u_{xx}, \quad j = 0,1,2. \tag{4}$$

Equation (3) has a lower order damping term and describes stress waves in a nonlinear Maxwell rod, while Eq. (4) has a spherical or cylindrical term besides those in (2). Both these equations have more general nonlinear convective terms than (2). We shall mention other generalised Burgers equations (GBE's) in the sequel.

It does not seem possible to linearise GBE's of the type (3) and (4) through Hopf-Cole like transformations. The question arises whether we can unify this class of equations in some manner. Since soliton behaviour and exact linearisation do not seem possible, the one option open is their characterisation through the ODE's obtained by similarity transformations. This turns out to be possible. We find that there is a class of nonlinear ODE's, which we refer to as Euler-Painlevé equations, which describes the self-similar single hump (and sometime other) type solutions of several GBE's. A very special case of this equation was studied by Euler and Painlevé [3], hence the name Euler-Painlevé transcendents.

In the present paper, we summarise our work on GBE's and Euler-Painlevé transcendents [4,5]. We first study the Burgers equation and its self-similar single hump solution. The self-similar solutions of (3) are then studied with reference to a connection problem with respect to the 'reduced' nonlinear ODE's on the whole real line. This is followed by a similar study for (4). Some current work on GBE's with variable viscosity coefficients is then briefly discussed. Finally, we refer to a large number of DE's, which have been categorised by Kamke [3] according to their formal appearance and appended with geometrical or physical significance. These are special cases of the Euler-Painlevé equations if the coefficients in the form of the latter are made to vary with the independent variable. A reference to the connection problem for the K-dV equation as represented by the second Painlevé transcendent is made again to show how K-dV type of equations and GBE's behave somewhat analogously in their self-similar forms. However, we caution that Euler-Painlevé equations are free from singularities on the finite real line, unlike Painlevé equations, which, in general, have an infinite number of poles.

2. SINGLE HUMP SOLUTION OF BURGERS EQUATION

We study in the following a specific self-similar solution which has a single hump form and vanishes at $x = \pm\infty$. For the Burgers equation (2), this solution can be written out as

$$u = (\delta/t)^{1/2} f(\eta), \quad \eta = x(2\delta t)^{-1/2} \tag{5}$$

where

$$f(\eta) = \left[\frac{(2\pi)^{1/2}}{e^R - 1} \exp(\eta^2) + (\pi/2)^{1/2} \exp(\eta^2).\mathrm{erfc}\eta\right]^{-1}$$

$$= \frac{1}{H_B(\eta)}, \text{ say,} \tag{6}$$

where the function $f(\eta)$ satisfies the DE

$$f'' + 2\eta f' + 2f - 2^{3/2} ff' = 0 . \tag{7}$$

The 'inverse' function $H_B(\eta)$ plays an important role for GBE's, as we discusss subsequently. In the present case, it satisfies the DE,

$$HH'' - 2H'^2 + 2\eta HH' - 2H^2 - 2^{3/2}H' = 0. \tag{8}$$

We note a few properties of this single hump solution, which, in a sense, is closest to the solitary wave. We emphasise that this solution is not symmetric, in contrast to the solitary wave. The solution (5) arises out of singular initial conditions $u(x,o) = c_o + A\delta(x)$, where c_o=const., $\delta(x)$ is the delta function—singular initial data are common to most self-similar solutions. By integrating (2) from $x = -\infty$ to $x = +\infty$, one may easily conclude that the Reynolds number

$$R = \frac{1}{\delta} \int_{-\infty}^{\infty} u \, dx \tag{9}$$

is constant. Here, we use the conditions that u and its x-derivatives vanish at $x = \pm\infty$. The solution (5) vanishes exponentially as $x \to +\infty$ and algebraically as $x \to -\infty$. One may further verify that in the limit $\delta \to 0$, that is, $R \to \infty$, the solution reduces to

$$u = x/t, \tag{10}$$

the solution of the inviscid Burgers equation with $\delta = 0$. For large but finite R, the single hump consists of the inviscid solution for $\xi < \sqrt{2}$, followed by a thin shock ending in a diffusive tail. For $R \to 0$, it easily follows from (6) that the solution (5) is essentially diffusive; the nonlinear convective term plays no essential role. Hopf [12] pointed out that the solution (5) represents a 'stationary' solution, to which an infinite number of solutions, with essentially similar behaviour at $x \to \pm\infty$ in their initial conditions, converge as $t \to \infty$. Such self-similar solutions in Soviet literature are now referred to as intermediate asymptotics [1].

3. SINGLE HUMP SOLUTIONS FOR GBE WITH DAMPING--THE CONNECTION PROBLEM

Now we seek self-similar single hump solutions of (3) [4]. We find that the solution has the form

$$u = t^{1/(1-\alpha)} f(\eta), \tag{11}$$

$$\eta = x(2\delta t)^{-1/2},$$

provided

$$\beta = (\alpha - 1)/2. \tag{12}$$

Equation (3) then reduces to the ODE

$$f'' + 2\eta f' - \frac{4}{1-\alpha} f - 4(2\delta)^{-1/2} f^{\frac{1}{2}(\alpha-1)} f' - 4\lambda f^{\alpha} = 0. \tag{13}$$

The solutions (11) decay (explicitly with time) if $\alpha > 1$ and grow if $\alpha < 1$. We employ the 'inverse' transformation (cf. (6))

$$H = \delta^{1/2} f^{(1-\alpha)/2} \tag{14}$$

and obtain the form

$$HH'' - 2(1+\alpha_1)H'^2 + 2\eta HH' - 2H^2 - 2^{3/2}H' - 2\lambda_1 = 0 \tag{15}$$

of (13), free from fractional powers of f. Here,

$$\alpha_1 = \frac{1}{2}\frac{3-\alpha}{\alpha-1}, \qquad \lambda_1 = \lambda\delta(1-\alpha). \tag{16}$$

Equation (15) generalises (8) in two ways: the coefficient of H' becomes $-2(1+\alpha_1)$ instead of -2, and a constant $-2\lambda_1$ gets added on to (8). These 'minor' changes represent a fairly general GBE (3). Before we pose a connection problem for (13), we note two special solutions of (13) and (15). First, (13) has a constant solution,

$$f = [\lambda(\alpha-1)]^{1/(1-\alpha)} = f_m, \text{ say.} \tag{17}$$

It is easy to check that f_m is the maximum value of f that the maxima of the single hump solutions can attain. The other exact solution is

$$f = \begin{cases} (A_+\eta)^{2/(1-\alpha)}, & \eta > 0, \\ (-A_-\eta)^{2/(1-\alpha)}, & \eta < 0, \end{cases} \tag{18}$$

where

$$A_+ = (2/\delta)^{1/2}\left(\frac{\alpha-1}{\alpha+1}\right)[(1+\lambda\delta(1+\alpha))^{1/2}+1],$$

$$A_- = (2/\delta)^{1/2}\left(\frac{\alpha-1}{\alpha+1}\right)[(1+\lambda\delta(1+\alpha))^{1/2}-1]. \tag{19}$$

The corresponding solutions for $H(\eta)$ can be written with the help of (14). Equation (15) has the Taylor series solution

$$H = \sum_{n=0}^{\infty} a_n\eta^n \tag{20}$$

for the decaying case $\alpha > 1$, where

$$a_2 = \frac{1}{a_o}\left\{(a_o^2 + 2^{1/2}a_1 + a_1^2) + \lambda_1 + \alpha_1 a_1^2\right\},$$

$$a_3 = \frac{1}{3a_o}\left\{(a_o a_1 + 2^{3/2}a_2 + 3a_1a_2) + 4\alpha_1 a_1 a_2\right\}, \text{ etc.} \tag{21}$$

Here a_o and a_1 are two arbitrary constants. It is easy to check that the solution H_B in (6), when $\exp(\eta^2)$ and erfc η are suitably expanded, is a special case of (20) with $\alpha = 3$, $\lambda = 0$, and $a_1 = -2^{\frac{1}{2}}$. For this choice of a_1, we obtain a single parameter family of solutions of Burgers equation, the single parameter a_o representing the amplitude at 'infinity' of the various curves, or their Reynolds number. In general, it is difficult to identify the ranges of parameters a_o and a_1 in (20), for which the series converges for $-\infty < \eta < \infty$. For this purpose we go back to the self-similar function f governed by (13), and pose for the latter a connection problem. Since we require the solutions to vanish at $\eta = \pm\infty$, we linearise (13) for its asymptotic behaviour:

$$f'' + 2\eta f' - \frac{4}{1-\alpha} f = 0. \tag{22}$$

This equation has the solution

$$f = A \exp(-\eta^2) H_\nu(\eta) \sim A \exp(-\eta^2)(2\eta)^{2\alpha_1} \text{ as } \eta \uparrow \infty, \tag{23a}$$

and

$$f \sim O(\eta^{-2\alpha_1 - 1}) \text{ as } \eta \downarrow -\infty, \tag{23b}$$

where H_ν is the Hermite function of order ν and A is the amplitude parameter. Thus, the asymptotic behaviour is similar to that for Burgers equation--exponential and algebraic for $\eta \sim \infty$ and $\eta \sim -\infty$, respectively.

Equation (13) was solved numerically starting with the asymptotic behaviour (23a) for large positive η. The main result of this study was that in the decaying range $\alpha > 1$, the self-similar form of single hump type exists if $1 < \alpha \le 3$; the amplitude parameter A for each admissible value of (α, β) could at most reach a maximum value of $A = A_{max}$. For this value of A, the solution starting with (23a) would rise to approach the special constant solution $f = f_m$ (see Eq. (17)). For $A > A_{max}$, there is no single hump solution. The solution grows to become unbounded at $\eta = -\infty$. Sachdev et al. [4] carried out a thorough investigation of the intermediate asymptotic nature of these self-similar solutions by integrating the PDE (3) by pseudo-spectral and implicit schemes. Here, we emphasize that the self-similar form of (3) is governed by the equation (15) for H, the inverse function. Equation (15) is a special case of Euler-Painlevé equation which we introduce in Sec. 6.

4. SINGLE HUMP SOLUTIONS OF NON-PLANAR GBE'S

Now we consider the self-similar solutions of (4) in the form

$$u = t^{-1/2\alpha} f(\eta), \quad \eta = x(2\delta t)^{-1/2}, \tag{24}$$

so that $f(\eta)$ is governed by the ODE

$$f'' - 2^{3/2} \delta^{-1/2} f^\alpha f' + 2\eta f' + \frac{2(1 - \alpha j)}{\alpha} f = 0, \tag{25}$$

where α is a parameter. The 'inverse' transformation

$$H = \delta^{1/2} f^{-\alpha} \tag{26}$$

changes (25) to

$$HH'' - \frac{\alpha+1}{\alpha} H'^2 + 2\eta HH' - 2(1 - \alpha j)H^2 - 2^{3/2} H' = 0. \tag{27}$$

Equation (27) may be compared with (8) corresponding to the standard Burgers equation (2): only the numerical coefficients in (27) differ from those in (8) and the former reduces to the latter when $j = 0$ and $\alpha = 1$. Equation (27) belongs to the Euler-Painlevé class (see Sec. 6). Equation (27) has a series solution (20) with two arbitrary parameters, and the recurrence relation for the coefficients can be written out [5]. To identify the single hump solutions for (25) and the inverted hump solutions for (27), we again pose the connection problem for (25). The linearised form of the latter,

$$f'' + 2\eta f' + \frac{2(1 - \alpha j)}{\alpha} f = 0, \tag{28}$$

has the solution

$$f(\eta) = A e^{-\eta^2} H_\nu(\eta), \quad \eta > 0$$

$$f(\eta) \sim \frac{B \pi^{1/2}}{\sqrt{(-\nu)}} |\eta|^{j-1/\alpha}, \quad \eta \ll 0, \tag{29}$$

provided $\alpha j < 1$. Here $\nu = 1/\alpha - (j+1)$, $H_\nu(\eta)$ is the Hermite function of order ν, and A and B are the amplitude parameters. Thus, the linear solution decays exponentially as $\eta \to \infty$ and algebraically as $\eta \to -\infty$. The connection problem may thus be stated as follows:

$$f'' - 2^{3/2} \delta^{-1/2} f^\alpha f' + 2\eta f' + \frac{2(1-\alpha j)}{\alpha} f = 0, \tag{30}$$

$$f \sim A \exp(-\eta^2) H_\nu(\eta), \quad \eta \uparrow \infty , \tag{31}$$

$$f \to 0 \quad (\eta \downarrow -\infty), \tag{32}$$

and

$$|f| < \infty, \quad -\infty < \eta < \infty . \tag{33}$$

Before we give the numerical results for the above problem, we discuss a class of special exact solutions which provides some clues to the general case. If $\alpha = 1/(j+1)$, Eq. (30) reduces to

$$f + \eta f' + \tfrac{1}{2} f'' = (2/\delta)^{1/2} f^\alpha f'. \tag{34}$$

Integrating (34) and using vanishing conditions (31) at $\eta = \infty$, we get

$$\eta f + \frac{1}{2} f' = \frac{1}{(\alpha+1)} (2/\delta)^{1/2} f^{\alpha+1}. \tag{35}$$

The transformation $G = f^{-\alpha}$ changes (35) to

$$G' - 2\alpha\eta G = - \frac{2\alpha}{\alpha+1} (2/\delta)^{1/2}. \tag{36}$$

This equation can be integrated. We, thus, have

$$f(\eta) = \exp(-\eta^2) \left\{ c - \frac{2}{\alpha+1} (2\alpha/\delta)^{1/2} \int_0^{\alpha^{\frac{1}{2}}\eta} e^{-t^2} dt \right\}^{-1/\alpha}, \tag{37}$$

where $c = f^{-\alpha}(0)$. This solution generalises the solution (6) for Burgers equation to other geometries, provided the parameter α equals $1/(j+1)$; thus for $j = 1$, $\alpha = 1/2$ and $j = 2$, $\alpha = 1/3$, we have explicit single hump solutions of (4). We can infer more about the solutions of (4) by deriving relations involving some integrals. Writing $F = f^{\alpha}$, Eq.(25) can be transformed as

$$\frac{1}{2} FF' - \frac{\alpha-1}{2\alpha} F'^2 + (1 - \alpha j)F^2 + \eta FF' - (2/\delta)^{1/2} F^2 F' = 0. \tag{38}$$

Integrating (38) from $\eta = -\infty$ to $\eta = \infty$, and using vanishing conditions there for F and F', we get

$$(2\alpha j - 1) \int_{-\infty}^{\infty} F^2 d\eta = \frac{1-2\alpha}{\alpha} \int_{-\infty}^{\infty} F'^2 d\eta. \tag{39}$$

Equation (39) implies the following:

(i) $j = 0$. The ratio

$$r \equiv \frac{\int_{-\infty}^{\infty} F^2 d\eta}{\int_{-\infty}^{\infty} F'^2 d\eta} = - \frac{(1 - 2\alpha)}{\alpha} > 0 \text{ if } \alpha > \frac{1}{2}. \tag{40}$$

Therefore, the single hump solutions in this case exist only if $\alpha > \frac{1}{2}$.

(ii) $j = 1$. In this case (39) holds if $\alpha = \frac{1}{2}$. This corresponds to the exact solution (37).

(iii) $j = 2$. In this case, the ratio r of the integrals in (40) is $(1 - 2\alpha)/\alpha(4\alpha - 1)$. This is positive if $1/4 < \alpha < 1/2$. This is the range for which the single hump solutions may exist.

The numerical study of the connection problem (30)-(33) indicated the existence of solutions for $\alpha < 1/(j+1)$, $j = 0,2$, which go to zero at a finite point $\eta = \eta_0$, say, instead of $\eta = -\infty$. Indeed, one can easily derive the relation

$$\frac{\alpha(j+1) - 1)}{\alpha} \int_{\eta_0}^{\infty} f \, d\eta = -\frac{1}{2} f'(\eta_0) \qquad (41)$$

for this case. Since, evidently, $f'(\eta_0) > 0$ and $f > 0$, (41) implies that $\alpha < 1/(j+1)$. Thus, single hump solutions of (30) vanishing at $\eta = \infty$ and $\eta = \eta_0$, a finite point, exist only if $\alpha < 1/(j+1)$. Combining these results with those in (i)-(iii) and the inferences from the numerical solution of the connection problem (30)-(33), we arrive at the qualitative nature of the solutions, as summarised in Table 1.

Table 1 Single hump, monotonic and diverging solutions of Eq. (30)

Behaviour at left boundary	$j = 0$	$j = 1$	$j = 2$
Solutions vanishing at $\eta = -\infty$	$\alpha = 1$	$\alpha = 1/2$	$1/3 \leq \alpha < 1/2$
Solutions vanishing at $\eta = \eta_0$	$1/2 < \alpha < 1$	-	$1/4 < \alpha < 1/3$
Solutions monotonically approaching a constant at $\eta = -\infty$	-	$\alpha = 1$	$\alpha = 1/2$
Solutions diverging to infinity at $\eta = -\infty$	-	$\alpha > 1$	$\alpha > 1/2$

The details of the numerical study of the problem (30)-(33) may be seen in Sachdev and Nair [5]. Apart from the results summarised in Table 1, we note that, unlike the single hump solutions of GBE (3) with damping, the nature of the solution depends on the parameters α and j. The solutions exist for the permissible values of α and j for all values of the amplitude parameter. The particular value $\alpha = 1/j$ seems to bifurcate the generality of solutions. For this value of α, $f = f_c$, a constant, is a solution of (30) so that, for a given value of A, the solution starting at $\eta = -\infty$, tends to a constant nonzero value at $\eta = -\infty$ instead of vanishing there or at $\eta = \eta_0$. For example, for $j = 1$, $A = 1$, $f_c = 0.41187$, and for $j = 2$, $A = 1$, $f_c = 0.10197$. The numerical solution of the connection problem agrees very closely with the series solution for H, governed by (27), when the values H(0) and H'(0) are suitably identified from the numerical solution. The intermediate asymptotic nature of the self-similar solutions was verified by solving the PDE[4] numerically for the relevant values of the parameters α and j and 'appropriate' initial conditions vanishing at $\eta = \pm\infty$.

5. GENERALISED BURGERS EQUATION WITH VARIABLE VISCOSITY

We consider the GBE

$$u_t + u^{\beta}u_x = \frac{\delta}{2}\, g(t)\, uu_x, \tag{42}$$

where g(t) is a smooth positive function, representing the dependence of viscosity on time, Scott [6] considered a special case of (42) with $\beta = 1$; the role of t and x, however, was interchanged in his study. He considered a piston problem for this equation and proved the intermediate asymptotic nature of the self-similar solution $u = \Omega(t/x)$ for cylindrical and sub-cylindrical cases. The latter case was defined such that $\frac{g(x)}{x} \to 0$ as $x \to \infty$. Conversely, if $\frac{g(x)}{x} \to \infty$, the GBE is referred to as super-cylindrical. Here, we consider the cylindrical and sub-cylindrical form

$$u_t + u^{\beta}u_x = \frac{\delta}{2}(1+t)^n\, u_{xx}, \tag{43}$$

where n is a parameter and $\beta > 0$. We shall find that self-similar solutions either decaying or oscillating at $x = \pm\infty$ exist for (43) only when $-1 \leq n \leq 1$, i.e., when (43) is either cylindrical or sub-cylindrical. We easily check that the self-similar form of the solutions of (43) is

$$u = [(1+t)^{n-1}\delta]^{1/2\beta} f(\eta),\ \eta = [(1+t)^{-(n+1)}\ \delta^{-1}]^{1/2} x, \tag{44}$$

so that it becomes

$$f'' - 2f^{\beta}f' + (n+1)\eta f' - \frac{(n-1)}{\beta}\, f = 0. \tag{45}$$

The inverse function

$$H = f^{-\beta} \tag{46}$$

is, therefore, governed by

$$HH'' - \frac{\beta+1}{\beta} H'^2 + (n+1)\eta\, HH' - 2H' + (n-1)H^2 = 0. \tag{47}$$

For single hump solutions of (45), we enquire when f has a maximum: $f' = 0, \quad f'' = \frac{n-1}{\beta} f < 0$. A necessary condition for single hump solutions for the case $\beta > 0$ is that $n < 1$. Equation (47), as we shall see, again belongs to the class of Euler-Painlevé equations. Before we pose a connection problem for (45), we note that it has a single parameter family of exact solutions. Assuming $\beta = \frac{1-n}{1+n}$, we write (45) in the form

$$f'' - 2f^{\beta}f' + \frac{2}{\beta+1}(\eta f' + f) = 0. \qquad (48)$$

Integrating (48) and assuming that f and f' tend to zero (such that $\eta f \to 0$) as $\eta \to \pm\infty$, we have

$$f^{-(\beta+1)}f' + \frac{2}{\beta+1}\eta f^{-\beta} = \frac{2}{\beta+1}. \qquad (49)$$

Integrating (49), we get

$$f = [\exp(\tfrac{a}{2}\eta^2 \{A - \sqrt{2a} \ \operatorname{erf}(\sqrt{a/2}\ \eta)\}]^{-1/\beta}, \qquad (50)$$

where $a = 2\beta/(\beta+1)$ and A is the constant of integration equal to $f(0)^{-\beta}$. For decaying solutions, we require that $a > 0$ so that $n < 1$. We also note that the special case $\beta = 1$, $n = 1$, which was treated by Scott [6] can be solved in a parametric form. The linearised form of (45) is

$$f'' + (n+1)\eta f' - \frac{n-1}{\beta} f = 0. \qquad (51)$$

Its solution is simply the confluent hypergeometric function $\Phi(\frac{1-n}{2\beta(1+n)}, \frac{1}{2}; z)$, where $z = -\frac{n+1}{2}\eta^2$. The asymptotic form of Φ immediately suggests that the solutions vanish at $\eta = \pm\infty$ provided $\frac{1}{2\beta}\frac{1-n}{1+n} > 0$, i.e., $-1 < n \le 1$. The case $n = 1$ leads to the erfc(η) solution of (51). The series solution for (47) exists in the permissible range of n and describes either the single hump solution or shock-like solution or the solutions with oscillatory tail and/or front. The results of the connection problem for (45) and (47) and other details will be published elsewhere [7].

6. EULER-PAINLEVÉ TRANSCENDENTS

In the compendium of nonlinear differential equations, compiled by Kamke [3], the equation

$$yy'' + ay'^2 + f(x)yy' + g(x)y^2 = 0 \qquad (52)$$

is attributed to Euler and Painlevé; the motivation for the study of this equation is not clear. This equation is, in fact, exactly linearised by the transformation

$$y = v^{1/(1+a)} \tag{53}$$

to

$$v'' + fv' + (a+1)gv = 0. \tag{54}$$

We have generalised Eq. (52) to the form

$$yy' + a(x)y'^2 + f(x)yy' + g(x)y^2 + b(x)y' + c(x) = 0, \tag{55}$$

where the constant a in (52) has been made to vary with x and a linear term in y' and a function c(x) have been added to (52). Our claim is that Eq. (55) characterises the GBE's in their self-similar form in the same manner as the Painlevé type equations describe Korteweg-de Vries and other model dispersive equations. Indeed, the three equations corresponding to the GBE's (3), (4) and (42), namely (15), (27) and (47) easily follow from (55) if special choices of the functions a(x), f(x), g(x), b(x) and c(x) are made. We must, however, emphasize the difference between the nature of Eq. (55) and the Painlevé type of equations. For the physically realistic cases of Eq. (55) which we have studied, the solutions are either single hump type or shock-like or have oscillatory tails, and there are no singularities of any kind in the finite part of the real line. In contrast, the Painlevé equations are typified by the property that their only movable singularities are poles. That Euler-Painlevé equations should be 'nicer' than Painlevé equations follows also from the nature of the BGE's and K-dV type of equations; roughly speaking, diffusion is 'smoother' than dispersion. We must also point out that Eq. (55) for general (smooth) coefficients would need more extensive investigation than we have carried out for the special cases, arising directly from GBE's.

It is remarkable that 65 nonlinear DE's in the compendia of Kamke [3] and Murphy [8] are special cases of (55) directly or by simple transformations. These equations are either autonomous or are linearisable by a power law or logarithmic transformation. Alternatively, they are reducible to first order equations of Riccati or Bernoulli type. Most of these equations are either integrable explicitly or admit at least one quadrature allowing their treatment in the phase plane. Here we give a listing of these equations and refer the reader to Kamke and Murphy for their physical importance and solution. In Kamke's book, these are listed in Sec. 6, p. 542 as 104-111, 117, 122, 124-127, 129,

131, 133-4, 136-9, 150-2, 155-8, 164, 166, 168-70, 173-9. In Murphy's book, these are enumerated as 133, 140, 142, 150, 190, 195, 199, 201, 203-4, 219-22, 227-31, 233-4 in Part 2, Chapter 4. It would thus appear that Eq. (55) has a quite ubiquitous nature and deserves close study and investigation.

7. CONCLUSIONS

We have shown with the help of 3 GBE's--(3), (4) and (42) that Eq.(55), which we have referred to as Euler-Painlevé equation, seems to characterise this class of equations. We have drawn an analogy with Painlevé type of equations which unify the study of K-dV equation and its kindred class. We have summarised the results of the connection problems for some GBE's. Similar study for Painlevé second equation has been carried out by Miles [9] and Rosales [10]. It is conceivable that not all GBE's will be typified by (55). Indeed, self-similar form of solutions does not always represent intermediate asymptotics. Besides, sometimes it is the self-similar form of the linearised PDE which may represent intermediate asymptotics to which a large class of solutions arising from a certain set of initial conditions approach [11]. This is the case, for example, for the super-cylindrical equation, as shown by Scott [6]. In this case, it is not the nonlinear Euler-Painlevé equation (55) which describes the asymptotic behaviour of a certain class of initial value problems but the corresponding self-similar solutions of the heat equation with a variable coefficient. To conclude, we would have to investigate other GBE's and study (55) more deeply to come to a firm understanding of the role of Euler-Painlevé equations.

REFERENCES

[1] P. L. Sachdev, **Nonlinear Diffusive Waves** (Cambridge University Press, London, 1987).

[2] G. B. Whitham, **Linear and Nonlinear Waves** (Wiley, New York, 1974).

[3] E. Kamke, **Differential Gleichungen: Losungsmethoden und Losungen** (Akademische Verlagsgesellschaft, Leipzig, 1943).

[4] P. L. Sachdev, K. R. C. Nair and V. G. Tikekar, Generalised Burgers equations and Euler-Painlevé transcendents-I, J. Math. Phys. **27** (1986) 1506.

[5] P. L. Sachdev and K. R. C. Nair, Generalised Burgers equations and Euler-Painlevé transcendents-II, J. Math. Phys. (1987), to appear.

[6] J. F. Scott, The long time asymptotics of solutions to the generalised Burgers equations, Proc. R. Soc. Lond. **A373** (1981) 443.

[7] P. L. Sachdev and K. R. C. Nair, Generalised Burgers equations and Euler-Painlevé transcendents-III (1987), to be published.

[8] G. M. Murphy, **Ordinary Differential Equations and Their Solutions** (Van Nostrand, New York, 1960).

[9] J. W. Miles, On the second Painlevé transcendent, Proc. R. Soc. Lond. **A361** (1978) 277.
[10] R. Rosales, The similarity solution for the Korteweg-de Vries equation and the related Painlevé transcendent, Proc. R. Soc. Lond. **A361** (1978) 265.
[11] G. I. Barenblatt, **Similarity, Self-Similarity and Intermediate Asymptotics** (Consultants Bureau, New York, 1979).
[12] E. Hopf, The partial differential equation $u_t + uu_x = \delta u_{xx}$, Communs. Pure Appl. Math. **3** (1950) 201.

Bäcklund Transformations and Soliton Wave Functions

J.A. Rao[1] *and A.A. Rangwala*[2]

[1]Department of Physics, Ramnarain Ruia College, Bombay 400019, India
[2]Department of Physics, University of Bombay, Vidyanagari, Bombay 400098, India

Using Darboux-Bargmann technique, we obtain (1) the Bäcklund transformations for any nonlinear evolution equation (NLEE) solvable by the inverse scattering method of Zakharov-Shabat—Ablowitz-Kaup-Newell-Segur (ZS/AKNS) and (2) the ZS/AKNS wave functions corresponding to the n-soliton solution of this NLEE.

1. BÄCKLUND TRANSFORMATIONS AND SOLITON WAVE FUNCTIONS

The ZS/AKNS scattering problem [1,2] is defined by the eigenvalue equation

$$v_{1x} + ikv_1 = q(x,t)v_2, \tag{1.1a}$$

$$v_{2x} - ikv_2 = r(x,t)v_1, \tag{1.1b}$$

where $v = \binom{v_1}{v_2}$ is the two-component ZS/AKNS wave function and q and r are functions of x and t satisfying the NLEE of interest. The eigenvalue is k. Equations (1.1) give the space evolution of the wave function. Let us distinguish the quantities, wave function and potentials q and r, referring to n solitons by primes and those referring to (n-1) solitons by unprimed ones. Thus:

$$v_1' = v_1(n), \quad v_2' = v_2(n), \quad v' = \binom{v_1'}{v_2'}; \; q' = q_n, \quad r' = r_n, \tag{1.2a}$$

and

$$v_1 = v_1(n-1), \quad v_2 = v_2(n-1), \; v = \binom{v_1}{v_2}; \; q = q_{n-1}, \; r = r_{n-1}. \tag{1.2b}$$

$v_{1,2}$ satisfy eqs. (1.1). $v_{1,2}'$ satisfy similar equations:

$$v_{1x}' + ikv_1' = q'v_2', \qquad v_{2x}' - ikv_2' = r'v_1'. \tag{1.3}$$

We appeal to Darboux's method [3] to expand the n-soliton wave function v' in terms of (n-1)-soliton solution v:

$$v_1' = Av_1 + Bv_2, \quad v_2' = Cv_1 + Dv_2, \tag{1.4}$$

where A, B, C, D are functions of x and t. From eqs. (1.1), (1.3) and (1.4), we see that A,...,D satisfy the differential equations:

$$A_x = -rB + q'C, \tag{1.5a}$$

$$B_x = -qA - 2ikB + q'D, \tag{1.5b}$$

$$C_x = r'A + 2ikC - rD, \tag{1.5c}$$

$$D_x = r'B - qC. \tag{1.5d}$$

We next invoke an idea due to Bargmann [4]. It is well known that the ZS/AKNS equation corresponding to the Korteweg-de Vries (KdV) equation is the Schrödinger equation [5]. In the context of the Schrödinger equation, Bargmann has shown that for a potential capable of giving n bound states, the solution of the Schrödinger equation can be written in the form $e^{ikx}\,\chi(k,x)$, where $\chi(k,x)$ is an n-th degree polynomial in k. On the other hand, we know that n-soliton solution of an NLEE envisaged by the ZS/AKNS eqs. (1.1) can be looked upon as a potential giving n bound states [5]. Thus the idea due to Bargmann suggests that v and v' will differ by a linear function of k. We therefore write:

$$A = a_1k + a_0, \quad B = b_1k + b_0, \tag{1.6a}$$

$$C = c_1k + c_0, \quad D = d_1k + d_0, \tag{1.6b}$$

where $a_i,\ldots,d_i$ are functions of x and t through (q,r) and (q',r'). The differential equations obtained for a,...,d by using eqs. (1.6) in eqs. (1.5) are easy to solve and a partial solution is:

$$a_1 = \alpha(t),$$

$$b_1 = 0, \quad b_0 = \frac{\alpha}{2i}[(\frac{\delta}{\alpha})q' - q],$$

$$c_1 = 0, \quad c_0 = -\frac{\alpha}{2i}[r' - (\frac{\delta}{\alpha})r],$$

$$d_1 = \delta(t), \quad d_0 = -(\frac{\delta}{\alpha})a_0 + \beta(t), \tag{1.7}$$

with a_0 satisfying the equation

$$a_{0x} = \frac{\alpha}{2i}(qr - q'r') \tag{1.8}$$

and, additionally, two equations of constraint need to be satisfied:

$$\frac{\alpha}{2i}[(\frac{\delta}{\alpha})q'_x - q_x] = -[q + (\frac{\delta}{\alpha})q']a_0 + \beta q', \tag{1.9a}$$

$$-\frac{\alpha}{2i}[r'_x - (\frac{\delta}{\alpha})r_x] = [r' + (\frac{\delta}{\alpha})r]a_o - \beta r. \tag{1.9b}$$

Here α, β, δ are constants of integration and in general depend on t. Subsequently we shall see that to be able to describe the soliton solutions they need to be taken as constants. From eqs. (1.7)-(1.8), we can see that we can obtain A,...,D in eqs. (1.4) provided we can solve eq. (1.8) for a_0 and be able to satisfy eqs. (1.9).

It is easy to obtain a_0 as a function of (q,r) and (q',r') and hence to show that eqs. (1.9) in fact represent the space part of Bäcklund transformations (BTs). To see this, we first notice that the choice of overall factor, α say, is at our disposal. We set $\alpha = -i$ and define $\varepsilon = -(\frac{\delta}{\alpha})$. We multiply eq. (1.9a) by r' and eq. (1.9b) by q, add the resulting equations and use eq. (1.8) to obtain

$$\varepsilon(qr')_x + \varepsilon^2(q'_xr' + qr_x) = 2(A_0^2)_x, \tag{1.10}$$

where

$$A_0 = \varepsilon a_0 + \beta. \tag{1.11}$$

Similarly multiplying eq. (1.9a) by r and eq. (1.9b) by q' and adding the resulting equations and once again using eq. (1.8), we get

$$\varepsilon(q'r)_x + q_xr + q'r'_x = 2(a_0^2)_x. \tag{1.12}$$

Adding eqs. (1.10) and (1.12), we see that the resulting equation is integrable only if

$$\varepsilon^2 = 1 \quad \text{or} \quad \varepsilon = \pm 1. \tag{1.13}$$

In all the subsequent discussion, we shall assume eq. (1.13) to hold. The sum of eqs. (1.10) and (1.12) when integrated gives a quadratic equation for a_0 whose solution is

$$a_0 = i\mu' \pm \frac{1}{2}[4\nu'^2+(q'+\varepsilon q)(r'+\varepsilon r)]^{\frac{1}{2}}H(\xi). \tag{1.14}$$

This solves eq. (1.8) consistent with eqs. (1.9). Here μ' and ν' are integration constants. They can in principle be functions of t and may even be complex. Soon, however, we shall identify them with soliton pole position which requires them not only to be constants but also real. We believe that the present method can be extended to include perturbed NLEEs and then this possibility of time dependence of pole position parameters becomes necessary. The function $H(\xi)$ is the Heaviside function

$$H(\xi) = \begin{cases} 1, & \xi > 0 \\ -1, & \xi < 0. \end{cases} \tag{1.15}$$

It can be shown that in the pure multisoliton case, the argument ξ is given by

$$\xi = x + \frac{t}{\nu'} \, \mathrm{Im}\omega(k') - x_0, \tag{1.16}$$

where $\omega(k)$ is the dispersion function and x_0 is a constant. The Heaviside function is required to ensure the continuity of both the soliton solutions and the wave functions [6].

Using a_0 from eq. (1.14) in eqs. (1.9), we get the space part of the BTs:

$$q'_x + \varepsilon q_x = -2i\mu'(q'+\varepsilon q) \pm (q'-\varepsilon q)[4\nu'^2 + (q'+\varepsilon q)(r'+\varepsilon r)]^{\frac{1}{2}} H(\xi), \tag{1.17a}$$

$$r'_x + \varepsilon r_x = 2i\mu'(r'+\varepsilon r) \pm (r'-\varepsilon r)[4\nu'^2 + (q'+\varepsilon q)(r'+\varepsilon r)]^{\frac{1}{2}} H(\xi). \tag{1.17b}$$

We see that the constants μ', ν' appearing in the above equations are in fact the parameters of the n-th soliton. Since soliton solution represents a pole in the complex k-plane at $k = k' \equiv \mu' + i\nu'$, we require (μ',ν') to be both real and independent of time. If the NLEE of interest is such that if (q,r) is a solution so is $(-q,-r)$ then both values of ε, ± 1, are permitted. Otherwise we would have either $\varepsilon=+1$ or $\varepsilon=-1$. The two possible signs accompanying the discriminant are linked with two possible directions of incidence; positive (negative) sign corresponds to the incidence from the left (right). These directions of incidence in their turn give rise to relevant analyticity structure of the scattering amplitude or the transmission coefficient in the upper half k-plane (right incidence) or in the lower half k-plane (left incidence).

It is easy to show that eqs. (1.17) reproduce the BTs in the standard cases of KdV, sine-Gordon (sG) and nonlinear Schrödinger equation (NLSE) [6,7] corresponding to the three classes r = constant, r=-q and r=-q* respectively. Our present treatment gives a unified treatment of the BTs for all NLEEs encompassed by the ZS/AKNS scheme.

Using a_0 from eq. (1.14) in eqs. (1.7) and substituting the results into eqs. (1.6), we get

$$A = -i(k-\mu')\pm \tfrac{1}{2}[4\nu'^2+(q'+\varepsilon q)(r'+\varepsilon r)]^{\frac{1}{2}}\, H(\xi), \tag{1.18a}$$

$$B = \tfrac{\varepsilon}{2}(q'+\varepsilon q), \tag{1.18b}$$

$$C = \tfrac{1}{2}(r' + \varepsilon r), \tag{1.18c}$$

$$D = \varepsilon[i(k-\mu')\pm \tfrac{1}{2}[4\nu'^2+(q'+\varepsilon q)(r'+\varepsilon r)]^{\frac{1}{2}}\, H(\xi)], \tag{1.18d}$$

which solve the differential equations in eqs. (1.5) and yield, when used in eqs. (1.4), the n-soliton wave function v' in terms of (n-1)-soliton solution v and the n-th and (n-1)-th soliton solutions (q',r') and (q,r) respectively. These results agree for the three cases (1) r = -1, (2) r = -q, (3) r = -q* discussed in our previous work [8]. The time evolution of ZS/AKNS wave functions is given by [2]

$$v_{1t} = \mathcal{A}(k;q,r)v_1 + \mathcal{B}(k;q,r)v_2, \tag{1.19a}$$

$$v_{2t} = \mathcal{C}(k;q,r)v_1 - \mathcal{A}(k;q,r)v_2. \tag{1.19b}$$

In the ZS/AKNS scheme one stipulates that $k_t = 0$. The functions $\mathcal{A},\mathcal{B},\mathcal{C}$ depend on particular NLEE of interest. Use of eqs. (1.4) leads to

$$A_t = A(\mathcal{A}'-\mathcal{A}) - B\mathcal{C} + C\mathcal{B}', \tag{1.20a}$$

$$B_t = B(\mathcal{A}'+\mathcal{A}) - A\mathcal{B} + D\mathcal{B}', \tag{1.20b}$$

$$C_t = -C(\mathcal{A}'+\mathcal{A}) + A\mathcal{C}' - D\mathcal{C}, \tag{1.20c}$$

$$D_t = -D(\mathcal{A}'-\mathcal{A}) + B\mathcal{C}' - C\mathcal{B}, \tag{1.20d}$$

where $\mathcal{A}' = \mathcal{A}(k;q',r')$, etc. From eqs. (1.18), we see that eqs. (1.20b,c) give the time part of the BTs.

2. RICCATI EQUATIONS

It is possible to extend the previous analysis further. To this end let us denote A,...,D given by eqs. (1.18) when evaluated at $k=k' \equiv \mu'+i\nu'$ by $\bar{A},\ldots,\bar{D}$. Thus, we have

$$\bar{A} = \nu' \pm \tfrac{1}{2}[4\nu'^2+(q'+\varepsilon q)(r'+\varepsilon r)]^{\frac{1}{2}}H(\xi), \tag{2.1a}$$

$$\bar{B} = \tfrac{\varepsilon}{2}(q' + \varepsilon q), \tag{2.1b}$$

$$\bar{C} = \tfrac{1}{2}(r' + \varepsilon r), \tag{2.1c}$$

$$\bar{D} = \varepsilon\{-\nu'\pm \tfrac{1}{2}[4\nu'^2+(q'+\varepsilon q)(r'+\varepsilon r)]^{\frac{1}{2}}H(\xi)\}, \tag{2.1d}$$

and we note that

$$\bar{A}.\bar{D} = \bar{B}.\bar{C}. \tag{2.2}$$

Also, we denote the wave functions in eqs. (1.4) evaluated at k=k' by $\bar{v}_1,\bar{v}_2$ and $\bar{v}_1',\bar{v}_2'$ so that

$$\bar{v}_1' = \bar{A}\bar{v}_1 + \bar{B}\,\bar{v}_2\,, \qquad \bar{v}_2' = \bar{C}\,\bar{v}_1 + \bar{D}\,\bar{v}_2. \tag{2.3}$$

If we define

$$\Gamma = \bar{v}_1/\bar{v}_2, \qquad \Gamma' = \bar{v}_1'/\bar{v}_2', \tag{2.4}$$

then it is easy to see from eqs. (1.1) and (1.19) that Γ satisfies the Riccati equations:

$$\Gamma_x = q - 2ik'\Gamma - r\Gamma^2, \tag{2.5a}$$

$$\Gamma_t = \mathcal{B} + 2\mathcal{A}\Gamma-\mathcal{C}\Gamma^2, \tag{2.5b}$$

with Γ' satisfying similar equations. If we now require that $\bar{v}_1'=0=\bar{v}_2'$ in eqs. (2.3), then we obtain

$$\Gamma = -\,\bar{B}/\bar{A} = -\,\bar{D}/\bar{C}. \tag{2.6}$$

The consistency of last equality coming from eq. (2.2). It also follows from eqs. (2.3) that $\bar{C}\bar{v}_1' - \bar{A}\,\bar{v}_2' = (\bar{B}\,\bar{C} - \bar{A}\,\bar{D})v_2$. The right-hand side of this equation vanishes in view of eq. (2.2) giving

$$\Gamma' = \bar{A}/\bar{C} = \bar{B}/\bar{D}. \tag{2.7}$$

Γ and Γ' are particular solutions of the Riccati eqs. (2.5) and their primed counterparts.

Two useful relations between Γ and Γ' can be easily established. Using eqs. (2.1) in eq. (2.6) and in eq. (2.7), it is seen that $\varepsilon\Gamma$ and Γ' both satisfy the quadratic equation $(r'+\varepsilon r)\gamma^2-4\nu'\gamma - (q'+\varepsilon q) = 0$. It therefore follows that

$$\varepsilon\Gamma + \Gamma' = 4\nu'/(r'+\varepsilon r), \tag{2.8}$$

and

$$(\varepsilon\Gamma)\Gamma' = -(q'+\varepsilon q)/(r'+\varepsilon r). \tag{2.9}$$

They can alternatively be written as

$$q'+\varepsilon q = -4\nu'(\varepsilon\Gamma)\Gamma'/(\varepsilon\Gamma+\Gamma'), \tag{2.10}$$

and

$$r'+\varepsilon r = 4\nu'/(\varepsilon\Gamma+\Gamma'). \tag{2.11}$$

These equations are generalizations of those given by Konno,Sanuki and Wadati [7] and were employed by us in Ref. 8. In the present general case, it is unfortunately not possible to go beyond this stage. In special situations, however, we can make further predictions. As an illustration, consider the situation where we stipulate a relationship between q and r like

$$\text{(a)}\quad r = r' = -1 \text{ (with } \varepsilon=+1); \quad \text{(b)}\quad r = -q,\ r'=-q'; \quad \text{(c)}\ r=-q^*, r'=-q'^*. \tag{2.12}$$

These then imply relationships between $\varepsilon\Gamma$ and Γ'. For instance, corresponding to choice (a) in eqs. (2.12), we get from eq. (2.8) that

$$\Gamma' = -\Gamma - 2\nu'; \tag{2.13a}$$

for the choice (b), we obtain from eq. (2.9) that

$$\Gamma' = 1/\varepsilon\Gamma\ ; \tag{2.13b}$$

and for the choice (c), we obtain from eq. (2.9) that

$$\Gamma' = 1/\varepsilon\Gamma^*. \tag{2.13c}$$

Equations (2.13) form important ingredients in the methods of Konno-Sanuki-Wadati [7] and Chen [6]. The above process is reversible. If we stipulate relationships in eqs. (2.13),we obtain from eqs. (2.9) and (2.10) relationships between q and r given in eqs. (2.12).

We now briefly outline the method of obtaining multi-soliton solutions and corresponding wave functions in the cases where r is stipulated as some known function of q as, for instance, in eqs. (2.12). The procedure is recursive and algebraic except for the starting zero-soliton case where it requires solution of very simple differential equations. We illustrate the procedure for the case of NLSE. Discussion for other cases will be found in Ref. 8. For the NLSE case, on using eq.(2.12c) in eq. (2.10), we get

$$q' = -\varepsilon[q + 4\nu' \frac{\Gamma}{1+|\Gamma|^2}], \tag{2.14}$$

which relates the n-soliton solution, q', to the (n-1)-soliton quantities, q, $\bar{v}_1$ and $\bar{v}_2$. This leads to the following procedure. We begin with the zero-soliton solution q(0) of NLSE which is q(0) = 0. This is substituted in eq. (1.1) to obtain the zero-soliton wave function v(0). The time dependence of v(0) is obtained from eqs. (1.19) with $\mathcal{A}, \mathcal{B}, \mathcal{C}$ appropriate to the NLSE case. After this the procedure is purely algebraic: We use these known q(0) and v(0) in eq. (2.14) to obtain q(1). These q(1), q(0) and v(0) are now used in eqs. (1.4) together with eqs. (1.18) to get v(1). This procedure can obviously be continued and furnishes in a simple manner all the higher order soliton solutions and wave functions.

References

[1] V. B. Zakharov and A. B. Shabat, Sov. Phys. JETP **34** (1972) 62.

[2] M. J. Ablowitz, D. J. Kaup, A. C. Newell and H. Segur, Phys. Rev. Lett. **31** (1973) 125;
M. J. Ablowitz, D. J. Kaup, A. C. Newell and H. Segur, Stud. Appl. Math. **53** (1974) 249.

[3] G. Darboux, **Leçons sur la Théorie Générale des Surfaces et les Applications Géometriques du Calcul Infinitesimal**, Vol. 2, 2d ed (Paris: Gauthier-Villars, 1915), p. 210.

[4] V. Bargmann, Rev. Mod. Phys. **21** (1949) 488.

[5] See, for instance, G. L. Lamb, Jr., **Elements of Soliton Theory** (New York: John Wiley, 1980) and R. K. Dodd, J. C. Eilbeck, J. D. Gibbon and H. C. Morris, **Solitons and Nonlinear Wave Equations** (New York: Academic Press, 1982).

[6] H. H. Chen, Phys. Rev. Lett. **33** (1974) 925.

[7] K. Konno and M. Wadati, Prog. Theor. Phys. **53** (1975) 1652;
M. Wadati, H. Sanuki and K. Konno, Prog. Theor. Phys. **53**(1975)419.

[8] A. A. Rangwala and J. A. Rao, Phys. Lett. **112A** (1985) 188.

Comparison of Some Numerical Schemes for the K-dV Equation

A. Hasan and M.S. Kalra

Nuclear Engineering Program, Indian Institute of Technology, Kanpur 208016, India

A numerical solution for the KdV equation has been obtained using a finite element Galerkin scheme based on cubic splines in the space variable. The results are compared with the finite difference schemes, and the finite element Galerkin and Petrov-Galerkin schemes reported in literature. It is found that the use of smoother trial and test functions leads to much less L_2 and L_∞ errors. It is also seen that quintic boundary polynomials used earlier are not necessary for obtaining an accurate solution.

1. INTRODUCTION

It is well known that the initial and boundary value problems associated with nonlinear partial differential equations are very difficult to handle in a general way. The nonlinear evolution equations have received particular attention over the past two decades or so. This is due to the fact that they arise in a natural way in a large number of physical problems and in many cases possess special types of solutions which may be of great practical use.

Our interest in the present work is in the numerical study of one such evolution equation known as Korteweg-deVries (KdV) equation

$$u_t + uu_x + K^2 u_{xxx} = 0. \qquad (1)$$

This equation and its generalizations play a major role in the study of nonlinear dispersive waves. Examples range from water waves and lattice waves to plasma waves [1].

The numerical solution of (1) has been the subject of many papers over the last few years. Zabusky and Kruskal were the first to study the KdV equation numerically through a leap-frog finite difference scheme [2]. Greig and Morris [3] proposed a Hopscotch finite difference method and compared it with the original scheme of Zabusky and Kruskal.

Fornberg and Whitham [4] used spectral methods for x-variable and leap-frog in t. The finite element Galerkin schemes and its modifications have been used to solve the KdV equation by Alexander and Morris [5], Sanz-Serna and Christie [6], and Mitchell and Schoombie [7], among others.

2. DETAILS OF THE PRESENT SCHEME AND COMPUTATIONAL RESULTS

In this paper we present some numerical results obtained for the KdV equation through a finite element Galerkin scheme. For this purpose we approximate u(x,t) in (1) as follows:

$$u(x,t) \cong U(x,t) = \sum_{i=2}^{N-2} U_i(t)\, \varphi_i(x), \tag{2}$$

where

$$\varphi_i(x) = \varphi\left(\frac{x-x_o}{h} - i\right), \quad h = (x_N - x_o)/N, \tag{3}$$

$$\begin{aligned}
\varphi(x) &= (2+x)^3, & -2 \le x \le -1,\\
&= 1+3(1+x)+3(1+x)^2-3(1+x)^3, & -1 \le x \le 0,\\
&= 1+3(1-x)+3(1-x)^2-3(1-x)^3, & 0 \le x \le 1,\\
&= (2-x)^3, & 1 \le x \le 2.
\end{aligned} \tag{4}$$

The function $\varphi(x)$ above is the basic or cardinal cubic spline function [8]. Here we have considered the range of interest of the x-variable from x_0 to x_N. Since $\varphi_1(x)$ as well as $\varphi_{N-1}(x)$ extend beyond this range, they have been omitted.

Now if we choose the test functions in the standard Galerkin scheme [9] to be the same as the trial functions $\varphi_i(x)$, the problem of approximately solving (1) reduces to obtaining a solution for the following system of ordinary differential equations:

$$A_{ik} \frac{dU_i}{dt} + B_{ijk} U_i U_j + C_{ik} U_i = 0, \tag{5}$$

where summation over repeated indices is implied and

$$A_{ik} = (\varphi_i, \varphi_k),$$

$$B_{ijk} = (\varphi_i \frac{d\varphi_j}{dx}, \varphi_k),$$

$$C_{ik} = (\frac{d^2\varphi_i}{dx^2}, \frac{d\varphi_k}{dx}).$$

Here we have used the standard notation for the L_2 inner product, viz.,

$$(f,g) = \int f(x)\, g(x)\, dx,$$

and also used integration by parts in writing C_{ik}.

For a given initial value $u_o(x)$, the initial values of $U_i(t)$ are obtained from the following:

$$(U_o, \varphi_k) = (u_o, \varphi_k), \quad k = 2,3,\ldots,N-2. \tag{6}$$

Equations (5) are solved by using IMSL subroutine DREBS.

In order to compare the results we define the following errors in the computed solution, U(x,T):

$$L_\infty \equiv \max_{[x_0,x_N]} |U(x,T) - u(x,T)|,$$

$$L_2 \equiv \left[\int_{x_0}^{x_N} (dx\ U(x,T) - u(x,T))^2 \right]^{1/2}, \tag{7}$$

where T is the time for which the solution is evolved and u(x,T) is the exact solution. For the comparison of different schemes, we use the same initial data and other parameters as used in the papers cited in the Introduction.

Table 1 gives the L_∞ and L_2 errors in the present scheme and compares them with four other methods we have referred to previously.

3. CONCLUSION

It is seen from Table 1 that the L_2 error in the Galerkin scheme used here is almost an order of magnitude less than that in the other schemes. L_∞ error is also found to be somewhat less. This can be attributed to the use of smoother trial and test functions. We have used the smoothest functions of degree 3. Here we may point out that Alexander and Morris [5] used quintic boundary polynomials especially constructed to maintain C_2-continuity in addition to cubic splines as trial and test functions. The use of these boundary functions does not appear to be strictly necessary in view of the fact that the solution goes to zero at the boundaries. In fact Table 1 shows that, if anything,

Table 1

Comparison of Different Numerical Schemes for the KdV Equation
Error × 10^3

T	Zabusky-Kruskal		Greig-Morris		Alexander-Morris (q = 0)*	Sanz-Serna-Christie (Petrov-Galerkin)		Present Method	
	L_∞	L_2	L_∞	L_2	L_∞	L_∞	L_2	L_∞	L_2
0.5†	63.5	122.7	67.4	122.4	57.0	51.9	102.5	37.0	18.1
1.0	161.4	298.2	141.6	228.1	---	100.4	150.6	52.7	24.7
1.5	---	---	---	---	---	---	---	50.7	30.4
2.0	---	---	---	---	---	---	---	53.6	31.6

*q is a parameter used by Alexander-Morris.

†Alexander-Morris L_∞ error is for T = 0.3958.

lesser L_∞ error results if we omit these higher order boundary functions. Finally we note that the computation time for the above scheme was 37 to 120 seconds for T = 0.5 to 2.0 seconds respectively.

REFERENCES

[1] R. K. Dodd et al., **Solitons and Nonlinear Wave Equations** (Academic, New York, 1982).
[2] N. J. Zabusky and M. D. Kruskal, Phys. Rev. Lett. 15 (1965) 240.
[3] I. S. Greig and J. L. Morris, J. Comp. Phys. 20 (1976) 64.
[4] B. Fornberg and G. B. Whitham, Phil. Trans. Roy. Soc. 289(1978)373.
[5] M. E. Alexander and J. L. Morris, J. Comp. Phys. 30 (1979) 428.
[6] J. M. Sanz-Serna and I. Christie, J. Comp. Phys. 39 (1981) 94.
[7] A. R. Mitchell and S. W. Schoombie, in **Numerical Methods in Coupled Systems**, ed. R. W. Lewis (Wiley, New York, 1984).
[8] J. T. King, **Introduction to Numerical Computations** (McGraw-Hill, New York, 1984).
[9] C. A. J. Fletcher, **Computational Galerkin Methods** (Springer, Berlin, 1984).

K-dV Like Equations with Domain Wall Solutions and Their Hamiltonians

Bishwajyoti Dey
Institute of Physics, Bhubaneswar 751005, India

We consider K-dV like equations with higher order nonlinearity and show that these have domain wall (kink) solutions for particular values of the coefficient of the nonlinear terms. The solutions are compared with the domain wall solutions of relativistic field theories. The exact Hamiltonian densities are also evaluated for these equations, using Dirac's constrained Hamiltonian formalism. The conservation of the Hamiltonians is explained in terms of the contribution of the corresponding fields from spatial infinities.

We consider certain nonlinear partial differential equations which are Korteweg-de Vries (K-dV) like equations with higher order nonlinearity. These equations are [1,2]

$$u_t + a(1+bu^n)u^n u_x + \delta u_{xxx} = 0, \qquad a,\delta > 0;\ n = 1,2,3,... \tag{1}$$

$$u_t + bu^2 u_x - \delta u_{xxx} = 0, \qquad b,\delta > 0, \tag{2}$$

which can be derived respectively from the Lagrangian density

$$\mathcal{L} = \frac{1}{2}\Theta_x\Theta_t + \frac{a}{(n+1)(n+2)}\Theta_x^{n+2} + \frac{ab}{(2n+1)(2n+2)}\Theta_x^{2n+2} - \frac{1}{2}\delta\Theta_{xx}^2 \tag{3}$$

and

$$\mathcal{L} = \frac{1}{2}\Theta_x\Theta_t + \frac{1}{12}b\Theta_x^4 + \frac{1}{2}\delta\Theta_{xx}^2, \tag{4}$$

where $u = \Theta_x$, and the subscripts denote partial derivatives.

In order to look for travelling wave solutions we make the simple transformation

$$\xi = x - ct \tag{5}$$

where c is the velocity of the solitary waves. Integrating equations (1) and (2) we get respectively the domain wall solutions of these equations as [1,2]

$$u(x,t) = \left\{ \frac{c(n+1)(n+2)}{2a} \right\}^{1/2} [1 \pm \tanh(c/\delta)^{\frac{1}{2}} \frac{n\xi}{2}]^{1/n}, \text{ for } n = 1,3,5,... \quad (6a)$$

$$u(x,t) = \pm\left\{ \frac{c(n+1)(n+2)}{2a} \right\}^{1/2} [1 \pm \tanh(c/\delta)^{\frac{1}{2}} \frac{n\xi}{2}]^{1/n}, \text{ for } n=2,4,6,... \quad (6b)$$

for

$$b = -\frac{a(2n+1)}{c(n+1)(n+2)^2} \quad (7)$$

and

$$u(x,t) = \pm(3c/b)^{\frac{1}{2}} \tanh[(c/2\delta)^{\frac{1}{2}}(\xi+k)]. \quad (8)$$

It can be noted that the solutions (equations (6)) of equation (1) resembles the solution of relativistic field theories with potential [3]

$$V(\varphi) = C\varphi^{2n+2} + B\varphi^{n+2} + A\varphi^2 + D. \quad (9)$$

Similarly equation (8) (solutions of equation (2)) resemble the kink/antikink solutions of $\lambda\varphi^4$ relativistic field theory.

A conservation law associated with a K-dV like equation is expressed by an equation of the form

$$T_t + X_x = 0, \quad (10)$$

where T the conserved density and -X, the flux of T, are functions of u(x,t). The K-dV equation (b = 0 and n = 1 case of equation (1)) has infinite number of conservation laws associated with it. However for equation (1) we could write only first two conservation laws (see [1] for n = 1 and 2 cases) and we are currently trying to find other conservation laws (if any) which are not obvious. On the other hand, equation (2) is a more interesting case, as one can write many conservation laws for this equation [1]. The third conservation law associated with K-dV type equations, usually describes the conservation of Hamiltonian. For example the Hamiltonian for equation (2) is conserved by the third conservation law associated with this equation. However for equation (1) we could not write the third conservation law. This led us to investigate the Hamiltonian nature of this type of K-dV like equations. It may be noted that the Hamiltonian density for K-dV like equations is not obvious, as these belong to degenerate Lagrangian system [4]. There are constraints in the system and one has to use Dirac's theory of constraints [5], for evaluating the correct Hamiltonian density.

Now to avoid higher derivatives in the Lagrangian we introduce a new field $\psi(x,t)$ and write the Lagrangian for equations (1) and (2) respectively as

$$\mathcal{L} = \frac{1}{2}\Theta_x\Theta_t + \frac{a}{(n+1)(n+2)}\Theta_x^{n+2} + \frac{ab}{(2n+1)(2n+2)}\Theta_x^{2n+2} + \delta\Theta_x\psi_x + \frac{1}{2}\delta\psi^2 \tag{11}$$

and

$$\mathcal{L} = \frac{1}{2}\Theta_x\Theta_t + \frac{1}{12}b\,\Theta_x^4 - \frac{1}{2}\delta\psi^2 - \delta\Theta_x\psi_x, \tag{12}$$

where as before $u(x,t) = \Theta_x$.
Considering independent variations with respect to $\Theta(x,t)$ we get from equations (11) and (12) respectively

$$\Theta_{xt} + a\Theta_x^n\Theta_{xx} + ab\Theta_x^{2n}\,\Theta_{xx} + \delta\psi_{xx} = 0 \tag{13}$$

and

$$\Theta_{xt} + b\Theta_x^2\Theta_{xx} - \delta\psi_{xx} = 0 \tag{14}$$

while independent variations with respect to $\psi(x,t)$ give for both equations (11) and (12)

$$\Theta_{xx} - \psi = 0. \tag{15}$$

Thus equations (13) and (14) together with $u = \Theta_x$ and equation (15) gives equations (1) and (2) respectively. The canonical momenta are

$$\pi_\psi = 0 \tag{16a}$$

$$\pi_\Theta = \frac{1}{2}\Theta_x. \tag{16b}$$

Thus these are degenerate Lagrangian systems as the canonical momenta for the field ψ is zero. So we use Dirac's theory [5] of constraints for evaluating the exact Hamiltonian density for these systems. The primary constraints

$$C_1 = \pi_\psi \approx 0 \tag{17a}$$

$$C_2 = \pi_\Theta - \tfrac{1}{2}\Theta_x \approx 0 \tag{17b}$$

(where $\approx$ denotes weak inequality) satisfy the relations

$$\{C_1(x),\ C_1(x')\} = 0, \tag{18a}$$

$$\{C_1(x),\ C_2(x')\} = 0, \tag{18b}$$

and

$$\{C_2(x), C_2(x')\} = -\delta_x(x-x'), \tag{18c}$$

where $\delta(x)$ denotes Dirac delta function. The symbol $\{\ ,\ \}$ denotes the Poisson bracket. The fields have their usual relationship

$$\{\psi(x),\ \pi_\psi(x')\} = \delta(x-x') \tag{19a}$$

and

$$\{\Theta(x),\ \pi_\Theta(x')\} = \delta(x-x'). \tag{19b}$$

Equations (18) show that the constraints are second class. The total Hamiltonian is defined as

$$H_T = \int_{-\infty}^{\infty} (\mathcal{H}_0 + \mathcal{H}_1)dx, \tag{20}$$

where the free part of the Hamiltonian density is given by

$$\mathcal{H}_0 = \pi_\psi \psi_t + \pi_\Theta \Theta_t - \mathcal{L} \tag{21}$$

and

$$\mathcal{H}_1 = \lambda C_1 + \sigma C_2, \tag{22}$$

where the Lagrange multipliers λ and σ have to be determined from the condition that the constraints are maintained in time, i.e.,

$$\{C_1,\ H_T\} = \{C_2,\ H_T\} = 0. \tag{23}$$

This condition requires an extra constraint condition, which we denote by a secondary constraint χ, and $\mathcal{H}_1$ is thus modified as

$$\mathcal{H}_1 = \lambda C_1 + \sigma C_2 + \mu\chi \tag{24}$$

where μ is another Lagrange multiplier. The Lagrange multipliers are determined as [2],

$$\sigma = \left(-\frac{a}{(n+1)}\,\Theta_x^{n+1} - \frac{ab}{(2n+1)}\,\Theta_x^{2n+1} - \delta\Theta_{xxx}\right) \tag{25}$$

for Lagrangian in equation (11) and

$$\sigma = (-\tfrac{1}{3}\, b\, \Theta_x^3 + \delta\Theta_{xxx}) \tag{26}$$

for Lagrangian in equation (12), and the multipliers

$$\mu = \psi - \Theta_{xx} \tag{27}$$

and

$$\lambda = \sigma_{xx}. \tag{28}$$

The total Hamiltonian density is now obtained from equation (20) using equations (17), (21), (24)-(28) which when evaluated for the Lagrangions in equations (11) and (12) gives respectively, upto a surface term,

$$\begin{aligned}\mathcal{H}_T = {} & \frac{na}{2(n+1)(n+2)}\Theta_x^{n+2} + \frac{nab}{(2n+1)(2n+2)}\,\Theta_x^{2n+2} + \tfrac{1}{2}\delta\Theta_{xx}^2 + \tfrac{1}{2}\delta\psi^2 \\ & +\delta\Theta_x\psi_x - \pi_\Theta\,(\delta\Theta_{3x} + \frac{a}{(n+1)}\,\Theta_x^{n+1} + \frac{ab}{(2n+1)}\Theta_x^{2n+1}) \\ & -\pi_\psi(\delta\Theta_{5x} + na\Theta_x^{n-1}\,\Theta_{xx}^2 + a\Theta_x^n\,\Theta_{3x} \\ & + 2nab\,\Theta_x^{2n-1}\,\Theta_{xx}^2 + ab\,\Theta_x^{2n}\,\Theta_{3x}),\end{aligned} \tag{29}$$

where Θ_{3x} denote Θ_{xxx}, etc., and

$$\begin{aligned}\mathcal{H}_T = {} & \frac{1}{12}\, b\Theta_x^4 - \tfrac{1}{2}\delta\Theta_{xx}^2 - \delta\Theta_x\psi_x - \tfrac{1}{2}\delta\psi^2 + \pi_\Theta(\delta\Theta_{3x} - \tfrac{1}{3}\, b\Theta_x^3) \\ & + \pi_\psi\,(\delta\Theta_{5x} - 2b\Theta_x\Theta_{xx}^2 - b\Theta_x^2\,\Theta_{3x}).\end{aligned} \tag{30}$$

To check that the Hamiltonian densities obtained are correct ones, we obtain the field equations (1) and (2) using the Hamiltonian equations of motion

$$\dot{\Theta}(x,t) = \{\Theta\, ,\, H_T\}. \tag{31}$$

Now to examine whether the Hamiltonian is a constant of motion it is sufficient to consider only the free ($\mathcal{H}_0$) part of the total Hamiltonian density [5]. Thus the Hamiltonian H when expressed in terms of the original field u(x,t) gives for equation (1)

$$H = \int_{-\infty}^{\infty} dx \left[\frac{1}{2}\delta u_x^2 - \frac{a}{(n+1)(n+2)} u^{n+2} - \frac{ab}{(2n+1)(2n+2)} u^{2n+2} \right] \quad (32)$$

and for equation (2)

$$H = \int_{-\infty}^{\infty} dx[-\frac{1}{12} bu^4 - \frac{1}{2}\delta u_x^2] \quad (33)$$

It is known that for a system, represented by K-dV type equation

$$u_t = \delta u_{xxx} + F''(u)u_x \quad (34)$$

which can be derived from the Lagrangian density

$$\mathcal{L} = \frac{1}{2}\Theta_x\Theta_t + \delta\Theta_x\psi_x + \frac{1}{2}\delta\psi^2 + F'(u) \quad (35)$$

the Hamiltonian is given by

$$H = \int_{-\infty}^{\infty} dx[-\frac{1}{2}\delta u_x^2 + F(u)]. \quad (36)$$

The Hamiltonians (32) and (33) for our systems (equations (1) and (2)) agree with equation (36). However it is to be noted that for describing the correct dynamics of the systems one has to use the total Hamiltonian density given by equations (29) and (30).

Now, the Hamiltonian H is a constant of motion if

$$\frac{dH}{dt} = 0. \quad (37)$$

For the Hamiltonian in equation (32) we get

$$\frac{dH}{dt} = \int_{-\infty}^{\infty} dx \left[\frac{1}{(n+1)} au^{n+1}(\delta u_{3x} + a(1+bu^n)u^n u_x) + \frac{ab}{(2n+1)} u^{2n+1}(\delta u_{3x} + a(1+bu^n)u^n u_x) - \delta^2 u_x u_{4x} - a\delta n u_x^3 u^{n-1} - a\delta u_x u_{xx} - 2nab\delta u^{2n-1}u_x^3 - ab\delta u^{2n} u_x u_{xx} \right].$$

Integrating by parts and using the fact that derivatives of u(x,t) are zero at spatial infinities (see equations (6)) we get

$$\frac{dH}{dt} = \left[\frac{a^2}{2(n+1)^2} u^{2n+2} + \frac{a^2 b}{(n+1)(2n+1)} u^{3n+2} + \frac{a^2 b^2}{2(2n+1)^2} u^{4n+2} \right]_{-\infty}^{\infty} \tag{38}$$

which is not zero as the contribution from spatial infinities do not cancel, since the field configuration corresponding to equation (1) is asymmetric (equation (6)). Thus the Hamiltonian corresponding to equation (1) is not a constant of motion. This explains why we could not get more than first two conservation laws for this system. However, for the Hamiltonian (equation (33)) of equation (2) we get

$$\frac{dH}{dt} = \int_{-\infty}^{\infty} dx(\tfrac{1}{3} b^2 u^5 u_x - \tfrac{1}{3} b\delta u^3 u_{3x} + 2b\delta u u_x^3 + b\delta u^2 u_x u_{xx} - \delta^2 u_x u_{4x}).$$

Integrating by parts and using the fact that derivatives of u(x,t) are zero at spatial infinities we get (see equation (8))

$$\frac{dH}{dt} = \frac{b^2}{18} (u^6)_{-\infty}^{\infty} \tag{39}$$

which is zero, as the field configuration corresponding to equation (2) is antisymmetric. Thus the Hamiltonian corresponding to equation (2) is conserved. It should be noted that, this Hamiltonian is also conserved by virtue of the third [1] among the many conservation laws satisfied by equation (2).

In case of relativistic field theories however such problem does not arise, where it can be shown by a simple calculation, that, if the Lagrangian does not contain explicit time dependence (as in equations (3) and (4)) then the Hamiltonian is conserved, even if the field configuration is nonzero at spatial infinities (but its derivatives are zero).

REFERENCES

[1] B. Dey, J. Phys. A: Math. Gen. **19** (1986) L9.
[2] B. Dey, to be published.
[3] B. Dey, J. Phys. C: Solid State Physics **19** (1986) 3365.
[4] Y. Nutku, J. Math. Phys. **25** (1984) 2007.
[5] E. C. G. Sudarshan and N. Mukunda in **Classical Dynamics: A Modern Perspective** (Wiley Interscience, New York, 1974).

Part III

Lattice Solitons

Lattice Solitons and Nonlinear Diatomic Models

P.C. Dash

Department of Physics, College of Basic Science and
Humanities Orissa University of Agriculture and Technology,
Bhubaneswar PIN 751003, India

The Fermi-Pasta-Ulam problem together with the explanation by Zabusky and Kruskal can be rightly considered as the origin of lattice solitons. This problem is reviewed in some detail along with a nice integrable nonlinear lattice, the Toda lattice. The recurrence phenomenon in case of KdV system and FPU discrete limit is also discussed. Three diatomic nonlinear lattice models as well as their solutions are considered. These are the simplest cubic nonlinear model in continuum limit, diatomic Toda system and continuum model with nonlinear onsite potential at one of the mass points and harmonic potential at the other, connected by harmonic springs.

1. ORIGIN OF LATTICE SOLITON

Though solitary wave was first discovered in 1834, the present upsurge of interest in solitons is mainly due to the attempts made to explain a nonlinear lattice problem, the Fermi-Pasta-Ulam recurrence found from computer experiments in 1955. Its explanation by Zabusky and Kruskal signalled the birth of lattice solitons. I intend to discuss in some detail this recurrence phenomenon not only because it supplied a major impetus to the development of soliton-physics but also it combines past excitement with present vigour.

2. FERMI-PASTA-ULAM PUZZLE

The equation of motion for a chain of mass points interacting through a potential $\Phi(r)$ can be written as

$$m\ddot{y}_n = \Phi'(y_{n+1} - y_n) - \Phi'(y_n - y_{n-1}), \tag{1}$$

where the force $f(r) = -\Phi'(r)$ and $r_n = y_{n+1} - y_n$ and dot represents derivative with respect to time and prime stands for spatial differentiation with respect to the argument. Linear lattice: $\Phi(r) = \frac{1}{2}\gamma r^2$

so that

$$m\ddot{y}_n = \gamma(y_{n+1} - 2y_n + y_{n-1}). \tag{2}$$

Solutions of eq. (2) with fixed boundary conditions (for n = 0, n = N+1, $y_n = 0$) can be represented by

$$y_n^{(\ell)}(t) = C_n \sin(\frac{\pi\ell}{N+1}n) \cos(\omega_\ell t+\delta_\ell) \tag{3a}$$

with

$$\omega_\ell = 2\sqrt{\gamma/m} \sin\frac{\pi\ell}{2(N+1)}, \quad \ell = 1,2,\ldots,N. \tag{3b}$$

Now one can introduce normal coordinates

$$\eta_r = \frac{2}{N+1} \sum_{n=1}^{N} y_n \sin\frac{\pi r}{N+1} n \tag{4a}$$

with

$$\omega_r = 2\sqrt{\gamma/m} \sin\frac{\pi r}{2(N+1)} \tag{4b}$$

such that energy

$$E = \Sigma\varepsilon_r = \sum_{r=1}^{N} \frac{1}{2}(\dot{\eta}_r^2 + \omega_r^2\eta_r^2). \tag{4c}$$

The natural motion of the harmonic lattice can be expressed as a superposition of these normal modes. The energy of each normal mode remains always constant. It was widely believed that a mild nonlinearity should bring the system to a state of statistical equilibrium. To verify this widely believed conjecture Fermi, Pasta and Ulam [1] considered a chain of masses and varied their number from 16 to 64. The nonlinear potentials considered were

$$\Phi(r) = \frac{1}{2}\gamma r^2 + \frac{1}{3}\gamma\alpha r^3 \tag{5a}$$

$$\Phi(r) = \frac{1}{2}\gamma r^2 + \frac{1}{4}\beta\gamma r^4 \tag{5b}$$

$$\Phi(r) = \begin{cases} \frac{1}{2}\gamma_1 r^2 & |r| < d, \\ \frac{1}{2}\gamma_2 r^2 + \delta r & |r| > d. \end{cases} \tag{5c}$$

Their **aim** was to obtain the energy distribution among the normal modes of the systems in the presence of weak nonlinearity and to determine the time of relaxation to equilibrium. The **result** they found from computer experiments was least expected: if the initial data assigned all the energy to the lowest mode, only a few modes were excited as time went on and almost all energy was eventually given back to the lowest mode (**Recurrence** phenomenon). The aftermath of this finding is really an exciting chapter in the history of soliton-physics. I shall briefly discuss the two serious consequences of FPU puzzle:

(a) How to explain the recurrence?

(b) what happens to statistical equilibrium?

After FPU, extensive studies were carried out with the intention of verifying and explaining the recurrence phenomenon. The conclusions are as follows [2]: Recurrence time

$$t_R = 0.44 \frac{N^{3/2}}{\sqrt{\alpha\beta}} t_L \tag{6a}$$

with fixed boundary condition $y_0 = y_N = 0$ and

$$y_n\Big|_{t=0} = B \sin(\tfrac{\pi n}{N}), \quad \dot{y}_n\Big|_{t=0} = 0 \tag{6b}$$

and t_L is the linear period defined by

$$t_L = 2N/(\gamma/m)^{\frac{1}{2}}. \tag{6c}$$

3. EXPLANATION OF RECURRENCE

Zabusky in 1967 first showed that the continuum limit of FPU lattice was the Korteweg-de Vries (KdV) equation. This was a major breakthrough and signalled the birth of **lattice solitons**. It also became very useful for providing an explanation to FPU recurrence. Subsequently, the continuum approximation to lattice problems are used in many contexts because (i) continuum approximation is easier for analytical as well as numerical study than its discrete counterpart; (ii) results can be conveniently related to the discrete version in many cases; continuum approximation is physically acceptable when wavelength is very large compared to spacing of particles in a lattice. To illustrate a continuum case I choose the following example which is not only the first historical model but contains alll the essential features of any nonlinear lattice problem in long wavelength limit.

Let

$$\Phi(r) = \frac{1}{2}\gamma r^2 + \frac{1}{3}\gamma\alpha r^3, \quad r_n \equiv y_{n+1} - y_n , \tag{7}$$

then

$$(m/\gamma)\ddot{y}_n = (y_{n+1} - 2y_n + y_{n-1})[1 + \alpha(y_{n+1} - y_{n-1})]. \quad (8)$$

In continuum approximation wavelength being large compared to spacing of particles, wave is very smooth so that one can make the following Taylor expansion:

$$y_{n\pm 1} = y \pm h \frac{\partial y}{\partial x} + \frac{1}{2}h^2 \frac{\partial^2 y}{\partial x^2} \pm \frac{1}{6} h^3 \frac{\partial^3 y}{\partial x^3} + \ldots, \quad (9)$$

where h is natural length of a spacing in lattice and $y(x) = y_n$, $x = nh$. Substituting eq. (8) in (7) and keeping terms upto h^2

$$\frac{1}{c^2}\frac{\partial^2 y}{\partial t^2} = \frac{\partial^2 y}{\partial x^2}[1 + 2\alpha h \frac{\partial y}{\partial x} + 3\beta h^2 (\frac{\partial y}{\partial x})^2] + \frac{h^2}{12}\frac{\partial^4 y}{\partial x^4}, \quad (10)$$

where $c = h\sqrt{\gamma/m}$.

For $h \simeq 0$, $\frac{1}{c^2}\frac{\partial^2 y}{\partial t^2} = \frac{\partial^2 y}{\partial x^2}$: linear wave equation (10a)

Keeping terms $\sim h$,

$$\frac{1}{c^2}\frac{\partial^2 y}{\partial t^2} = (1 + 2\alpha h \frac{\partial y}{\partial x}) \frac{\partial^2 y}{\partial x^2}. \quad (10b)$$

This is hyperbolic equation, whose solutions become discontinuous after a time $\sim(\alpha h y_o c)^{-1}$ where y_o is the maximum amplitude at $t = 0$ (obtained numerically), that is, the solutions break down. Zabusky's conjecture is that this is not the solution in discrete case and hence not physical. So, considering terms $\sim h^2$, we have

$$\frac{1}{c^2}\frac{\partial^2 y}{\partial t^2} = (1 + \varepsilon \frac{\partial y}{\partial x}) \frac{\partial^2 y}{\partial x^2} + \frac{h^2}{12}\frac{\partial^4 y}{\partial x^4} \quad (10c)$$

with $\varepsilon = 2\alpha h$.

Let us consider waves travelling to the right:

$$u \equiv \frac{\partial y}{\partial \xi}, \quad \xi = x - ct, \quad \tau = \varepsilon^* t \text{ and } \varepsilon^* = \frac{1}{2}\varepsilon c$$

so that $\frac{\partial}{\partial x} \equiv \frac{\partial}{\partial \xi}$, $\frac{\partial}{\partial t} \equiv \varepsilon^* \frac{\partial}{\partial t} - c\frac{\partial}{\partial \xi}$.

Now eq. (10c) takes the following form,

$$\frac{\partial u}{\partial \tau} + u\,\frac{\partial u}{\partial \xi} + \mu\,\frac{\partial^3 u}{\partial \xi^3} = 0, \tag{11}$$

with $\mu = h/24\alpha$.

This is the Kortweg de Vries equation as obtained by Zabusky. It will not be out of place to mention that even when the potential is of exponential type, the continuum limit of the discrete case is also the KdV equation (Toda).

Recurrence phenomenon is now known to be due to the motion of solitons which carry energy but the first explanation in this line was advanced by Zabusky and Kruskal [4] using the continuum version of the lattice equation of motion (11). FPU considered the initial condition

$$y_n\Big|_{t=0} = B\,\sin\frac{\pi n}{N} \tag{12a}$$

$$\dot{y}_n\Big|_{t=0} = 0. \tag{12b}$$

From (4a) the corresponding normal coordinate

$$\eta_r = \frac{2}{N+1}\sum_{n=1}^{N} B\,\sin\frac{\pi n}{N}\,\sin\frac{\pi r}{N+1}n = B\delta_{1r}. \tag{13}$$

From (12b) there is no initial kinetic energy, so total energy from (4c)

$$E = \Sigma\varepsilon_r = \frac{1}{2}\,\Sigma\,\omega_r^2 B\delta_{1r}$$

$$= \frac{1}{2}B\omega_1^2 \quad \text{(lowest mode is only excited).}$$

Equation (12a) represents a sine-wave which remains confined in the lattice. This can be thought of as a stationary wave which can be approximated by the superposition of two progressive waves of half amplitude:

$$y_n\Big|_{t=0} = \frac{1}{2}\,B\,\sin\frac{\pi n}{N}. \tag{14}$$

Writing $\xi = nh = x$, $\tau = \frac{1}{2}\,\varepsilon ct$, $u = \partial y/\partial\xi$ and considering a cyclic lattice of 2N particles one can find in the continuum limit

$$u\Big|_{\tau=0} = A_o\,\cos\pi x \tag{15}$$

with boundary condition

$$u(x) = u(x+2) \tag{16}$$

where $A_o = B\pi/2$.

4. RECURRENCE IN KdV CONTINUUM

KdV equation (11) has a solution (soliton)

$$u = u_\infty + a\ \mathrm{sech}^2\beta(\xi - c\tau) \tag{17}$$

where u_∞ and a are constants ($a > 0$),

$$c = u_\infty + \frac{a}{3} \text{ and } \beta = \sqrt{a/12}\,\mu.$$

Further with KdV equation (11) is associated an eigenvalue equation

$$6\mu\frac{\partial^2\Phi}{\partial\xi^2} + (\lambda - U)\Phi = 0 \tag{18}$$

where $U = -u$, such that if u develops in time as in KdV, then λ is independent of time. Equation (18) is analogous to a Schrödinger eigenvalue equation with $\hbar^2$, the Planck's constant being replaced by 12μ. As λ is independent of time we can use $u|_{\tau=o} = 0$ and find out eigenvalues which will remain same for all time. For a single soliton

$$\lambda = u_\infty - \frac{a}{2} \text{ and } c = \text{constant} - \frac{2}{3}\lambda. \tag{19a,b}$$

For different eigenvalues, different 'a' values can be obtained and so, each eigenvalue of eq. (18) is associated with a soliton. Now from eq. (15) expanding near the bottom of the potential-well

$$U \simeq A_o + \frac{1}{2}A_o\pi^2\xi^2 \tag{20a}$$

and

$$\lambda_n = -A_o + (n+\tfrac{1}{2})\sqrt{12\mu A_o\pi^2} \tag{20b}$$

where $n = 0,1,2,...$

Using eq. (19b) the velocities of the solitons associated with λ_n form an arithmatic series with common difference

$$\Delta c = \frac{2}{3}(\lambda_n - \lambda_{n-1}) = \frac{2}{3}\sqrt{12\mu A_o\pi^2}. \tag{21}$$

These are the solitons which move independently because actually these solitons travel, interact as they collide and pass through one another (here we neglect acceleration during interaction). When solitons move in the periodic region (length = 2), the same configuration of solitons as the initial one will come back again after a time interval

$$\tau_R = \frac{2}{\Delta c} = 0.364/\sqrt{\mu A_o} \tag{22}$$

when $A_o = 1$, $\sqrt{\mu} = 0.0222$

$$\tau_R = 40/\pi \ . \tag{23a}$$

Numerical experiment with the same values of μ and A_o gives

$$\tau_R = \frac{30.4}{\pi}. \tag{23b}$$

The discrepancy in between (23a) and (23b) may be ascribed to change in velocities during interaction.

4.1 FPU Recurrence

From numerical experiments Zabusky obtained the recurrence time for FPU lattices as

$$t_R = 0.44 \frac{N^{3/2}}{\sqrt{\alpha\beta}} t_L \tag{24a}$$

$$t_L = 2N/\sqrt{\gamma/m} \tag{24b}$$

using expressions $\tau = \frac{1}{2} \varepsilon ct$, $\varepsilon = 2\alpha h$, $c = h\sqrt{\gamma/m}$, $A_o = B\pi/2$ and $\mu = h/24\alpha$ in eq. (22), theoretical estimate for FPU discrete case gives

$$t_R = 0.53 \frac{N^{3/2}}{\sqrt{\alpha\beta}} t_L \ . \tag{25}$$

The discrepancy between (25) and (24a) is still less.

This gives a very good account of the recurrence phenomena in both discrete and continuum limit.

4.2 Statistical Equilibrium

Ford et al. and Saito et al. [2] advanced reasonable explanation to the problem of statistical equilibrium, energy sharing, ergodicity and the equipartition. It is now clear that there exists a critical value α_c of the coupling constant α which determines ergodicity. For $\alpha > \alpha_c$, the system will be stochastic and energy sharing would take place. If $\alpha < \alpha_c$, the recurrence phenomena may occur and one may find nonlinear normal modes which is a consequence of Kolmogorov, Arnold and Moser theorem. Here again as suggested by Ford a resonance relation exists: $\Sigma\, n_k \omega_k = 0$, n_k being integers, all not equal to zero. When this condition is satisfied for small nonlinearity case there will be energy sharing, otherwise recurrence phenomena will occur. Further if the initial state is far from a normal mode the resonant nonlinear system exhibits rapid energy sharing and equipartition of energy is readily established.

5. TODA LATTICE

Previous sections give an account of the origin of lattice solitons but the most remarkable model for their study is the Toda monatomic chain. With the nearest neighbour interaction the Toda chain happens to be the only integrable nonlinear model [3]. Here the nearest neighbour interaction potential is given by

$$\Phi(r) = \frac{a}{b}\, e^{-br} + ar + \text{const.} \quad (ab > 0) \tag{26}$$

and the equation of motion becomes

$$m\ddot{s}_n/(1+\dot{s}_n) = s_{n-1} + s_{n+1} - 2s_n, \tag{27}$$

where

$$\dot{s}_n = -\,\partial\Phi(r_n)/\partial r_n = a(e^{-br_n} - 1) \tag{28}$$

or equivalently,

$$m(d^2 y_n/dt^2) = a\,\{e^{-b(y_n - y_{n-1})} - e^{-b(y_{n+1} - y_n)}\}\,. \tag{29}$$

The exponential potential (26) includes the linear harmonic case when $b \to 0$ and strongest nonlinear case of a system of hard spheres if $b \to \infty$. The equation of motion (27) or (29) admits exact M-pulse solutions whose form for M = 1 is given by

$$\exp(-br_n) - 1 = (m/ab)\beta^2 \operatorname{sech}^2(\alpha n \mp \beta t + \delta) \tag{30}$$

or with

$$m\dot{y}_n = S_n - S_{n-1} \qquad (31)$$

$$S_n = (m/b) \log \{1+\exp[2(\alpha n \mp \beta t+\delta)]\}, \qquad (32)$$

where $\beta = (ab/m)^{\frac{1}{2}} \sinh\alpha$. (β/α) gives the speed and eq. (30) is the single pulse solution. The M-pulse solution can be obtained in closed analytic form and is represented assymptotically (for $t \to \mp \infty$) by

$$\exp(-br_n) - 1 = (m/ab)\beta_j^2 \operatorname{sech}^2(\alpha_j n+\beta_j t+\delta_j^{\mp}) \qquad (33)$$

with $j = 1,2,\ldots,$ N and constants δ_j are related by

$$\sum_j \delta_j^- = \sum_j \delta_j^+. \qquad (34)$$

This last expression represents conservation of momentum.

These pulses move almost independently in the lattice. They emerge after collisions with same shapes and velocities. So they behave like particles and are called **solitons** or **lattice solitons.** Further it can be seen from eq. (32) that when the soliton moves in the lattice with a constant velocity it causes a contraction of the lattice.

In addition to this M-soliton solution the Toda lattice admits a nonlinear periodic solution known as Cnoidal solution:

$$\exp(-br_n) - 1 = (m/ab)(2k\nu)^2[\mathrm{dn}^2\{2(\tfrac{n}{\lambda} - \nu t)K - \tfrac{E}{k}\}], \qquad (35)$$

where K and E are elliptic integrals of the first and second kinds, and dn is an elliptic function (Jacobian).

If the modulus k is small, $k \simeq 0$, a cnoidal wave reduces to a sinusoidal wave,

$$r_n = -\frac{m\omega^2 k^2}{8ab^2} \cos(\omega t - 2\pi n/\lambda) \qquad (36)$$

$$\omega = 2(ab/m)^{1/2} \sin(\pi/\lambda) \cdot \qquad (37)$$

The cnoidal wave (eq. 35) can be written as

$$\exp(-br_n) - 1 = (m/ab)[\sum_{\ell=-\infty}^{\infty} \beta^2 \operatorname{sech}^2\{\alpha(n-\lambda\ell)-\beta t\} - 2\beta\nu] \qquad (38)$$

with $\alpha = \pi K/K'$, $\beta = \pi K\nu/K'$.

Equation (38) represents an infinite sequence of solitons at equal intervals λ and shifted downwards.

As a mathematical model for lattice-soliton Toda chain no doubt occupies a unique place but its applications in different fields (like wave propagation in nerve systems, ladder circuit, chemical reactions in atoms and molecules, and ecological systems) make it very important and interesting from physical point of view.

6. NONLINEAR DIATOMIC MODELS

Lattice solitons are extensively studied using monatomic models in both discrete and continuum limits. One of the applications of these studies is to explain certain important characteristics of structural phase transitions. As many of the solids undergoing displacive type phase transitions have a diatomic structure along (100) symmetry direction, the study of diatomic nonlinear models attracts much attention. Besides, some of these nonlinear diatomic lattices are very helpful in explaining nonlinear flow of heat in solids.

In the literature mainly two types of models are available for nonlinear diatomic cases: one deals with nonlinear interactions between nearest neighbours [4-6] and the other with nonlinear onsite potentials connected by harmonic springs [7,8]. Study of diatomic Toda chain [6] happens to be the first attempt in arriving at an exact solution to a discrete nonlinear diatomic model. Now it is becoming gradually clear that diatomic Toda chain represents a nonintegrable system [9]. However, very recently nonintegrable rational billiard systems are found to be analytically tractable in terms of Fourier expansion [10] and so the earlier study of diatomic Toda chain with the help of Fourier series deserves some special mention. On the other hand, Büttner and Biltz [7] reported exact solutions to a lattice with nearest and a next nearest interaction, having a nonlinear Φ^4 onsite potential. I wish to consider here the following three nonlinear lattices which will involve all the techniques and characteristics of the available models.

6.1 A Continuum Diatomic Model of First Kind

A large number of studies on diatomic models include nonlinear interaction between nearest neighbours. These are discussed in continuum or long wavelength limit using a procedure by which solutions can be separated into optic and accoustic modes. The following nonlinear diatomic model is chosen to illustrate the main features of this type of lattices. Here we consider a chain of alternate mass points m_1 and m_2 connected by nonlinear springs, with potential energy,

$$\Phi(r) = \frac{1}{2} k_2 r^2 + \frac{1}{3} k_3 r^3 \tag{39}$$

and equation of motion for displacements y_{1n} and y_{2n} of atoms m_1 and m_2 of the n-th cell can be written as

$$m_1 \ddot{y}_{1n} = k_2(y_{2n} - y_{1n}) + k_2(y_{2n-1} - y_{1n}) + k_3(y_{2n} - y_{1n})^2 - k_3(y_{2n-1} - y_{1n})^2, \tag{40}$$

$$m_2 \ddot{y}_{2n} = -k_2(y_{2n} - y_{1n+1}) - k_2(y_{2n} - y_{1n}) + k_3(y_{2n} - y_{1n+1})^2 - k_3(y_{2n} - y_{1n})^2. \tag{41}$$

To solve in the continuum limit the expansions used are

$$y_{pn\pm 1} = y_{pn} \pm 2hy'_{pn} + 2h^2 y''_{pn} \pm \frac{4}{3} h^3 y'''_{pn} + \ldots \tag{42}$$

$$y_{2n} = a(y_{1n} + \sigma_1 h y'_{1n} + \sigma_2 \frac{h^2}{2} y''_{1n} + \ldots). \tag{43}$$

Consideration of a harmonic lattice using these expansions suggests that $a = 1$, corresponds to accoustic mode and $a = -m_1/m_2$ to optical case. Now substitution of equations (42,43) in (41) yields (with appropriate choice of variables), the following KdV equation for a=1:

$$\frac{\partial u}{\partial \tau} + u \frac{\partial u}{\partial \xi} + \rho \frac{\partial^3 u}{\partial \xi^3} = 0 \cdot \tag{44}$$

Therefore in the accoustic region KdV type pulse solitons are obtained. Now for $a = -M_1/M_2$ as in [4] we obtain travelling wave solutions of the form $\exp(\pm i\Theta)$ whose amplitude satisfies the nonlinear Schrödinger equation:

$$i \frac{\partial A}{\partial \tau} + \frac{1}{2} \frac{d^2 w_o}{dk^2} \frac{\partial^2 A}{\partial \xi^2} + \frac{P_1}{2w}\left(\frac{P_2}{D_2} + \frac{N}{v_q^2 - 1}\right) |A|^2 A + \frac{P_1}{2w} A = 0. \tag{45}$$

It describes the motion of the amplitude as nonlinear wave modulation.

6.2 Diatomic Toda Lattice

Using the notations of Section 5 and putting constants a = b = 1, equations of motion for diatomic Toda chain becomes [6]

$$-m_1 m_2 \ddot{s}_{2n}/(1+\dot{s}_{2n}) = (m_1+m_2)s_{2n} - m_1 s_{2n-1} - m_2 s_{2n+1} \tag{46}$$

$$-m_1 m_2 \ddot{s}_{2n-1}/(1+\dot{s}_{2n-1}) = (m_1+m_2)s_{2n-1} - m_2 s_{2n-2} - m_1 s_{2n}. \tag{47}$$

We look for periodic solutions in the following form because this form exists in the harmonic limit as well as in the equal mass limit.

$$s_{2n} = \sum_{j=-\infty}^{\infty} a_j \exp[i2\pi j(\nu t + \frac{2n}{\lambda})] \tag{48}$$

$$s_{2n-1} = \sum_{j=-\infty}^{\infty} b_j \exp[i2\pi j(\nu t + \frac{2n-1}{\lambda})]. \tag{49}$$

Substituting expressions (48,49) in (47) and then integrating over a time period we get

$$\begin{pmatrix} m_1 m_2(2\pi\nu)^2 - m_1 - m_2 & e(1) \\ e(-1) & m_1 m_2(2\pi\nu)^2 - m_1 - m_2 \end{pmatrix} \begin{pmatrix} a_1 \\ b_1 \end{pmatrix} = \begin{pmatrix} M \\ N \end{pmatrix}, \tag{50}$$

where

$$M = i2\pi\nu(m_1+m_2) \sum_{j=-\infty}^{\infty} j a_j a_{1-j} - i2\pi\nu \sum_{j=-\infty}^{\infty} j a_j b_{1-j} e(1-j) \tag{51}$$

$$N = i2\pi\nu(m_1+m_2) \sum_{j=-\infty}^{\infty} j b_j b_{1-j} - i2\pi\nu \sum_{j=-\infty}^{\infty} j b_j a_{1-j}\, e(j-1), \tag{52}$$

with $e(j) = m_1 \exp(-i2\pi j/\lambda) + m_2 \exp(i2\pi j/\lambda)$. (53)

Equation (50) can be solved for getting the coefficients of the Fourier expansion, so that

$$s_{2n} = a_o - 2\pi\nu \frac{4m_1 m_2 \sin^2\frac{\pi}{\lambda} \sum_{j=1}^{\infty} \frac{2q^j}{1-q^{2j}} \sin 2\pi j(\nu t + \frac{2n}{\lambda})}{m_1\exp(i2\pi/\lambda) + m_2\exp(-i2\pi/\lambda) - m_1 - m_2}, \tag{54}$$

$$s_{2n-1} = b_o - 2\pi\nu \frac{4m_1 m_2 \sin^2\frac{\pi}{\lambda} \sum_{j=1}^{\infty} \frac{2q^j}{1-q^{2j}} \sin 2\pi j(\nu t + \frac{2n-1}{\lambda})}{m_1\exp(-i2\pi/\lambda) + m_2\exp(i2\pi/\lambda) - m_1 - m_2}. \tag{55}$$

6.3 Henry and Oitma Diatomic Model

Let us consider a diatomic chain of harmonically coupled nearest neighbour atoms m_1 and m_2 including a nonlinear Φ^4 potential on mass M_1. The Hamiltonian is given by

$$H = \Sigma[\frac{1}{2} m_1\dot{u}_i^2 + \frac{1}{2} m_2\dot{v}_i^2 + \frac{1}{2} (u_i - v_{i-1})^2 + \frac{1}{2} \gamma(u_i-v_i)^2+V(u_i)] , \tag{56}$$

where u_i, v_i are displacements of the two types of atom in the i-th unit cell. $V(u_i)$ is the nonlinear onsite potential at lattice points with mass m_1. With the following prescription, equations of motion can be obtained in continuum limit in the displacive regime:

$$\begin{aligned} u_i &\to u(x,t) \to u \\ v_i &\to v(x + \frac{a}{2}, t) \to v + \frac{a}{2}v' + \frac{1}{2}(\frac{a}{2})^2 v'' \\ v_{i-1} &\to v(x - \frac{a}{2}, t) \\ \sum_i &\to \int\frac{dx}{a} \end{aligned} \tag{57}$$

$$m_1\ddot{u} + 2\gamma(u-v) - \frac{1}{4}\gamma\, a^2v'' + \frac{\partial V}{\partial u} = 0 \tag{58}$$

$$m_2\ddot{v} + 2\gamma(v-u) - \frac{1}{4}\, \gamma\, a^2u'' = 0. \tag{59}$$

(a) If $V = 0$, the equations reduce to continuum equations of motion for a harmonic diatomic chain.

(b) If $V \neq 0$, but instead is given by a Φ^4 nonlinear single site potential then the following three types of excitations are obtained as exact solution to the field eqs. (58) and (59).

(i) Linearized periodic solutions

Here

$$V(u) = - \frac{A}{2}\, u^2 + \frac{B}{4}\, u^4 \quad (A, B > 0) \tag{60}$$

with potential minima at $\pm u_o$ $(= \pm A/B)$. These linearized solutions are low energy phonons and represent oscillation of m_1 atoms in one of the double well potential $(\pm u_o)$ and oscillation of m_2 atoms about $\pm u_o$ from their equilibrium position.

$$u = \pm u_o + u_L \sin(kx - \omega_L t + \Phi), \quad u_L << 1 \tag{61}$$

$$v = \pm u_o + v_L \sin(kx - \omega_L t + \Phi), \quad v_L << 1 \tag{62}$$

with the dispersion relation

$$\omega_L^2(\pm) = \frac{Am_2+\gamma(m_1+m_2)}{m_1m_2} \left(1 \pm \left[1 - \frac{(4A\gamma+\gamma^2a^2k^2 - \frac{\gamma^2a^4k^4}{16})m_1m_2}{[Am_2 + (m_1+m_2)]^2}\right]^{1/2}\right). \tag{63}$$

(ii) Solitary wave solutions (large amplitude solutions)

Equations (61) and (62) represent low amplitude solutions. Besides, the field equations (58) and (59) represent large amplitude solutions also. $v = u = \pm u_o$ is the simplest large amplitude solutions with lowest energy. It is the ground state and is taken as the reference level. Other solutions are regarded as excitations above this level:

$$E = \int_{-\infty}^{+\infty} \frac{dx}{a} \{ \tfrac{1}{2}m_1\dot{u}^2 + \tfrac{1}{2}m_2\dot{v}^2 + \gamma(u-v)^2 + \frac{\gamma a^2}{4} u'v' - \frac{A}{2}(u^2 - v_o^2) + \frac{B}{4}(u^4 - u_o^4)\}. \tag{64}$$

Seeking, now, solutions in the form $u(x,t) = f(s)$, $v(x,t) = g(s)$ with $s = x-ct$ equations (58) and (59) becomes

$$\begin{pmatrix} \frac{\gamma a^2}{4} & 2\gamma \\ m_2c^2 & 2\gamma \end{pmatrix} \begin{pmatrix} \frac{d^2g}{ds^2} \\ g \end{pmatrix} = \begin{pmatrix} m_1c^2 \frac{d^2f}{ds^2} + 2\gamma f + \frac{\partial V}{\partial f} \\ \frac{\gamma a^2}{4}\frac{d^2f}{ds^2} + 2\gamma f \end{pmatrix}.$$

For $c = c_o$, where $c_o = \sqrt{\gamma a^2/4m_2}$,

$$g(s) = f_o(s), \quad s = [c_o^2(m_1 - m_2)]^{\frac{1}{2}} \int^{f_o} \frac{df}{[c - 2V(f)]^{\frac{1}{2}}}.$$

For a Φ^4 potential depending upon the values of the integration constant c equal displacement field solutions can be obtained in the form of a tanh-kink or a sech-pulse.

(iii) Nonlinear periodic solutions

For $c = c_o$, the displacement fields are not equal and solutions may become extended instead of localized. The exact solutions can be written as

$$f = f_o \sin(ks), \quad k = [(2\gamma)/(9c^2 m_2)]^{1/2}$$

$$g = g_o \sin(ks) + g_1 \sin(3ks)$$

with

$$f_o = \{ \frac{A}{3B}[m_1 m_2 c^4 k^4 - (2\gamma(m_1+m_2)-Am_2)c^2k^2 - 2A\gamma + \gamma^2 a^2 k^2 - \frac{a^4\gamma^2 k^4}{16}]/(m_2 c^2 k^2 - 2\gamma) \}^{1/2}$$

$$g_o = (2\gamma - \frac{\gamma a^2 k^2}{4}) f_o/(2\gamma - m_2 c^2 k^2), \quad g_1 = (Bf_o^3)/(9\gamma a^2 k^2 - 8\gamma)$$

$$c = \{ \frac{2\gamma}{9k^2 m_2} \}^{1/2}, \quad \omega = ck = [(2\gamma)/(9m_2)]^{\frac{1}{2}} .$$

The linearized periodic solutions are low energy phonons whereas these nonlinear periodic solutions are high energy phonons. For discrete diatomic case, Büttner and Biltz [7] found "periodon" solutions and the above nonlinear periodic solutions are shown to be the long wavelength (continuum) limit of these solutions.

7. CONCLUSION

In conclusion, it may be stressed that the history and development of the soliton concept is intimately connected with the studies on lattice solitons. Now it is known that further researches will help understand among other things, some unexplained facts in the field of ferro-electrics, conformational change in biological molecules and chopping phenomenon. However, some challenging areas of research include mono and diatomic nonlinear lattices in higher dimensions [11], scattering and destruction of solitons by impurities (thermal conduction [12]), onset of chaotic behaviour (that is, how solitons and chaos compete or compromise) [13], effect of perturbation on solitons and quantization [14].

REFERENCES

[1] E. Fermi, J. Pasta and S. Ulam, **Collected Papers of Enrico Fermi**, Vol. II (University of Chicago Press, 1965), p. 978.

[2] A good review of the recurrence phenomena can be found in the review article by M. Toda, **Phys. Rept. 18C** (1975) 1; **Theory of Nonlinear Lattices** , Springer Series in Solid State Sciences, Vol. 20 (Springer, Berlin, 1981).

[3] K. Sawada and T. Kotera, Proc. Theo. Phys. Suppl. **59** (1976) 101.
[4] P. C. Dash and K. Patnaik, Proc. Nucl. Phys. and S. S. P. Symp. (India) **21C** (1978) 483; Prog. Theor. Physics **65** (1981) 1526, and references therein.
[5] H. Büttner, and H. Biltz, in **Solitons and Condensed Matter Physics**, ed. A. R. Bishop and T. Schneider (Springer, Berlin, 1978),p.162.
[6] P. C. Dash and K. Patnaik, Phys. Rev. **A23** (1981) 959.
[7] H. Büttner and H. Biltz, in **Recent Developments in Condensed Matter Physics**, ed. J. T. Devreese (Plenum, New York, 1981), Vol. 1, p. 49.
[8] B. I. Henry and J. Oitman, Aust. J. Phys. **36** (1983) 339.
[9] B. Kostat, Adv. Math. **34** (1979) 195.
[10] P. J. Richens and M. V. Berry, Physica **2D** (1981) 495.
[11] H. Büttner, H. Frosch, C. Behnke and H. Biltz, Springer Series in Solid-State Sciences, **47** (1983) 281, and references therein.
[12] G. Casati, J. Ford, F. Vivaldi and W. M. Visscher, Phys. Rev. Lett. **52** (1984) 1861 and ibid **53** (1984) 1120; F. Mokross and H. Büttner, J. Phys. **C16** (1983) 4539.
[13] Till now there are no serious attempts in lattice models for studying the coexistence of chaos and solitons. However in differential equations some studies are made in this direction. A good and brief review is given by A. Bishop in Springer Series in Solid-State Sciences **47** (1983) 197 and Y. Imry, ibid **47** (1983) 170.
[14] N. Theodorakopoulos and F. G. Mertens, Phys. Rev. **B28** (1983) 3512.

Recent Results in Toda Lattice

Z. Popowicz

Institute of Theoretical Physics, University of Wroclaw,
ul. Cybulskiego 36, PL-50-205 Wroclaw, Poland

It is shown how different generalizations of Toda lattice problems occur in the context of eighteenth and nineteenth century mathematics. A possible classification and applications of different generalizations of the Toda lattice are given, with special attention to gauge theory.

The topic of this paper will be the following system of equations

$$\frac{\partial^2}{\partial x \partial y} \varphi_n = e^{\varphi_{n+1} - \varphi_n} - e^{\varphi_n - \varphi_{n-1}} , \tag{1}$$

where $\varphi_n = \varphi_n(x,y)$ and $n \in Z$, and its different generalizations which appear in the physics and mathematics. For our purpose, it will be convenient to define the system (1) in a different manner as

$$\frac{\partial^2}{\partial x \partial y} r_n = e^{r_{n+1}} - 2e^{r_n} + e^{r_{n-1}}, \tag{2}$$

$$\frac{\partial^2}{\partial x \partial y} h_n = e^{\pm k_{nm} h_m} , \tag{3}$$

where $r_n = \varphi_{n+1} - \varphi_n$ and $h_n = h_n(x,y)$,

$$k_{nm} = \begin{cases} 2 & \text{for } n = m \\ -1 & \text{for } n = m \pm 1 , \\ 0 & \text{for rest} \end{cases} \tag{4}$$

and for $x = y$ as

$$\frac{d}{dx} M_n = M_n [N_n - N_{n-1}], \tag{5}$$

$$\frac{d}{dx} N_{n-1} = M_n - M_{n-1}, \tag{6}$$

where $M_n = \exp(\varphi_{n+1} - \varphi_n)$ and $N_n = d\varphi_n/dx$.

Equations (1-6) are known to the physicist as the Toda lattice. Toda defined equation (2) for $x = y$ in 1967 [1,2] considering a one-dimensional lattice, which consisted of N particles of unit mass interacting through the potential

$$\Phi(r) = e^{-r} + r. \qquad (7)$$

Toda considered such a potential because it admitted analytic solutions and it generalized the famous computer experiments of Fermi, Pasta and Ulam [3]. Fermi, Pasta and Ulam considered a finite number of pendula arranged in a line interacting with their nearest neighbour via anharmonic forces. The system was started by displacing the end pendulum with the other at rest. Soon all were moving but after a finite time, the initial situation recurred. This meant that the system was not ergodic. The same situation has been obtained for Toda lattice [4].

After introducing the Toda lattice its importance was quickly and widely recognized by physicists. It appears that the Toda lattice and its generalizations can describe different physical phenomena and are contained among the soliton equations. These facts have given a strong impetus for a deep investigation of these equations.

Surprisingly, recently [5-8] it has been noted that the Toda lattice in the forms (1)-(6) has been "known" for the mathematicians of the eighteenth and nineteenth centuries. Here one should specify the meaning "known" and distinguish those mathematicians who used that system consciously [9] and those who were close to define it [10,11]. I do not define here the meaning of "known". It is the problem for the historians and philosophers, similar to the question "Did ancient Greek know the differential calculus?" On the other hand, I would like to apply the Toda lattice to the eighteenth and nineteenth century mathematics using the ideas of Euler and Sylwester.

First let us note that Euler solved the following linear differential equation

$$g = a_o\ddot{g} + b_o\dot{g}, \qquad (8)$$

where $g = g(x)$, a_o, b_o are arbitrary functions of x, differentiable infinite times, and

$$\dot{g} = \frac{dg}{dx}, \quad \ddot{g} = [\dot{g}]^2, \ . \ . \ ., \ [\dot{g}]^n = \frac{d^n g}{dx^n}, \qquad (9)$$

without using the series expansion of g. His method known now as the Euler method is based on the following trick. Let us write down equation (8) as

$$g/\dot{g} = b_o + a_o(\dot{g}/\ddot{g})^{-1} . \qquad (10)$$

Using equation (8) we can compute

$$\dot{g}/\ddot{g} = b_1 + a_1(\ddot{g}/\dddot{g})^{-1} , \tag{11}$$

and similarly for

$$\frac{[\dot{g}]^n}{[\dot{g}]^{n+1}} = b_n + a_n \left(\frac{[\dot{g}]^{n+2}}{[\dot{g}]^{n+3}}\right)^{-1} , \tag{12}$$

where the functions a_n, b_n could be computed explicitly from (8), differentiating n times. Substituting (11)-(12) into (10) we obtain

$$\left(\frac{d}{dx} \ln g\right)^{-1} = b_o + \frac{a_o}{b_1+} \frac{a_1}{b_2+} \frac{a_2}{b_3+} \cdots \frac{a_n}{b_{n+1}+} \cdots \tag{13}$$

In this way, Euler was able to represent g as a continuous fraction [12], which is denoted by us by the use of the special notation in (13). On the other hand, the investigation of the convergence of the continuous fraction is equivalent to the investigation of the corresponding series. It may be that this was the reason why Euler stopped these investigations at this stage. Following the line proposed by Common and Roberts [8], let us apply the Euler method to the Riccati equation

$$\frac{dg_o(t)}{dt} = E_o(t) + g_o(t)F_o(t) + g_o^2(t), \tag{14}$$

where E_o, F_o are given functions of t. In [8], the authors proved the following theorem.

Theorem If g_{k-1} satisfies the Riccati equation for some E_{k-1}, F_{k-1}, then g_k , defined by

$$g_{k-1} = U_o + (N_k - g_k)^{-1} M_k, \tag{15}$$

where U_o is an arbitrary constant and $k = 1,2,\ldots,$satisfies the Riccati equation for E_k, F_k if

$$M_k = E_{k-1} + U_o F_{k-1} + U_o^2, \tag{16}$$

$$N_k = \dot{M}_k M_k^{-1} - (F_{k-1} + U_o), \tag{17}$$

$$E_k = \dot{N}_k + N_k U_o + M_k, \tag{18}$$

$$F_k = -\dot{N}_k - U_o. \tag{19}$$

Note that the initial functions M_1 and N_1 are determined from the coefficients of the original Riccati equation (14) by using (16) and (17) with k = 1. If we eliminate the functions E_k, F_k and the coefficient U_o from (16-17), we obtain M_k, N_k satisfying the Toda equations (5-6), where now n = 2,3,.... In this way, we have obtained the solution of the Riccati equation (14) in the form of continuous fraction (15) which is constructed from the solutions of the Toda lattice. It is surprising that the "old is so new."

Quite a different application of the Toda lattice, with reference to the nineteenth century mathematics, can be constructed using a work of Sylvester [11]. Note that Sylvester is also well known for coining many new words or jargons in mathematics, for example, the persymmetric determinant. He defined it as the determinant which possesses the same elements on the perpendicular line to the main diagonal. For example, the persymmetric determinant of third order is defined by

$$D_3(g) \overset{df}{=} \det \begin{pmatrix} g_o & g_1 & g_2 \\ g_1 & g_2 & g_3 \\ g_2 & g_3 & g_4 \end{pmatrix} . \tag{20}$$

Assuming that

$$g_n = \frac{d^n}{dx^n} g_o, \quad n = 1,2,.... \tag{21}$$

Sylvester proved that

$$D_2[D_3(g)] = D_4(g).D_2(g) \tag{22}$$

and in the general case that

$$D_2[D_n(g)] = D_{n+1}(g)D_{n-1}(g). \tag{23}$$

Putting

$$D_{N+1}(g) = 1, \tag{24}$$

for some N we obtain a complicated nonlinear differential equation for the function g. Probably, Sylvester would have fallen head over heels to know that equations (23) and (24) are equivalent with the finite one-dimensional Toda lattice with the free endpoints in the form (3). This connection can be easily seen due to the following correspondence. Let us rewrite the equation (3) for minus sign and for x = y = z as

$$\partial^2_{zz} h_1 = e^{-2h_1+h_2}, \tag{25}$$

$$\partial^2_{zz} h_2 = e^{-2h_2+h_1+h_3}, \tag{26}$$

$$\cdots\cdots\cdots\cdots$$

$$\partial^2_{zz} h_N = e^{-2h_N + h_{N-1}}, \tag{27}$$

from which we obtain

$$e^{h_1} \overset{dt}{=} D_1(e^{h_1}) \overset{dt}{=} D_1(g), \tag{28}$$

$$e^{h_2} = \partial^2_{zz} h_1 . e^{2h} = e^{h_1} \partial^2_{zz} e^{h_1} - \partial_z e^{h_1} \partial_z e^{h_1} = D_2(g), \tag{29}$$

$$e^{h_3} D_1(g) = D_2[D_2(g)]. \tag{30}$$

Using equation (23) we obtain

$$e^{h_3} = D_3(g) \tag{31}$$

and by recurrence

$$e^{h_n} = \frac{D_2[D_{n-1}(g)]}{D_{n-2}(g)} = D_n(g), \tag{32}$$

for $n = 3,4,\ldots,N$. In order to terminate this recurrence according to (27), we assume (24). Hence, we have established the above mentioned correspondence.

Surprisingly the Toda lattice (2) where $r_n = \ln f_n$ can be seen in the book of Darboux [9]. Darboux found the singular solutions of this lattice in the form

$$f_n = (n-a)(b-n)(x-y)^{-2}, \tag{33}$$

where a, b are arbitrary constants.

In this way, we showed that the different forms of the Toda lattice can be recovered in terms of the eighteenth and nineteenth century mathematics. Furthermore, from both physical and mathematical points of view, it is possible to generalize the Toda lattice in other ways also. We will try to introduce a possible classification of such generalized Toda lattices in the following. We call the one-dimensional system introduced by Toda as the standard Toda lattice, which has been thoroughly investigated. We have three possibilities: finite, infinite and periodic Toda lattices. Out of these infinite and periodic Toda lattices are completely integrable Hamiltonian systems [13,14] and can be solved by the inverse scattering transformation [15] or by the Hirota method [16] or by the Bäcklund transformation [4]. The finite Toda lattice has been considered by Kostant [17] and by Olshanetsky and Perelomov [18].

One can also generalize the standard Toda lattice in one-dimensional space to the two-dimensional space-time [19,20]. This model corresponds to a non-trivial relativistic invariant model in field theory, the so-called nonlinear Klein-Gordon equation and can be solved by the inverse scattering method [19] or by the Bäcklund transformation. Barbashov and Nesterenko in 1981 [21] showed that the relativistic string model in a space-time of constant curvature (de Sitter universe) is described by the system of equations

$$\varphi_{xt} = e^{\varphi} \cos\psi - e^{-\psi} , \tag{34}$$

$$\psi_{xt} = e^{\varphi} \sin \psi . \tag{35}$$

Interestingly, this system is equivalent with the N = 4 periodic two-dimensional Toda lattice. Indeed as was shown by Fordy and Gibbons [20] that in this case the Toda lattice can be reduced to (34-35) if we make the substitution

$$\Theta = \frac{1}{2} \varphi , \qquad \Phi = \frac{i}{2}\psi \tag{36}$$

after rescaling the variables x and t.

Ueno and Takasaki generalized [22] the standard Toda lattice to the multidimensional case using the idea of Kadomtsev-Petviashvilli hierarchies [23]. Its algebraic structure and bilinearization in terms of the τ function and some special solutions were investigated in detail in [24].

The fourth generalization is a purely theoretical approach where one generalizes the Toda lattice by including supersymmetry. Here, one can distinguish between the non-extended [25] or extended [26] supersymmetric Toda lattice.

In the next approach, one can utilize the connection, discovered by Bogoyavlensky [27], between the Toda lattice and simple Lie algebras.

It appears that for every simple Lie algebra one can associate a Toda lattice which bears the name of the Lie algebra. For example, the SU(N) one-dimensional finite, nonperiodic Toda lattice is exactly the standard finite, nonperiodic Toda lattice with the free endpoints. The standard periodic Toda lattice corresponds to contragradient Lie algebras. It appears that these equations have important applications in the gauge field theory, namely, in the construction of the spherically symmetrical instantons or monopoles [28-31]. Let me briefly present how one can recover the Toda lattice from the self-dual equation for the Yang-Mills field theory. The instantons are defined as the finite action self-dual of the Yang-Mills field theory [32]. The monopoles are defined as the static solutions with finite energy of the self-dual Yang-Mills-Higgs field assuming the so-called Bogomolny-Prasad-Sommerfeld limit [32]. Here the self-dual condition means that

$$F^{x}_{\mu\nu} = \frac{1}{2}\varepsilon_{\mu\nu\alpha\beta} F_{\alpha\beta} = F_{\mu\nu}, \tag{37}$$

where

$$F_{\mu\nu} = \partial_{\mu}A_{\nu} - \partial_{\nu}A_{\mu} + [A_{\mu},A_{\nu}], \tag{38}$$

and A_{μ} is a Lie algebra valued matrix function.

The self-dual conditions (37) are usually written down in the complexified space-time. This is achieved by introducing new coordinates

$$y = \frac{1}{\sqrt{2}}(x_{o}+ix_{1}), \qquad z = \frac{1}{\sqrt{2}}(x_{2}-ix_{3}) \tag{39}$$

$$\bar{y} = \frac{1}{\sqrt{2}}(x_{o}-ix_{1}), \qquad \bar{z} = \frac{1}{\sqrt{2}}(x_{2}+ix_{3}) \tag{40}$$

in terms of which (39) takes the form

$$F_{yz} = F_{\bar{y}\bar{z}} = 0, \tag{41}$$

$$F_{y\bar{y}} + F_{z\bar{z}} = 0. \tag{42}$$

Notice that due to the complexification of the space-time the gauge group is also complexified. Now let us consider the special ansatz [33] for A_{u}, $A_{\bar{u}}$ assuming that $u = y = z$, $\bar{u} = \bar{y} = \bar{z}$.

$$A_{u} = \sum_{\alpha\in M^{+}}(y_{\alpha} \exp[-\sum_{\beta} k_{\alpha\beta}\tilde{\psi}_{\beta}E_{-\alpha} + \partial_{u}\psi_{\alpha} H_{\alpha}), \tag{43}$$

$$A_{\bar{u}} = \sum_{\alpha\in M^{+}}(\tilde{y}_{\alpha} \exp[-\sum_{\beta} k_{\alpha\beta}\psi_{\beta}E_{+\alpha} + \partial_{\bar{u}}\tilde{\psi}_{\alpha} H_{\alpha}), \tag{44}$$

where y_α, $\tilde{y}_\alpha$, ψ_α, $\tilde{\psi}_\alpha$ are functions of z, $\bar{z}$ and $E_{\pm\alpha}$, M_α are the generators of the gauge group which we choose in the Cartan-Chevalle basis [34]. Hence they satisfy the following commutation relations

$$[H_\beta, E_{\pm\alpha}] = \pm k_{\beta\alpha} E_{\pm\alpha} , \tag{45}$$

$$[E_{+\alpha}, E_{-\beta}] = \delta_{\alpha\beta} \; H_\beta . \tag{46}$$

Here M^+ denotes the set of the simple roots of a given Lie algebra of a gauge group and $k_{\alpha\beta}$ is the Cartan matrix. For the SU(N) gauge group it has the form (4) with finite n. Substituting (43-46) after equating the coefficients corresponding to the same generators we obtain

$$y_\alpha = \tilde{y}_\alpha = \exp(-\sum_\beta k_{\alpha\beta}(\psi_\beta+\tilde{\psi}_\beta), \tag{47}$$

$$\partial_{u\bar{u}}(\psi_\alpha + \tilde{\psi}_a) = \exp(-\sum_\beta k_{\alpha\beta}(\psi_\beta+\tilde{\psi}_\beta)). \tag{48}$$

Equation (48) is the Toda lattice in the form (3) with the free endpoints for the SU(N) gauge group. For different groups, we have different Cartan matrices and hence different Toda lattices. These equations can be solved by pure algebraic method [31,33,35] or for the classical semisimple gauge group by the non-auto Liouville-Bäcklund transformation [36]. This name follows from the fact that for the SU(2) gauge group, equation (48) reduces to the Liouville equation which possesses the non-auto Bäcklund transformation which transforms the solutions of the two-dimensional Laplace equation onto the Liouville equation. The Liouville-Bäcklund transformation transforms N-1 solutions of the two-dimensional Laplace equations onto the SU(N) Toda lattice.

On the other hand, equations (41-42) can be solved in a different way. First, let us notice that equation (41) tells that the potentials A_y, $A_{\bar{y}}$, A_z, $A_{\bar{z}}$ are pure gauge. This means that

$$A_y = D^{-1}\partial_y D , \qquad A_{\bar{y}} = \bar{D}^{-1} \partial_{\bar{y}}\bar{D}, \tag{49}$$

$$A_z = D^{-1}\partial_z D , \qquad A_{\bar{z}} = \bar{D}^{-1}\partial_{\bar{z}}\bar{D}, \tag{50}$$

where D, $\bar{D}$ are arbitrary matrices. Introducing the new matrix $R = \bar{D}^{-1}D$ equation (42) reduces to

$$(R^{-1}R_y)_{\bar{y}} + (R^{-1}R_z)_{\bar{z}} = 0. \tag{51}$$

In the special case when $z = \bar{z}$ one can show [32] that equation (51) describes the monopole solutions for the Yang-Mills-Higgs field theory. Surprisingly if we now consider the following equation

$$[R_n^{-1}R_{ny}]_{\bar{y}} = R_{n+1}^{-1} R_n - R_n^{-1}R_{n-1}, \tag{52}$$

where R_n is the matrix-valued function of y, $\bar{y}$ belonging to SL(N,C), then it can be considered as the lattice approximation of the equation (51) for $z = \bar{z}$, what can be easily seen [37] using the Taylor expansion of

$$R_{n\pm 1} = J \pm \varepsilon\partial_z J + \frac{\varepsilon^2}{2!} \pm \ldots . \tag{53}$$

In the special case when

$$R_n = e^{\varphi_n}.1 \tag{54}$$

we obtain the equation (1) the Toda lattice. This generalization of the Toda lattice is known as the nonabelian Toda lattice. Perk and Capel [38] were the first to introduce this concept to physics. They showed that the correlation between x and y components of the spin in the inhomogenous X-Y model can be described by the one-dimensional nonabelian Toda lattice. The present author has shown that this one-dimensional nonabelian Toda lattice can be considered as the lattice approximation of the chiral models [39]. This nonabelian infinite Toda lattice possesses the inverse scattering transformation as well as a Bäcklund transformation [37,39,40]. The multisoliton solutions for the periodic nonabelian Toda lattice can be obtained from the inverse scattering transformation [41] by the use of the so-called "soliton correlation matrix" [42].

In order to show the basic difference between abelian and nonabelian solutions of Toda lattice let us present a straightforward method [43] of constructing the solutions of Toda lattice in both cases. Let us define the Bäcklund transformation and the inverse scattering transformation [37] respectively as

$$R_n^{-1} \partial_y R_n - R'^{-1}_{n-1} \partial_y R'_{n-1} = \alpha \{R'^{-1}_n R_n - R'^{-1}_{n-1} R_{n-1}\}, \tag{55}$$

$$\partial_{\bar{y}} R'_n R'^{-1}_n - \partial_{\bar{y}} R_n R_n^{-1} = \frac{1}{\alpha} \{R'_n R_{n+1} - R'_{n-1} R_n^{-1}\} , \tag{56}$$

where α is an arbitrary parameter, R_n and R'_n are respectively the known and unknown solutions of equation (52),

$$\partial_y \psi(n) = \psi(n+1) - R^{-1}(n)\partial_y R(n)\psi(n), \tag{57}$$

$$\partial_{\bar{y}} \psi(n) = R_n^{-1} R_{n-1} \psi(n-1), \tag{58}$$

$$\partial_{y\bar{y}} \psi(n) = R_{n+1}^{-1} R_n \psi(n) - \partial_{\bar{y}}(R_n^{-1}\partial_y R_n) \psi(n) - R_n^{-1}\partial_y R_n \partial_{\bar{y}} \psi(n), \tag{59}$$

where equation (59) is obtained from (57) and (58) by differentiating (57) by $\partial_{\bar{y}}$ and using (58). These formulae are obtained by the generalization of the corresponding formulae in the abelian Toda lattice. Notice that in (57) and (58) the absence of the so-called spectral parameters. We can introduce these by scaling the matrix function

$$\psi \rightarrow \exp(\mu y + \mu' \bar{y}) , \tag{60}$$

where μ and μ' are quite arbitrary parameters. We are now ready to prove the following.

Theorem: If R_n and $\psi(n)$ satisfy (52) and (57-59) respectively, then

$$R_n' = \alpha R_n \psi(n) \psi^{-1}(n+1), \tag{61}$$

where α is an arbitrary parameter, satisfies (52). The proof is elementary; it is enough to substitute (61) into (52) and use (57-58). In order to generate new solutions from (57-59), we should specify the seed solution R_n and then solve (57-59). First let us consider the abelian case for which we choose

$$R_n = \text{const. } 1, \tag{62}$$

then equations (57-59) read as

$$\partial_y \psi(n) = \psi(n+1), \tag{63}$$

$$\partial_{\bar{y}} \psi(n) = \psi(n-1), \tag{64}$$

$$\partial_{y\bar{y}} \psi(n) = \psi(n), \tag{65}$$

The system (63-65) is well known in physics because it is the special case of the telegraph equation. Kametake [44] studied this system in the context of the soliton solutions of the Toda lattice and the present author has shown [45] that the Kametake method is equivalent to the Bäcklund transformation. Kametake showed that the base of the solutions of (63-65) can be chosen as

$$Z_{n,\lambda} = \lambda^n \exp(\lambda y + \frac{1}{\lambda}\bar{y}), \tag{66}$$

where $\lambda \neq 0$ is an arbitrary constant. The one soliton solution is obtained from (61) where

$$\psi(n) = Z_{n,\lambda} \pm Z_{n,1/\lambda}, \tag{67}$$

and are reduced to the well-known one (anti) soliton solutions for the one-dimensional infinite Toda lattice for $y = \bar{y}$. For the non-abelian case we made the following choice of R_n for the SL(2,C) group

$$R_n = \begin{pmatrix} 1 & i \\ -nS & 1-inS \end{pmatrix} ; \quad \det R_n = 1, \quad s \in R, \tag{68}$$

which satisfies the nonabelian Toda lattice. Indeed

$$R_{n+1}^{-1} R_n = R_n^{-1} R_{n-1} = \begin{pmatrix} 1-is & s \\ s & 1+is \end{pmatrix}. \tag{69}$$

Substituting (68) and (69) into (57-59) one can quickly realize that these equations constitute a system of linear partial differential equations with constant coefficients. We can solve that system using its characteristic equation. In our case this equation is

$$\det \begin{pmatrix} 1-is-\mu & s \\ s & 1+is-\mu \end{pmatrix} = (1-\mu)^2 = 0, \tag{70}$$

and possesses a two-fold root only. According to the general theory of linear partial differential equations the solution can be represented as

$$\psi_{i,j}(n) = (A^0_{i,j}(n)y + A^1_{i,j}(n)\bar{y} + A^2_{i,j}(n))e^{\lambda_i y + \frac{1}{\lambda_i}\bar{y}}, \tag{71}$$

where λ_i are arbitrary constants and $A^k_{i,j}$, $k = 0,1,2$, $i,j = 1,2$ are coefficients which can be determined from the assumption that $\psi(n)$ satisfies (57-58), from the fact that $R'_n \in SL(2,C)$ and from the boundary conditions. Let us compare the solutions (71) with (66). We see that they coincide if $s = 0$. For $s \neq 0$ we obtain typical nonabelian solutions. Similarly to the abelian case we can construct the linear combinations of (71). The matrix R'_n obtained in this way I named in [43] as the nonabelian one (anti) solitons.

Let us return once more to the equations (1-6). From the soliton theory point of view this equation can be considered as the model of solitons in one discrete and in one or two continuous variables. One can ask whether it is possible to construct solitons in two discrete variables. Recently, Ablowitz et al. [46] considered the discretization of the so-called "∂" (DBAR) problem. They considered the finite-differential analogue of the Schrödinger equation in the form

$$\psi(n-1,m) + B(n,m)\psi(n,m) + A(n,m)\psi(n+1,m) = \lambda\psi(n,m+1), \qquad (72)$$

where $(n,m) \in Z^2$ and the "potentials" B, A-1 vanish sufficiently fast as n and/or m goes to infinity. Due to it they obtained the Toda lattice equation in two discrete variables and in one continuous as the compatibility conditions between (72) and

$$\frac{d}{dt}\,\psi(n,m) = -\lambda^{-1}\,\psi(n,m) + g^{o}(n,m)\psi(n+1,m), \qquad (73)$$

$$g^{o}(n,m) = \prod_{i=0}^{\infty} A(n-j,m+j)/A(n-1-j,m+j), \qquad (74)$$

which reads as

$$\frac{d}{dt}A(n,m) = B(n+1,m)g^{o}(n,m+1) - B(n,m)g^{o}(n,m), \qquad (75)$$

$$\frac{d}{dt}B(n,m) = g^{o}(n,m+1) - g^{o}(n-1,m). \qquad (76)$$

This is the next possible way of generalization of the Toda lattice. From (75-76) we recover the Toda lattice (5-6) by assuming that A and B do not depend on m. Recently, Ragnisco and the present author used in [47] the concept of Darboux transformation and obtained some new formulae for the solutions of the equation (75-76). Let us briefly demonstrate how it works. The concept of Darboux transformation can be applied for the explicit integration of linear evolution equations with scalar or matrix-valued coefficients. We define the Darboux transformation for (75-76) by

$$\tilde{\psi}(n,m) = \psi(n,m) - \frac{f(n,m)}{f(n+1,m)}\,\psi(n+1,m) \qquad (77)$$

$$\tilde{B}(n,m) = B(n,m) - \frac{f(n,m+1)}{f(n+1,m+1)} + \frac{f(n-1,m)}{f(n,m)} \qquad (78)$$

$$\tilde{A}(n,m) = \frac{f(n,m+1)f(n+2,m)}{f(n+1,m)f(n+1,m+1)}\,A(n+1,m), \qquad (79)$$

where $f(n,m)$ is a particular solution of equations (72-73) with $\lambda = \lambda_1$.

To see that $\tilde{\psi}$, $\tilde{A}$ and $\tilde{B}$ satisfy the equations (72-73) let us substitute the formula (77-79) to this equation and use the fact that f, A, B, satisfy (72-73) also. Notice that in the special case when we assume that A, B, ψ do not depend on m we recover the Darboux transformation for the Toda lattice (5-6). This Darboux transformation is different from the transformation in the paper of Matveev [48] because he used a different representation for the so-called L-A pair. Notice that the concept of Darboux transformation is similar to the concept of Bäcklund transformation. Indeed, from the knowledge of the Darboux transformation one can recover the Bäcklund transformation for the Toda lattice (5-6) assuming additionally that

$$\frac{f(n)}{f(n+1)} = - \alpha \quad \exp[\varphi'(n+1) - \varphi(n)], \tag{80}$$

where α is an arbitrary constant and $\varphi'(n)$ and $\varphi(n)$ are old and new solutions of the Toda lattice. To see this, it is enough to substitute (80) with (77) into (72-73) assuming that m dependence does not appear and equate the coefficient standing on the same $\psi(n)$.

Concluding this brief review let us recapitulate what we considered here. Firstly, we considered the applications of the concept of the Toda lattice in the context of the eighteenth and nineteenth century mathematics. Secondly, we tried to classify the different generalizations of the Toda lattice which appear in mathematics and physics. We paid special attention to the application of Toda lattice to the gauge theory. However, our classification does not exhaust this subject and the scope of this paper does not allow us to consider other interesting generalizations of the Toda lattice. Interested readers may refer to the papers [49-53] where this problem is considered.

References

[1] M. Toda, J. Phys. Soc. Japan **22** (1967) 431.
[2] M. Toda, Phys. Rep. **18C** (1975) 1.
[3] E. Fermi, J. Pasta, S. Ulam, **Collected Papers of Enrico Fermi**, Vol. II (University of Chicago Press, 1965), p. 978.
[4] M. Wadati, Prog. Theor. Phys. Supp. **59** (1976) 36.
[5] Y. Kametaka, Proc. Japan Acad. **60A** (1984) 145.
[6] Y. Kametaka, RIMS Kokyuroko **554** (1985) 26.
[7] H. Au-Yang and J. H. H,Perk,Physica**18D** (1986) 365.
[8] A. Common & D. Roberts, "Solutions of the Riccati equation and their relations to the Toda lattice," (1986) Preprint.
[9] G. Darboux, **Lecons sur La theorie generale des surfaces et les applications geometrique du calcul infinitesimal** II (Chelsea,1972).

[10] E. Ince, **Ordinary Differential Euqations** (New York: Dover, 1956).
[11] J. Sylvester, Compt. Rend. Acad. Sci. **54**, No. 129 (1862) 170.
[12] H. Wall, **Analytic Theory of Continued Fraction** (New York: Van Nostrand, 1948).
[13] S. Manakov, Soviet Phys. JETP **40** (1974) 269.
[14] H. Flaschka, Prog. Theor. Phys. **51** (1974) 703.
[15] V. Zakharov, S. Manakov, P. Novikov, and A. Pytaevsky, **The Theory of Solitons** (Moscow: Nauka, 1980).
[16] R. Hirota, J. Phys. Soc. Japan **35** (1973) 286.
[17] B. Kostant, Adv. Math. **34** (1979) 159.
[18] M. Olshanetsky and A. Perelomov, Theor. Math. Phys. **45** (1980) 3.
[19] A. Mikhailov, JETP Letters **30** (1979) 443.
[20] A. Fordy and J. Gibbons, Commun. Math. Phys. **77** (1980) 21.
[21] B. Barbashov and V. Nesterenko, Commun. Math. Phys. **78** (1981) 499.
[22] K. Ueno, K. Takasaki, Proc. Japan Acad. **59** (1983) 168, 216.
[23] M. Jimbo, and T. Miwa, Publ. RIMS. Kyoto **19** (1983) 943 .
[24] K. Ueno, and K. Takasaki, "Toda Lattice Hierarchy" RIMS 425 Kyoto, Japan (1983).
[25] M. Olshanetsky, Commun. Math. Phys. **88** (1983) 63.
[26] Z. Popowicz, J. Phys. A. Math. Gen. **19** (1986) 1495.
[27] O. Bogoyavlensky, Commun. Math. Phys. **51** (1979) 201.
[28] A. Leznov and M. Saveliew, Lett. Math. Phys. **3** (1979) 489.
[29] A. Leznov and M. Saveliew, Phys. Elem. Part A: Nuclei. **11** (1980)40.
[30] A. Leznov, Theor. Math. Phys. **42** (1980) 343.
[31] A. Leznov, and M. Saveliew, **The Group's Method of Integrations of the Nonlinear Dynamical Systems** (Moscow: Nauka, 1985).
[32] M. Prasad, Physica **2D** (1980) 167.
[33] R. Farwell and M. Minami, J. Phys. A. Math. Gen. **15** (1982) 25.
[34] J. Humphreys, **Introduction to Lie Algebras and Representation Theory** (New York: Springer, 1972).
[35] N. Ganoulis, P. Goddard and D. Olive, Nucl. Phys. **B205 FS5** (1982) 601.
[36] Z. Popowicz, J. Math. Phys. **25** (1984) 2212.
[37] Z. Popowicz, Z. Phys. Part and Fields **19** (1983) 79.
[38] J. Perk and M. Capel, Physica **92** (1978) 1563.
[39] Z. Popowicz, Phys. Lett. **81A** (1981) 235.
[40] D. Levi, L. Pinolli and P. Santini, J. Phys. A. Math. Gen. **14** (1981) 1567.
[41] A. Mikhailov, Physica **3D** (1981) 73.
[42] J. Harnad, Y. Saint-Aubin, and S. Shnider, Commun. Math. Phys. **93** (1984) 33.
[43] Z. Popowicz, "The nonabelian solitons for the SL(2, C) nonabelian Toda lattice" to appear in Inverse Problems.
[44] Y. Kametaka, Proc. Japan Acad. **60** (1984) 79.
[45] Z. Popowicz, Phys. Lett. **116A** (1986) 255.
[46] M. Ablowitz, S. Chitlaru Briggs, O. Ragnisco and P. Santini, "An example of $\bar{\partial}$ problem in a finite difference context: Direct and inverse problem for the discrete analogue of the equation $\psi_{xx} + u\psi = \sigma\psi_y$, to appear in J. Math. Phys.
[47] Z. Popowicz, and O. Ragnisco, "Darboux transformation and one-soliton solution for the two-dimensional Toda lattice" preprint Universita di Roma "La Sapienza" (1987) (submitted).
[48] V. Matveev, and M. Salle, Lett. Math. Phys. **3** (1979) 425.
[49] A. Nakamura, J. Phys. Soc. Japan **52** (1979) 425.
[50] N. Saitoh, E. Takizawa and S. Takeno, J. Phys. Soc. Japan **54** (1985) 3701, 4525.
[51] N. Saitoh, J. Phys. Soc. Japan **54** (1985) 3261.
[52] M. Leo, R. Leo, G. Solani and L. Solombrino, Lett. Math. Phys. **8** (1984) 267.
[53] Li Yi-Shen, Cheng Yi, "Infinite dimensional Lie Algebra and a theory on the symmetries and constants of motion in 1+1 dimensional systems" preprint.

Construction of Exact Invariants for One- and Two-Dimensional Classical Systems

R.S. Kaushal

Department of Physics, Ramjas College, University of Delhi,
Delhi 110007, India

A general method to construct the invariants of both time independent and time dependent one- and two-dimensional classical system is outlined by complexifying the two dimensions.

1. INTRODUCTION

If all possible constants of motion (briefly called invariants) of a dynamical system are known or can be constructed, then it becomes easier to study or to predict the behaviour of the system rather completely, and the system is said to be integrable [1]. Since there exists scarcity of integrable systems in contrast to nonintegrable systems, in recent years, there has been considerable interest in exploring the methods of construction of these invariants for both time-dependent (TD) [2-12] and time independent (TID) systems [13-16,18].

These invariants play an important role in various disciplines, particularly in plasma physics, hydrodynamics, astrophysics, biophysics and in quantum field theory. In fact, nonlinear equations involving higher degrees of freedom can be reduced to quadrature from the knowledge of these invariants. Besides their mathematical usefulness in solving a problem, the physical interpretations of these invariants have not yet been found for most of the systems. The most-studied example [2-7] is that of one-dimensional TD harmonic oscillator. For this system, the Hamiltonian

$$H = \frac{1}{2} p^2 + \frac{1}{2} \omega^2(t) q^2, \tag{1}$$

is not the constant of motion, and the invariant is found to be

$$I = \frac{1}{2}\left[\frac{q^2}{\rho^2} + (\rho p - \dot{\rho} q)^2\right], \tag{2}$$

where ρ satisfies an auxiliary equation

$$\ddot{\rho} + \omega^2(t)\rho = \rho^{-3}. \tag{3}$$

This is perhaps the only example for which some plausible physical interpretations of I have been sought. Eliezer and Gray [6] argue that I is equivalent to the angular momentum in a projected two-dimensional plane in which the radial equation is given by the auxiliary Eq. (3). Takayama [6] relates I and subsequently the auxiliary Eq. (3) to the betatron oscillations about the equilibrium orbits. In the present talk, my main emphasis will be on the construction aspects of these invariants for one- and two-dimensional classical systems.

2. ONE-DIMENSIONAL TD SYSTEMS

After the work of Lewis [2] and Lewis and Riesenfeld [2] several other methods have been developed for the construction of invariants for one-dimensional TD systems. Although most of these methods deal with the invariants upto second order in momenta (this is perhaps to maintain an analogy with the Hamiltonian of the system), the existence of higher order invariants cannot be ruled out.

2.1 Second Order Invariants

Various approaches used for the construction of second order invariants for the system

$$L = \frac{1}{2}\dot{q}^2 - V(q,t) \tag{4}$$

can broadly be classified as

(a) Ermakov approach as used by Ray and Reid [7];
(b) Lutzky's approach using Noether's theorem [5];
(c) Transformation group approach of Burgan et al. [3]; and
(d) Dynamical Lie algebraic approach [4].

While the approach (a) is more heuristic than the other three, it assumes the form of auxiliary equation in advance. The approaches (b) and (c) are rather involved and also used for solving a class of nonlinear differential equations [5,17]. The approach (d), based on the closure property of Lie algebra of phase-space functions has relatively more transparent connection with the corresponding quantum systems. More general cases in which V depends on momentum and also being non-separable can easily be dealt within this approach.

2.2 Higher Order Invariants

Recently, we have carried out [9] the construction of invariants of third and fourth order in momenta for the system (4), by assuming the form of I as

$$I = b_0 + b_1 q + \frac{1}{2!} b_2 q^2 + \frac{1}{3!} b_3 q^3 + \frac{1}{4!} b_4 q^4, \tag{5}$$

where b_i's are the functions of q and t. The fact that I satisfies $\frac{dI}{dt} = 0$, leads to a set of coupled partial differential equations

$$\left.\begin{aligned} &\frac{\partial b_4}{\partial q} = 0 \; ; \quad 4\frac{\partial b_3}{\partial q} + \frac{\partial b_4}{\partial t} = 0, \\ &3\frac{\partial b_2}{\partial q} + \frac{\partial b_3}{\partial t} = b_4 \frac{\partial V}{\partial q} \; ; \quad 2\frac{\partial b_1}{\partial q} + \frac{\partial b_2}{\partial t} = b_3 \frac{\partial V}{\partial q}, \\ &\frac{\partial b_0}{\partial q} + \frac{\partial b_1}{\partial t} = b_2 \frac{\partial V}{\partial q} \; ; \quad \frac{\partial b_0}{\partial t} = b_1 \frac{\partial V}{\partial q} . \end{aligned}\right\} \tag{6}$$

By method of elimination of the coefficients in (6) nonlinear differential equations in V (henceforth called 'potential' equations) can be obtained separately for third and fourth order invariants. Further assuming the separability of V in q and t variables, these potential equations are reduced to simpler forms. The system,

$$V(q,t) \sim t^{-4/3} q^{1/2}, \tag{7}$$

is found to be integrable and admits [9] a third order invariant of the form

$$I = -\frac{1}{2}[f - (\tfrac{1}{3}c_3 q - c_4)^{1/2} q]^2 + \frac{1}{6}\psi_1 q^3 . \tag{8}$$

Other integrable systems can be found by solving these potential equations.

3. TWO-DIMENSIONAL SYSTEMS

Several methods have been developed [10-16] and applied to the study of two-dimensional systems. While the third and fourth order invariants in momenta are constructed [13] for TID systems, construction of invariants for TD systems beyond the second order seems to be not pursued that vigorously. Here we outline a method which can uniformly be used for the study of both TID and TD systems by complexifying [16] the two-dimensions. The method has not only led to some simplifications in the construction of invariants but also provided with several new integrable systems in two-dimensions.

3.1 Time Independent Systems

We consider a dynamical system described by the Lagrangian,

$$L = \frac{1}{2} \dot{z} \dot{\bar{z}} - V(z,\bar{z}) \tag{9}$$

with $z = q_1+iq_2$, $\bar{z} = q_1-iq_2$, $\ddot{z} = -2 \frac{\partial V}{\partial z}$, $\ddot{\bar{z}} = -2 \frac{\partial V}{\partial \bar{z}}$.

We assume the second constant of motion (the first being the Hamiltonian itself for the TID case) upto fourth order in momenta in the form

$$I = a_0 + a_i\xi_i + \frac{1}{2!} a_{ij}\xi_i\xi_j + \frac{1}{3!} a_{ijk}\xi_i\xi_j\xi_k + \frac{1}{4!} a_{ijkl}\xi_i\xi_j\xi_k\xi_l, \tag{10}$$

where $i,j,k,l = 1,2$; $\xi_1 = \dot{z}$ and $\xi_2 = \dot{\bar{z}}$ and the coefficients a_0, a_i, a_{ij}, a_{ijk}, and a_{ijkl} are functions of z and $\bar{z}$ only. These coefficients are symmetric with respect to any interchange of their indices. The fact that $\frac{dI}{dt} = 0$, implies an expression in $\dot{z}$, $\dot{\bar{z}}$ which must vanish identically, viz.,

$$a_{0,i}\xi_i + a_{i,j}\xi_i\xi_j + a_i\dot{\xi}_i + \frac{1}{2} a_{ij,k}\xi_i\xi_j\xi_k + \frac{1}{2} a_{ij}(\dot{\xi}_i\xi_j + \xi_i\dot{\xi}_j)$$
$$+ \frac{1}{6}a_{ijk,\ell}\xi_i\xi_j\xi_k\xi_\ell + \frac{1}{6} a_{ijk}(\dot{\xi}_i\xi_j\xi_k + \xi_i\dot{\xi}_j\xi_k + \xi_i\xi_j\dot{\xi}_k)$$
$$+ \frac{1}{24} a_{ijk\ell,m}\, \xi_i\xi_j\xi_k\xi_\ell\xi_m + \frac{1}{24} a_{ijk\ell}(\dot{\xi}_i\xi_j\xi_k\xi_\ell + \xi_i\dot{\xi}_j\xi_k\xi_\ell + \xi_i\xi_j\dot{\xi}_k\xi_\ell +$$
$$\xi_i\xi_j\xi_k\dot{\xi}_\ell) = 0. \tag{11}$$

Now, after accounting for the proper symmetrisation of the coefficients, we obtain the following conditions on a_{ijkl}, a_{ijk}, a_{ij}, a_i and a_0:

$$a_{ijkl,m} + a_{jklm,i} + a_{klmi,j} + a_{lmij,k} + a_{mijk,l} = 0, \tag{12a}$$

$$a_{ijk,l} + a_{jkl,i} + a_{kli,j} + a_{lij,k} = 0, \tag{12b}$$

$$a_{ij,k} + a_{jk,i} + a_{ki,j} + a_{ijkl}\dot{\xi}_1 = 0, \tag{12c}$$

$$a_{i,j} + a_{j,i} + a_{ijk}\dot{\xi}_k = 0, \tag{12d}$$

$$a_{0,i} + a_{ij}\dot{\xi}_j = 0, \tag{12e}$$

$$a_i\dot{\xi}_i = 0. \tag{12f}$$

From these equations it can be seen that the coupling between the coefficients corresponding to even and odd powers of momenta in (10) does not arise. As a result, the number of equations to be solved simultaneously for the third and fourth order invariants is nine and twelve, respectively. To solve these equations we adopt the following procedure:

(a) Equations which are free from potential terms in the set (12) are solved explicitly for the coefficients involved and with some arbitrary constants;

(b) Equations involving the potential terms are reduced to a couple of 'potential' equations by method of elimination;

(c) Most of the arbitrary constants which arise in the step (a), can be fixed either by solving or by rationalising these potential equations.

(d) Using these values of the arbitrary constants, various coefficients in (10) can be determined which in turn provide the explicit structure of I.

The method is successfully applied [15] to a large class of potentials (including the Toda, Holt and Fokas potentials and also a class of Toda-type potentials) which admit the third order invariants. The potential equation corresponding to the fourth order invariants requires further investigation.

3.2 Time-Dependent Systems

If we restrict upto the third order invariants ($a_{ijkl} = 0$), and the potential V and the coefficients a_0, a_i, a_{ij}, a_{ijk} in Eqs. (9) and (10) are allowed to depend explicitly on t, then the expression (11) takes the form,

$$\left(\frac{\partial a_0}{\partial t} + a_i\dot{\xi}_i\right) + \left(a_{0,i} + \frac{\partial a_i}{\partial t} + \frac{1}{2} a_{ij}\dot{\xi}_j + \frac{1}{2} a_{ji}\dot{\xi}_j\right)\xi_i +$$

$$\left(a_{i,j} + \frac{1}{2}\frac{\partial a_{ij}}{\partial t} + \frac{1}{6} a_{kji}\dot{\xi}_k + \frac{1}{6} a_{ikj}\dot{\xi}_k + \frac{1}{6} a_{ijk}\dot{\xi}_k\right)\xi_i\xi_j +$$

$$\left(\frac{1}{2} a_{ij,k} + \frac{1}{6}\frac{\partial a_{ijk}}{\partial t}\right)\xi_i\xi_j\xi_k + \frac{1}{6} a_{ijk,\ell}\xi_i\xi_j\xi_k\xi_\ell = 0. \tag{13}$$

This expression, after accounting for the proper symmetrisation of the coefficients as before, will now yield [12] the following restrictions on a_0, a_i, a_{ij} and a_{ijk}:

$$a_{ijk,l} + a_{jkl,i} + a_{kli,j} + a_{lij,k} = 0, \tag{14a}$$

$$a_{ij,k} + a_{jk,i} + a_{ki,j} + \frac{\partial a_{ijk}}{\partial t} = 0, \quad a_{i,j}+a_{j,i}+ \frac{\partial a_{ij}}{\partial t} + a_{ijk}\dot{\xi}_k = 0, \tag{14b}$$

$$a_{0,i} + \frac{\partial a_i}{\partial t} + a_{ij}\dot{\xi}_j = 0; \quad \frac{\partial a_0}{\partial t} + a_i\dot{\xi}_i = 0. \tag{14c}$$

In contrary to the TID case, here the coupling between the coefficients corresponding to the even and odd powers of momenta in (10) does exist. Subsequently, the number of equations to be satisfied simultaneously for a given order of invariant is much more as compared to the corresponding TID case. This complicates the construction of invariants for TD systems. Note that for the second order case ($a_{ijk} = 0$), the set (14) reduces to ten equations whereas for the third order it provides fifteen equations. To solve these equations we follow the same procedure as described before. However, the aribtrary constants which arise in the step (a) now turn out to be the arbitrary functions of t and the same can be fixed in terms of potential parameters.

Using the complexification method in the context of TD systems a large number of integrable systems which admit the second order invariants are found [11]. Recently, Eqs. (14) are used [12] to obtain a third order invariant for a TD Henon-Heiles system,

$$V(q_1,q_2,t) = \tfrac{1}{2}\alpha(t)(q_1^2 + q_2^2) + \beta(t)(q_1^2 q_2 - \tfrac{1}{3}q_2^3). \tag{15}$$

The invariant turns out to be

$$I = \tfrac{1}{2}c_4\dot{\alpha}q_1(q_2^2 - \tfrac{1}{3}q_1^2) + \tfrac{1}{2}c_4\alpha(p_1q_1^2 - p_1q_2^2 - 2p_2q_1q_2)$$
$$+ (q_2p_1-q_1p_2)[c_4\beta_0(q_1^2+q_2^2) + \tfrac{4}{3}c_3p_1p_2] + c_4p_1(\tfrac{1}{3}\, p_1^2-p_2^2),$$

where c_3, c_4 and β_0 are arbitrary real constants and α satisfies, $\ddot{\alpha} = -3\alpha^2$.

A part of this work has been carried out in collaboration with Prof. K. C. Tripathy and Dr S. C. Mishra.

REFERENCES

[1] E. T. Whittaker, **A Treatise on the Analytical Dynamics of Particles and Rigid Bodies** (Cambridge Univ. Press, London, 1965).
[2] H. R. Lewis Jr., J. Math. Phys. **9** (1968) 1976; H. R. Lewis and W. B. Riesenfeld, J. Math. Phys. **10** (1969) 1458.
[3] J. R. Burgan et al., Phys. Lett. **74A** (1979) 11.
[4] H. J. Korsch, Phys. Lett. **74A** (1979) 294; R. S. Kaushal and H.J. Korsch, J. Math. Phys. **22** (1981) 1904.
[5] J. R. Ray and J. L. Reid, J. Math. Phys. **20** (1979) 2054; Phys. Lett. **74A** (1979) 23.
[6] C. J. Eliezer and A. Gray, SIAM J. Appl. Math. **30** (1976) 463; K. Takayama, Phys. Lett. **88A** (1982) 55.
[7] J. R. Ray and J. L. Reid, J. Math. Phys. **23** (1982) 503.
[8] P. G. L. Leach, J. Math. Phys. **22** (1981) 465; Phys. Lett. **98A** (1983) 89.
[9] R. S. Kaushal, Pramana J. Phys. **24** (1985) 663.
[10] G. N. Katzin and J. L. Levine, J. Math. Phys. **24** (1983) 1761; B. Grammaticos and B. Dorizzi, J. Math. Phys. **25** (1984) 2194.
[11] R. S. Kaushal, S. C. Mishra and K. C. Tripathy, Phys. Lett. **102A** (1984) 7; S. C. Mishra, R. S. Kaushal and K. C. Tripathy, J. Math. Phys. **25** (1984) 2217.
[12] R. S. Kaushal, "Third Order Invariants for Two-Dimensional Time-Dependent Classical Systems: Henon-Heiles System," to be published.
[13] O. R. Holt, J. Math. Phys. **23** (1982) 1037; L. S. Hall, Physica **D8** (1983) 90; T. Sen, Phys. Lett. **111A** (1985) 97; B. Dorizzi et al., J. Math. Phys. **25** (1984) 2200.
[14] M. J. Ablowitz, A. Ramani and H. Segur, J. Math. Phys. **21** (1980) 715; 1006; A. S. Fokas and P. A. Lagerstrom, J. Math. Ana. Appln. **74** (1980) 325.
[15] R. S. Kaushal, S. C. Mishra and K. C. Tripathy, J. Math. Phys. **26** (1985) 420; R. S. Kaushal and S. C. Mishra, Pramana J. Phys. **26** (1986) 109.
[16] S. C. Mishra, Ph.D. Thesis, Delhi University (1985).
[17] See e.g.,P.B.Burt and J.L.Reid,J.Math.Ana.Appln. **55** (1976) 43.
[18] R. Sahadevan and M. Lakshmanan, Phys. Rev. **33A** (1986) 3563; R. Sahadevan, Ph.D. Thesis, University of Madras (1986).

Nonlinear Chains and Kink-Impurity Interactions

Bishwajyoti Dey
Institute of Physics, Bhubaneswar 751 005, India

One-dimensional monatomic and diatomic chains with harmonic coupling between neighbouring sites and an on-site anharmonic potential $V(\varphi) = C\varphi^{2n+2} + B\varphi^{n+2} + A\varphi^2 + D$ are examined in the displacive limit, which serves as a model for a structural phase transition. It is shown that these systems admit kinks, nonlinear phonons and periodon solutions. The eigenfunctions of the small oscillations about the kink solutions are also obtained. This complete set of eigenfunctions has been used to investigate the influence of impurities on the amplitude kink for $n = 1$ case of the potential. The kink solution change near the impurity is evaluated with the help of linear perturbation theory.

In recent years one-dimensional field theoretic models exhibiting solitary-wave solutions (kink, solitons, non-linear phonons) have played an important role in condensed matter physics, as these provide a non-perturbative approach to strongly anharmonic systems. These models have been used to describe many systems, especially those that undergo a structural phase transition. For example, the domain wall (kink) solutions of $\lambda\varphi^4$ field theoretic models are identified with the central peak phenomena observed in ferroelectric crystals [1].

Here we consider a model of a one-dimensional lattice with harmonic coupling between neighbouring lattice sites and an on-site potential

$$V(\varphi) = C\varphi^{2n+2} + B\varphi^{n+2} + A\varphi^2 + D \tag{1}$$

with adjustable non-linearity for $n = 1,2,3,\ldots$. This potential has triply degenerate minima for even values of n and doubly degenerate minima for odd values of n, and for

$$B^2 = 4AC \quad \text{with} \quad A, C > 0 \text{ and } B < 0. \tag{2}$$

This model describes a first-order structural phase transition at zero temperature. We consider the model Hamiltonian for the monatomic chain [2],

$$H = \sum_i \left(\frac{1}{2} m\dot{\varphi}_i^2 + V(\varphi_i)\right) + \frac{1}{2} \sum_{i,j} C_{ij}(\varphi_i - \varphi_j)^2, \tag{3}$$

where the on-site potential $V(\varphi)$ is given by equation (1). The continuum limit is

$$H = \int_{-\infty}^{\infty} \frac{dx}{a}\left[\frac{1}{2}m\left(\frac{\partial\varphi}{\partial t}\right)^2 + \frac{1}{2}mC_0^2\left(\frac{\partial\varphi}{\partial x}\right)^2 + V(\varphi)\right] \tag{4}$$

and the equation of motion is

$$m\left(\frac{\partial^2\varphi}{\partial t^2}\right) - mC_0^2\left(\frac{\partial^2\varphi}{\partial x^2}\right) + \frac{dV}{d\varphi} = 0 , \tag{5}$$

where a is lattice constant and C_0, the velocity of sound. This equation can be readily integrated to give the kink and anti-kink solutions as

$$\varphi_k = 2^{-1/n} \varphi_0\left(1 \pm \tanh \frac{n\xi}{2}\right)^{1/n} \;;\; \text{for } n = 1,3,5,... \tag{6a}$$

$$= \pm 2^{-1/n} \varphi_0\left(1 \pm \tanh \frac{n\xi}{2}\right)^{1/n};\; \text{for } n = 2,4,6,... \tag{6b}$$

where $\varphi_0 = (2A/B)^{1/n}$; $\xi = (x-vt)/\xi_0$; $\xi_0^2 = m(C_0^2 - v^2)/2A$.
The solutions are stable and have finite energy.

For the diatomic chain we take a model Hamiltonian

$$H = \sum_i \left[\frac{1}{2}M_1\dot{u}_i^2 + \frac{1}{2}M_2\dot{v}_i^2 + \frac{1}{2}\gamma(u_i-v_{i-1})^2 + \frac{1}{2}\gamma(u_i-v_i)^2 + V(u_i)\right] , \tag{7}$$

where u_i and v_i are the displacements of masses M_1 and M_2 respectively, γ is the harmonic force constant between neighbouring atoms and V(u) is the on-site potential represented by equation (1). The coupled equations of motion are given by

$$M_1\ddot{u}_n = -\gamma(2u_n - v_{n-1} - v_n) - V'(u_n) \tag{8a}$$

$$M_2\ddot{v}_n = -\gamma(2v_n - u_n - u_{n+1}) \tag{8b}$$

The continuum limit of the Hamiltonian is

$$H = \int_{-\infty}^{\infty} \frac{dx}{a}\left[\frac{1}{2}M_1\dot{u}^2 + \frac{1}{2}M_2\dot{v}^2 + \gamma(u-v)^2 + \frac{1}{4}\gamma a^2 u'v' + V(u)\right] \tag{9}$$

and gives the coupled fields equations

$$\begin{pmatrix} \frac{1}{4}\gamma a^2 & 2\gamma \\ M_2 C^2 & 2\gamma \end{pmatrix} \begin{pmatrix} \frac{d^2 g}{dS^2} \\ g \end{pmatrix} = \begin{pmatrix} M_1 C^2 \frac{d^2 f}{dS^2} + 2\gamma f + \frac{dV}{df} \\ \frac{1}{4}\gamma a^2 \frac{d^2 f}{dS^2} + 2\gamma f \end{pmatrix}, \qquad (10)$$

where $u(x,t) = f(S)$; $v(x,t) = g(S)$ and $S = x - Ct$.

For a particular velocity

$$C = C_0 = (\gamma a^2/4M_2)^{1/2} \qquad (11)$$

we get kink and antikink solutions ($M_2 > M_1$ and $D = 0$) as

$$u_k(x,t) = v_k(x,t) = 2^{-1/n} u_0 [1 \pm \tanh(\frac{nS}{2\xi_k})]^{1/n}, \text{ for } n = 1,3,5,... \qquad (12a)$$

$$= \pm 2^{-1/n} u_0 [1 \pm \tanh(\frac{nS}{2\xi_k})]^{1/n}, \text{ for } n = 2,4,6,... \qquad (12b)$$

where $u_0 = (2A/B)^{1/n}$ and $\xi_k = [C_0^2(M_2 - M_1)/2A]^{1/2}$.

These solutions are also stable and have finite energy.

For $C \neq C_0$ (equation (11)) we can obtain the nonlinear periodic solutions. For $n = 1$ case the solutions of equation (10) are [2],

$$f(S) = \frac{1}{2}(1 + f_0 \sin kS) \qquad (13a)$$

and

$$g(S) = \frac{1}{2} + g_1 \sin kS + g_2 \sin^3 kS, \qquad (13b)$$

where the amplitudes f_0, g_1 and g_2 are determined in terms of the parameters and $k = (2\gamma/9M_2C^2)^{1/2}$. The periodon solutions which solve the discrete equations of motion (equations (8)), can be expressed as (for $n = 1$ case of potential)

$$u_n = \frac{1}{2}[1 + F_0 \sin(\omega t - n\,a)], \qquad (14a)$$

$$v_n = \frac{1}{2} + G_1 \sin[\omega t - (n+\tfrac{1}{2})\,a] + G_2 \sin^3[\omega t - (n+\tfrac{1}{2})\,a]. \qquad (14b)$$

One can also obtain kink and antikink solutions for a model Hamiltonian generalised by including a single well on-site potential on the M_2 species of atoms [2].

The eigenfunctions and eigenvalues of the small oscillations about the static kink solutions (equations (6)) is obtained from the equation [4]

$$mC_0^2 \frac{d^2\psi(x)}{dx^2} + \left(m\omega^2 - \frac{d^2V(\varphi)}{d\varphi^2}\bigg|_{\varphi=\varphi_k} \right)\psi(x) = 0 , \tag{15}$$

where

$$\frac{d^2V}{d\varphi^2}\bigg|_{\varphi=\varphi_k} = A[n^2 + 1 - \tfrac{1}{2}(n+1)(2n+1)\mathrm{sech}^2 \frac{nx}{2\xi_0} + (n+1)(n-1)\tanh \frac{nx}{2\xi_0}]. \tag{16}$$

The number of bound states is given by the relation [5]

$$n_b = 0,1,2,\ldots < \frac{1}{n}[n+1 - \sqrt{n^2 - 1}]. \tag{17}$$

Thus for n = 1, there are two bound states corresponding to $n_b = 0$ and 1. There is only one bound state for all other values of n. The eigenfunctions of the bound states are given in terms of the hypergeometric series $F(a,b,|c|p)$ as $(K = nx/2\xi_0)$

$$\psi_m = N \frac{\exp(-a_m^K)}{(e^K+e^{-K})^{b_m}} F(-m, 2[\frac{(n+1)(2n+1)}{n^2} + \tfrac{1}{4}]^{\frac{1}{2}} - m|a_m+b_m+1|\frac{e^{-K}}{e^K+e^{-K}}), \tag{18}$$

where

$$a_m = \frac{2(n+1)(n-1)}{n(3n+2)-2(m+\frac{1}{2})} \tag{19}$$

and

$$b_m = \frac{(3n+2)}{2n} - (m + \tfrac{1}{2}). \tag{20}$$

The frequency of the bound states is given by

$$\omega_m^2 = \frac{An^2}{2m}\left[\frac{2(n^2+1)}{n^2} - \left(\frac{3n+2}{2n} - (m+\tfrac{1}{2})\right)^2 - \frac{(n+1)^2(n-1)^2}{n^4\left(\frac{3n+2}{2n} - (m+\frac{1}{2})\right)^2}\right]. \tag{21}$$

The continuum states are

$$\psi_{k_+} = N \exp[\tfrac{1}{2}i(k_+ - k_-)K]\,(e^K+e^{-K})^{\frac{1}{2}i(k_+ + k_-)} \times$$

$$F(-\tfrac{1}{2}ik_+ - \tfrac{1}{2}ik_- + \tfrac{1}{2} - \gamma, -\tfrac{1}{2}ik_+ - \tfrac{1}{2}ik_- + \tfrac{1}{2} + \gamma|1-ik_+|\frac{e^{-K}}{e^K+e^{-K}}), \tag{22}$$

where $k_-^2 = k_+^2 + 4 - 4/n^2$ and the frequency of the continuum states can be expressed in terms of the wave vector k_+ as

$$\omega_{k_+}^2 = (2 + \frac{1}{2}k_+^2)\frac{An^2}{m}. \tag{23}$$

For n = 1 case only the hypergeometric case can be terminated and the continuum states can be expressed in terms of hypergeometric functions. We use this complete set of states for n = 1 case of the potential to calculate the kink impurity interaction, using the linear perturbation theory.

The model Hamiltonian for kink-impurity interaction is taken as[6],

$$H = \int_{-\infty}^{\infty}\frac{dx}{a}[\frac{1}{2}m(\frac{\partial\varphi}{\partial t})^2 + \frac{1}{2}mC_0^2(\frac{\partial\varphi}{\partial x})^2 + V(\varphi) + \frac{1}{2}V(x)\varphi^2], \tag{24}$$

where

$$V(x) = \alpha\delta(x-x_p) \tag{25}$$

describes the short range kink-impurity interaction potential. The equation for $\varphi(x,t)$ is

$$m\frac{\partial^2\varphi}{\partial t^2} - mC_0^2\frac{\partial^2\varphi}{\partial x^2} + 4C\varphi^3 - 3|B|\varphi^2 + 2A\varphi + \gamma\frac{\partial\varphi}{\partial t} = -V(x)\varphi, \tag{26}$$

where the damping constant γ describes phenomenologically the stochastic character of kink motion between their collisions. The solution of equation (26) is given by

$$\varphi(x,t) = \varphi_K(x) + u(x,t), \tag{27}$$

where $\varphi_K(x)$ denotes the static kink solution (equation (6a)) and u(x,t) the fluctuation in $\varphi_K(x)$ due to its interaction with impurity. Substituting equation (27) in equation (26) we get the equation for u(x,t) as

$$m\frac{\partial^2 u}{\partial t^2} - mC_0^2\frac{\partial^2 u}{\partial x^2} + A(2 - 3\,\mathrm{sech}^2\frac{x}{2\xi_0})u + \gamma\frac{\partial u}{\partial t} = -V(x)(\varphi_K+u). \tag{28}$$

Using linear perturbation theory, the fluctuation u(x,t) can be written in terms of the basis functions $\{\psi\}$ (equations (18) and (22), for n = 1) as [6],

$$u(x,t) = \beta_0(t)\psi_0(x) + \beta_1(t)\psi_1(x) + \int_{-\infty}^{\infty}dk\ \beta_k^{(t)}\ \psi_k^{(x)}. \tag{29}$$

Substituting equation (29) in equation (28) we obtain coupled equations for the coefficients $\{\beta\}$, the solutions of which determine u(x,t) completely and hence $\varphi(x,t)$ in equation (27). The coefficients $\{\beta\}$ is determined to be

$$\beta_j(t) = \beta_j^o[1 - e^{-\gamma t/m} (\cos\Omega_j t + \frac{\gamma}{2m\Omega_j} \sin \Omega_j t)]; \; j = 0,1,k \quad (30)$$

where

$$m\Omega_j^2 = m\omega_j^2 + A_{jj} - \gamma^2/4m, \quad \text{for } j = 0,1 \quad (31a)$$

$$m\Omega_k^2 = m\omega_k^2 - \gamma^2/4m, \quad \text{for } j = k \quad (31b)$$

$$A_{jj} = \int_{-\infty}^{\infty} dx[V(x)\psi_j^*\psi_j], \quad (31c)$$

$$\beta_0^o = - \frac{A}{|B|}(\tfrac{8}{3}\xi_0)^{\frac{1}{2}} \cosh^2 \frac{X_0-X_p}{2\xi_0}(1 - \tanh \frac{X_0-X_p}{2\xi_0}), \quad (31d)$$

$$\beta_1^o = \alpha\frac{A}{|B|} (\tfrac{3}{4}\xi_0)^{\frac{1}{2}} \frac{(1 - \tanh \frac{X_0-X_p}{2\xi_0}) \operatorname{sech} \frac{X_0-X_p}{2\xi_0} \tanh \frac{X_0-X_p}{2\xi_0}}{\frac{3A}{2m} + \alpha(\frac{3}{4}\xi_0) \operatorname{sech}^2 \frac{X_0-X_p}{2\xi_0} \tanh^2 \frac{X_0-X_p}{2\xi_0}}, \quad (31e)$$

$$\beta_k^o = - \frac{\alpha A}{\omega_k^2|B|}(1 - \tanh \frac{X_0-X_p}{2\xi_0}) \psi_k^*(X_0-X_p). \quad (31f)$$

X_0 and X_p denote the initial position of the kink and position of impurity respectively and ω_j^2 denote the frequency of the eigenstates (equations (21) and (23), for n = 1). This solution u(x,t) can be used for calculating the kink-impurity binding (pinning) energy [6].

REFERENCES

[1] J. A. Krumhansl and J. R. Schrieffer, Phys. Rev. **B11** (1975) 3535.
[2] B. Dey, J. Phys. C: Solid State Physics **19** (1986) 3365.
[3] B. I. Henry and J. Oitmaa, Aust. J. Phys. **36** (1983) 339.
[4] B. Dey, Proc. SSP Symp. (India) **29C** (1986) 137 and unpublished (1986).
[5] P. M. Morse and H. Feshbach in **Methods of Theoretical Physics** (McGraw-Hill, New York, 1953), Vol. 2, p. 1651.
[6] V. L. Aksenov, A. Yu. Didyk and R. Zakula, Preprint E17-84-483, JINR, DUBNA (1984).

Part IV

Statistical Mechanics and Quantum Aspects

Quantum Solitons: An Overview*

R. Rajaraman

Centre for Theoretical Studies, Indian Institute of Science,
Bangalore 560012, India

The quantisation of the classical solutions of nonlinear field equations by semiclassical methods is explained.

1. INTRODUCTION

It is well known that nonlinear relativistic field equation permit localised classical solutions, called solitary waves and solitons. Here we discuss the relevance of such classical solutions to the corresponding quantum field theories and show how these solutions can be associated with quantum, extended, non-pertubrative-particle states. To obtain the quantum particle-states we quantise the classical static solutions. There is a variety of techniques for executing such quantisation. The idea of associating these classical soliton-solutions with particle states in the corresponding quantum theory comes from the fact that these classical solutions resemble extended particles, as these are localised (finite size) and have finite energy. However, this connection between the classical solution and the quantum particle is not trivial. For example, it is not correct to identify the extended classical solution with say, something like the wave function of the quantum particle. It should be noted that though the classical soliton solutions resemble particles, the concept of particles exists only in quantum field theory and not in classical field theory. In classical field theory, the fields are continuous c-number functions of space-time and these functions specify the states of the classical system. The dynamics of these fields is governed by nonlinear partial differential field equations. In quantum field theory however, the fields are operator functions of space-time acting on vectors in a Hilbert space. The states of the quantum system are specified by vectors in that Hilbert space and not by the fields. In particular, 'particles' are special state-vectors of the quantum system. They are simultaneous eigenstates of the Hamiltonian and momen-

*Notes taken by Bishwajyoti Dey.

tum operators, with eigenvalues forming a discrete hyperboloid in the $E-\vec{P}$ space. Additional requirements are that the form-factors in these states should be localised to reflect their particle-nature.

Thus we see that the classical solutions of field equations on the one hand, and the concept of a particle in the corresponding quantum field theories on the other are quite different entities. However the two entities are related. Certain properties of the quantum states, e.g., energy or form factors can be expanded in semiclassical series and the leading terms in these series are found to be related to the corresponding classical solutions. These relations between the two entities can be easily established by generalising to field theory certain concepts of quantum mechanics such as the correspondence principle and the semiclassical expansion (which relate quantum levels to classical orbits). To develop this generalisation, we recall the analogy between objects in quantum mechanics and in quantum field theory.

Consider a non-relativistic, unit mass particle in one dimension under the influence of a potential V(x). Classically, the particle is described by giving x as a function of time, and this is obtained by solving Newton's equation

$$\frac{d^2x}{dt^2} = -\frac{dV}{dx} . \tag{1}$$

In quantum mechanics, the particle is described, not by giving the value of x, but of the wave function $\psi(x,t)$. x now becomes an operator which operates on the space of all wave functions (Hilbert space). Similarly momentum $p = \dot{x}$ is also an operator which satisfies

$$[p,x] = -i\hbar \tag{2}$$

where [,] denotes the commutator. The Hamiltonian which is also an operator is given in terms of the operators x and p by

$$H = \frac{p^2}{2} + V(x). \tag{3}$$

The energy eigenstates $\psi_n(x,t)$ obey the time independent Schrödinger equation

$$H\psi_n(x,t) = E_n\psi_n(x,t). \tag{4}$$

Now consider the theory of a scalar field governed by the Lagrangian

$$L = \int[\tfrac{1}{2}(\partial_\mu\varphi\partial^\mu\varphi) - U(\varphi)]d\mathbf{x}.$$

The analogue of x(t) in field theory is the field $\varphi(\mathbf{x},t)$ which is classically obtained by solving the field equation. To quantise the system we consider the commutator

$$[\varphi(\mathbf{x},t),\ \dot{\varphi}(\mathbf{y},t)] \quad = \quad i\hbar\delta(\mathbf{x}-\mathbf{y}), \tag{5}$$

where $\dot{\varphi}(\mathbf{x},t)$ is momentum conjugate to the field $\varphi(\mathbf{x},t)$ and they both are operators. Corresponding to wave function $\psi(\mathbf{x},t)$ in quantum mechanics here we have $\psi[\varphi(\mathbf{x},t)]$ whose square gives the probability of finding the field in the configuration $\varphi(\mathbf{x},t)$. The energy eigenstates are again determined by

$$H|\psi_n\rangle \quad = \quad E_n|\psi_n\rangle\ , \tag{6}$$

where the Hamiltonian operator H is given by

$$H \quad = \quad \int d\mathbf{x}\left(\frac{\dot{\varphi}^2}{2} + \frac{(\nabla\varphi)^2}{2} + U[\varphi]\right). \tag{7}$$

The lowest eigenstate $|\psi_o\rangle$ is called the 'vacuum' of the quantum field theory and by a suitable choice one can make $E_o = 0$ (and the momentum = 0) for the vacuum state. The Hamiltonian and momentum operators commute with each other, and a particle in quantum field theory corresponds to a discrete hyperboloid in the spectrum, i.e., a family of simultaneous eigenvectors of the Hamiltonian and momentum operators, with eigenvalues E and **P** obeying

$$E^2 \quad = \quad \mathbf{P}^2 + m^2 \tag{8}$$

with fixed m. We are interested in the relationship of these particle-states in quantum field theory to localised solutions of the classical field equation

$$\Box\varphi + \frac{\partial U}{\partial\varphi} \quad = \quad 0. \tag{9}$$

Now, recall the relation between quantum energy eigenstates determined by Eq. (4) and the static solution of Eq. (1). Let us consider a potential V(x) which has an absolute minimum at x = a, and a local minimum at x = c. Then clearly there is a classical static solution,

$$x(t) \quad = \quad a.$$

It may be called the 'classical ground state' as it represents, the lowest-energy classical solution. Its energy is

$$E_o^{cl} \quad = \quad V(a). \tag{10}$$

A particle placed at the point x = a will stay there classically. In quantum theory such a state is not allowed. The uncertainty principle will not permit the particle to have both zero momentum and a fixed position. Consequently, even in the ground state, the particle will fluctuate around x = a leading to a ground state energy

$$E_o = E_o^{cl} + \Delta_o = V(a) + \Delta_o , \tag{11}$$

where Δ_o is the energy of zero-point motion. Furthermore, if the potential is harmonic (approximately) near x = a, one can make a Taylor series expansion of V(x) near x = a

$$V(x) = V(a) + \frac{1}{2}\omega^2(x-a)^2 + \frac{1}{3!}\lambda_3(x-a)^3 + \frac{1}{4}\lambda_4(x-a)^4 + \ldots . \tag{12}$$

Then, for those wave functions that satisfy

$$\lambda_r \langle(x-a)^r\rangle \ll \omega^2 \langle(x-a)^2\rangle , \quad r = 3,4,\ldots \tag{13}$$

where <...> denotes expectation value, the effect of the anharmonic terms of V(x) will be small. For λ_r sufficiently small, the potential will be dominantly that of a harmonic oscillator and the low-lying energy eigenstates whose spread is localised in the vicinity of x = a will have energies

$$E_n = V(a) + (n + \frac{1}{2})h\omega + 0(\lambda_r). \tag{14}$$

In particular the ground state energy is given by

$$E_o = V(a) + \frac{1}{2}h\omega + 0(\lambda_r) \cdot \tag{15}$$

Note that the first term in the expansion (15) is just the energy V(a) of the classical solutions. Equation (15) represents the simplest relation between some quantum states and corresponding classical solution. Apart from this, the classical solution also gives other features of the ground state wave function. For instance the position expectation value

$$\langle x\rangle \equiv \int|\psi_o(x)|^2 x\, dx = a + \ldots . \tag{16}$$

The quantity on the r.h.s. is the classical solution whereas the l.h.s. is a property of the quantum ground state.

Repeating the same procedure for the local minimum of the potential at x = c, we have a classical solution

$$x(t) = c. \tag{17}$$

It has the classical energy

$$E^{cl} = V(c), \tag{18a}$$

which is higher than V(a). This solution is interesting, as it is analogous to the classical static soliton in field theory. The solitons too are static solutions, but with higher energies than the classical vacua in the corresponding field theories. As before we can make a Taylor series expansion about x = c (local minimum) giving

$$V(x) = V(c) + \frac{1}{2}(\omega')^2(x-c)^2 + \frac{\lambda_3'}{3!}(x-c)^3 + \frac{\lambda_4'}{4!}(x-c)^4 + \ldots \tag{18b}$$

and if λ_r' (r=3,4,...) are sufficiently small, then the approximate harmonic oscillator states centred at x = c will have energies

$$E_n = V(c) + \omega'(n' + \frac{1}{2}) + 0(\lambda_r') \tag{18c}$$

and expectation values

$$\langle x \rangle = c + 0(\lambda_r'). \tag{18d}$$

Thus again we have a set of energy eigenstates whose energy is related to V(c), the classical energy of a static solution and whose $\langle x \rangle$ is related to the classical solution itself.

However while deriving results in Eq.(18c,d) we ignored the effect of the deeper potential well at x = a. Actually the wave packets built in the potential well around x = c will tunnel into the well around x = a and vice versa. But, if the λ_r and λ_r' are all small, the two minima x = a and x = c will be widely separated with a large potential barrier between them. Therefore, the tunneling will be small. Thus to any finite order in the weak coupling expansion, the set of levels around x = c can be considered separately from the set around x = a.

In field theory on the other hand we have examples of both situations. The ones where such tunneling takes place, are associated with "instantons" and the ones where tunneling does not take place, are associated with "quantum solitons." The states built around most of the soliton solutions can be shown not to decay at all through tunneling. This is related to the non-zero topological index carried by these solutions, which effectively places an infinite energy barrier between them and the vacuum solutions.

Now in order to generalise these ideas to field theory let us first consider a system with a large (but finite) number of degrees of freedom.

Let a bunch of unit-mass non-relativistic particles be collectively described by a cartesian coordinate $\mathbf{x} = (x_1,..x_N)$ with a potential $V(x_1...x_N)$. Let $\mathbf{x} = \mathbf{a}$ be a (local or absolute) minimum of V in the N-dimensional space. Then $\mathbf{x}(t) = \mathbf{a}$ will be a stable static classical solution. Expanding $V(x_1,..x_N)$ around $x_i = a_i$,

$$V(\mathbf{x}) = V(\mathbf{a}) + \frac{1}{2}(x_i-a_i)(x_j-a_j)\left[\frac{\partial^2 V}{\partial x_i \partial x_j}\right]_{\mathbf{x}=\mathbf{a}}$$

$$+ \frac{1}{3!}(x_i-a_i)(x_j-a_j)(x_k-a_k)\left[\frac{\partial^3 V}{\partial x_i \partial x_j \partial x_k}\right]_{\mathbf{x}=\mathbf{a}} + \ldots \quad (19)$$

In the weak coupling limit we can construct a set of low lying states around $\mathbf{x} = \mathbf{a}$ by diagonalising the matrix of second derivatives $[\partial^2 V/\partial x_i \partial x_j]_{\mathbf{x}=\mathbf{a}}$ by changing variables to some normal modes ξ_i. Then in terms of ξ_i, the problem reduces to a set of oscillators of frequencies ω_i, with ω_i^2, the eigenvalues of the matrix. The lowest-energy state constructed around $\mathbf{x} = \mathbf{a}$ will have energy

$$E_o = V(a) + \sum_{i=1}^{N} \frac{1}{2}h\omega_i + \text{higher order terms} \quad (20)$$

and the higher excitations will be given by

$$E_{\{n_i\}} = V(\bar{a}) + \sum_{i=1}^{N} (n_i + \frac{1}{2})\omega_i + \text{higher order terms.} \quad (21)$$

These higher order terms are to be obtained by perturbation theory. As can be seen these results are the generalisation of the one-degree of freedom problem, with the set of eigenvalues $(\omega_i)^2$ of $[\partial^2 V/\partial x_i \partial x_j]_{\mathbf{x}=\mathbf{a}}$ replacing the single number ω^2.

2. QUANTISATION OF SOLITON SOLUTIONS

In field theory, instead of a finite number of degrees of freedom $x_1,...,x_N$, we have a continuous infinity of degrees of freedom $\varphi(\mathbf{x})$, i.e., the value of φ at each space point $\mathbf{x}$. We consider the example of the sine-Gordon field $\varphi(x,t)$ here. The action is

$$S = \int dt\, d\mathbf{x}\, \{\frac{1}{2}(\partial_\mu\varphi\partial^\mu\varphi) + \frac{m^4}{\lambda}(\cos\frac{\sqrt{\lambda}}{m}\varphi - 1). \quad (22)$$

Changing variables to

$$x \to mx, \quad t \to mt \quad \text{and} \quad \frac{\sqrt{\lambda}}{m}\varphi \to \varphi \quad , \quad (23)$$

we get the Lagrangian in the standard form

$$L = T[\varphi] - V[\varphi] \tag{24}$$

as

$$L = \int dx[\tfrac{1}{2}(\dot{\varphi})^2] - \int dx(\tfrac{1}{2}(\varphi')^2 + 1 - \cos\varphi) = T[\varphi] - V[\varphi]. \tag{25}$$

The stable static solutions of the field equation

$$\frac{\partial^2\varphi(x,t)}{dt^2} = - \frac{\partial V[\varphi]}{\partial\varphi(x,t)} \tag{26}$$

are once again the minima of the potential, satisfying

$$\frac{\delta V(\varphi)}{\delta\varphi(x)} = 0 = \varphi'' - \sin\varphi \quad . \tag{27}$$

These are given by

$$\text{(i)}\ \varphi(x) = 0,\ \pm 2\pi,\ \pm 4\pi,\ .\ .\ . \tag{28a}$$

$$\text{(ii)}\ \varphi = \varphi_{sol}(x) \tag{28b}$$

$$\text{(iii)}\ \varphi = \varphi_{anstisol}(x), \tag{28c}$$

$$\text{where}\quad \varphi_{\substack{sol\\(antisol)}}(x) = 4\tan^{-1}(e^{\pm x}) \tag{29}$$

(the ± sign describes soliton and antisoliton solutions respectively). In quantum field theory, as mentioned before, the field φ , π_φ (the conjugate momentum) and the Hamiltonian are operators. The generalisation of the matrix of second derivative of V occurring in Eq. (19) is the operator $(-\partial^2/\partial x^2 + \cos\varphi)$ evaluated at $\varphi(x) = \varphi_o(x)$ which is any one of the solutions in Eq. (27). Its eigenvalues and eigenfunctions are given by the differential equation

$$(-\frac{\partial^2}{\partial x^2} + \cos\varphi_o(x))\eta_i(x) = \omega_i^2\eta_i(x) \quad , \tag{30}$$

where $\eta_i(x)$ are the orthonormal 'normal modes' of fluctuations around $\varphi_o(x)$. Thus in quantum theory (weak coupling approximation), one can construct a set of approximate harmonic-oscillator states, spread in field space around $\varphi_o(\mathbf{x})$. The energies of these states would be

$$E_{\{n_i\}} = V[\varphi_o] + h\sum_i(n_i + \tfrac{1}{2})\omega_i + \text{higher order terms} \tag{31}$$

For the solution given by Eq. (28a) we have as the eigenvalue equation

$$\left(\frac{-\partial^2}{dx^2} + \cos\varphi\right)_{\varphi=0} \eta_i(x) = \left(-\frac{\partial^2}{\partial x^2} + 1\right)\eta_i(x) = \omega_i^2\,\eta_i(x). \qquad (32)$$

The eigenvalues are $\omega_i^2 = k_i^2 + 1$

$$\text{with } k_i = 2\pi N_i/L \text{ and } L \to \infty. \qquad (33)$$

The set of approximate harmonic oscillator states built around $\varphi = 0$ have energies

$$E_{n_i} = V[\varphi=0] + \sum_i (n_i + \tfrac{1}{2})\omega_i + \text{higher order terms} \qquad (34)$$

(we have used the units $\hbar = c = 1$). The vacuum state has energy

$$E_o = E_{vac} = \frac{1}{2}\sum_i (k_i^2 + 1)^{\frac{1}{2}} + \text{higher order terms.} \qquad (35)$$

The next excited state (one particle state) has energy

$$E_1 = E_o + (k_1^2 + 1)^{\frac{1}{2}} + \text{higher order terms.} \qquad (36)$$

$E_1 - E_o$ represents the energy of the single-quantum particle state. Its mass, in terms of the original parameters in (22), is

$$\text{mass} = m + 0(\lambda) \qquad (37)$$

(in terms of original variable $(k^2+1)^{\frac{1}{2}} \to (k^2+m^2)^{\frac{1}{2}}$, etc.). Similarly for the soliton solution in Eq. (28b) the eigenvalue equation is

$$\left\{-\frac{\partial^2}{\partial x^2} + \cos\varphi(x)_{sol}\right\}\eta_i(x) = \tilde{\omega}_i^2\,\eta_i(x). \qquad (38)$$

This equation is exactly solvable and has one discrete eigenvalue

$$\tilde{\omega}_o^2 = 0 \qquad (39a)$$

followed by a continuum,

$$\tilde{\omega}_N^2 = (q_N^2 + 1) \qquad (39b)$$

where $\frac{L}{\sqrt{2}}q_N = 2\pi N + \delta(q_N)$

One can now associate a set of energy levels around $\varphi(x) = \varphi_{sol}(x)$ with energies

$$E_{\{n_i\}} = V[\varphi_{sol}] + \sum_i (n_i + \tfrac{1}{2})\tilde{\omega}_i + \text{higher order terms.} \qquad (40)$$

Of these, the lowest state has energy

$$E = V[\varphi_{sol}] + \frac{1}{2} \sum_i \tilde{\omega}_i + \text{higher order terms}. \tag{41}$$

This is the basic soliton-particle state in quantum field theory. Higher states are soliton-plus-meson states. Together they form the soliton sector of states. Finally we make some comments:

i) All we are doing is quantising small fluctuations about any given stable static classical solution, whether it be $\varphi = 0$ or $\varphi = \varphi_{sol(antisol)}$.

ii) The method is 'semiclassical' but no less quantum mechanical than standard perturbation quantum field theory, which amounts to expanding about $\varphi = 0$.

iii) The states in the soliton sector carry a nonzero topological quantum number

$$Q = \frac{1}{2} \int_{-\infty}^{\infty} \frac{d\varphi}{dx}\, dx = \frac{1}{2}[\varphi(\infty) - \varphi(-\infty)]. \tag{42}$$

The associated current is

$$j_\mu = \varepsilon_{\mu\nu} \partial_\nu \varphi \tag{43}$$

which is conserved, i.e.,

$$\partial^\mu j_\mu = 0 \tag{44}$$

and gives rise to the quantum number

$$Q = \int j_o\, dx. \tag{45}$$

iv) The quantisation of time-dependent solutions can be carried out by extending the standard WKB method to field theory.

v) The occurrence of 'ultraviolet' divergences in quantum field theory due to short-distance behaviour of products of field operators is well known. It is removed by adding suitable 'counter terms' to the Hamiltonian. When this is done, ultraviolet divergences in the sum $\sum_i \omega_i$ over modes in expressions like (41), (40), (35), cancel out.

Fuller details and references are given in the reviews listed below:

The method for quantising solitons was first developed systematically by

R. F. Dashen, B. Hasslacher and A. Neveu, Phys. Rev. **D10** (1974) 4130
and independently by
J. Goldstone and R. Jackiw, Phys. Rev. **D11** (1975) 1486.
The extension of W. K. B. method to field theory to quantise time-dependent soliton solutions (such as the sine-Gordon breather solutions) was developed by
R. F. Dashen, B. Hasslacher and A. Neveu, Phys. Rev. **D11** (1975) 3424.
A review of the above work and other developments is given in
R. Rajaraman, Phys. Reports **21C** (1975) 227.

Soliton Statistical Mechanics: Statistical Mechanics of the Quantum and Classical Integrable Models

*R.K. Bullough, D.J. Pilling, and J. Timonen**

Department of Mathematics, U.M.I.S.T., P.O. Box 88,
Manchester M60 1QD, United Kingdom

It is shown how the Bethe Ansatz (BA) analysis for the quantum statistical mechanics of the Nonlinear Schrödinger Model generalises to the other quantum integrable models and to the classical statistical mechanics of the classical integrable models. The bose-fermi equivalence of these models plays a fundamental role even at classical level. Two methods for calculating the quantum or classical free energies are developed: one generalises the BA method the other uses functional integral methods. The familiar classical action-angle variables of the integrable models developed for the real line $\mathbb{R}$ are used throughout, but the crucial importance of periodic boundary conditions is recognized and these are imposed. Connections with the quantum inverse method for quantum integrable systems are established. The R-matrix and the Yang-Baxter relation play a fundamental role in the theory. The lectures draw together the quantum BA method, the quantum inverse method, and the generalised BA and functional integral methods introduced more recently.

1. INTRODUCTION TO THE CLASSICAL AND QUANTUM INTEGRABLE MODELS

The simplest integrable model with real physical interest is probably the 'repulsive' nonlinear Schrödinger (NLS) model with equation of motion

$$-i\varphi_t = \varphi_{xx} - 2c\varphi|\varphi|^2 ; \tag{1.1}$$

φ is a complex scalar field. This classical system is a Hamiltonian system with Hamiltonian

$$H[\varphi] = \int [\varphi_x\varphi_x^* + c\varphi^2\varphi^{*2}]dx ; \tag{1.2}$$

---

*On leave from Department of Physics, University of Jyväskylä, SF-40100 Jyväskylä, Finland the parameter c is a real-valued coupling constant, and, for the 'repulsive' NLS model, $c > 0$. Evidently this is consistent with the repulsive character of the nonlinear self-potential $c\varphi^2\varphi^{*2}$, which is positive when $c > 0$.

By imposing the Poisson bracket

$$\{\varphi(x,t),\ \varphi^*(x',t)\} \quad = \quad i\delta(x - x') \tag{1.3}$$

one finds that Hamilton's equation of motion

$$\varphi_t \quad = \quad \{H[\varphi],\ \varphi\} \tag{1.4}$$

is exactly (1.1). There is now an obvious extension to a quantum theory in which φ, φ^* become operators $\varphi, \varphi^\dagger$ with commutation relation

$$[\varphi(x,t),\ \varphi^\dagger(x',t)] \quad = \quad \delta(x - x') \tag{1.5}$$

and (1.4) becomes Heisenberg's equation of motion. For a well defined ground state of the quantum model φ, $\varphi^\dagger$ must be placed in normal order:

$$\hat{H}[\varphi] \quad = \quad \int [\varphi_x^\dagger\varphi_x + c\varphi^{\dagger 2}\varphi^2]dx\ . \tag{1.6}$$

The 'hat' on $\hat{H}[\varphi]$ indicates it is now an operator.

We shall see that $\hat{H}[\varphi]$, (1.6), describes N bosons on a line with repulsive δ-function interactions. Indeed there is evidently a number operator $\hat{N} = \int\varphi^\dagger\varphi\, dx$ which commutes with $\hat{H}$. Moreover, since (1.1) is now

$$-i\varphi_t \quad = \quad \varphi_{xx} - 2c\varphi^\dagger\varphi^2 \tag{1.7}$$

(in normal order) in second quantisation, there is the equivalent Schrödinger problem in first quantisation

$$\left\{- \sum_{i=1}^{N} \frac{\partial^2}{\partial x_i^2} + 2c \sum_{i>j} \delta(x_i - x_j) \right\}\Psi \quad = \quad E\Psi\ , \tag{1.8}$$

and this displays the repulsive δ-function interactions. The 'on a line' means that the model is a one space-dimensional model defined on (say) $b \leq x \leq d$. The cases of most interest are $b \to -\infty$, $d \to +\infty$ (i.e., $x \in \mathbb{R}$) with appropriate decaying boundary conditions, or periodic boundary conditions of period L (say) on $-\frac{1}{2}L \leq x \leq \frac{1}{2}L$. The quantum model with periodic boundary conditions we have just described was the model of a pioneering paper in quantum statistical mechanics by Yang and Yang [1]. Their analysis was based on the quantum Bethe Ansatz (BA) method which we very briefly describe later. There is an ansatz for Ψ which solves (1.8) exactly [2]. However, the fact that a quantum model with a Schrödinger problem such as (1.8) is exactly solvable has much wider implications. The purpose of these lectures is to show how the Yangs' result for the free energy of this quantum model apparently generalises to all of the integrable models including the classical integrable models.

The history of the classical integrable models is that they were first solved with φ, defined for $x \in \mathbb{R}$, vanishing 'fast enough' at $|x| = \infty$. Typically, this is exponential decay at $x = \pm\infty$ [3]. However, we shall find that it is *periodic* boundary conditions, rather than these decaying boundary conditions, which play a fundamental role in the statistical mechanics.

At first sight the 'attractive' classical NLS ($c < 0$) is more interesting than the repulsive case ($c > 0$); for the former has soliton solutions. For $c = -1$ the 1-soliton solution is

$$\varphi(x,t) = \frac{2\eta \exp\{-4i(\xi^2-\eta^2)t + 2i\xi x + i\delta\}}{\cosh[8\eta\xi t - 2\eta(x - x_o)]} \; ; \tag{1.9}$$

ξ, η, δ and x_o are real free parameters and (for $\eta > 0$) decays as $e^{-2\eta|x|}$ as $|x| \to \infty$. However when $c < 0$ the quantum NLS has no lower bound to its ground state. Thus starting with the simpler cases we should first of all be concerned with the *repulsive* case of the quantum or classical NLS.

Still, in a sense which becomes plain once we transform to action-angle variables (§2), the two NLS models are non-relativistic forms of two covariant models: the attractive NLS corresponds to the sine-Gordon (s-G) model

$$\varphi_{xx} - \varphi_{tt} = m^2 \sin\varphi \, , \tag{1.10}$$

the repulsive NLS to the sinh-Gordon (sinh-G) model

$$\varphi_{xx} - \varphi_{tt} = m^2 \sinh\varphi \, . \tag{1.11}$$

The m's are "masses", so the units are such that $\hbar = c$ (= velocity of "light") = 1. The classical s-G has the kink and antikink solutions

$$\varphi(x,t) = 4\tan^{-1}\exp\{\pm m(x-Vt)\gamma\} \;; \tag{1.12}$$

$\gamma \equiv (1-V^2)^{-1/2}$. It also has the "breather" (bound kink-antikink pair) solutions

$$\varphi(x,t) = 4\tan^{-1}[\tan\eta\,\sin\{\gamma m(\cos\eta)(t-Vx)\}\,\mathrm{sech}\{\gamma m(\sin\eta)(x-Vt)\}]. \tag{1.13}$$

These are the soliton solutions of the s-G model. Evidently the boundary conditions (b.c.s.) are now $\varphi \to 0$ (mod 2π) 'fast enough' as $|x| \to \infty$; and $\varphi_x \to 0$ fast enough. On the other hand the sinh-G, like the repulsive NLS has no soliton solutions. We shall find in consequence that we can develop the statistical mechanics of sinh-G wholly in parallel with that of the repulsive NLS.

Both the s-G and sinh-G are Hamiltonian: for s-G

$$H[\varphi] = \gamma_0^{-1}\int[\tfrac{1}{2}\Pi^2\gamma_0^2 + \tfrac{1}{2}\varphi_x^2 + m^2(1-\cos\varphi)]dx \tag{1.14}$$

and γ_0 (the real valued coupling constant) > 0. Then sinh-G is found by $\varphi \to i\varphi$, $\gamma_0 \to -\gamma_0$. The bracket is

$$\{\Pi(x,t), \varphi(x',t)\} = \delta(x-x') \tag{1.15}$$

and (1.4) for s-G is

$$\varphi_t = \gamma_0\Pi, \quad \Pi_t = \gamma_0^{-1}[\varphi_{xx} - m^2\sin\varphi] \tag{1.16}$$

which is (1.10). The canonical transformation $\varphi \to \gamma_0^{\frac{1}{2}}\varphi, \Pi \to \gamma_0^{-\frac{1}{2}}\Pi$ means γ_0 appears only as $-m^2\gamma_0^{-\frac{1}{2}}\sin(\gamma_0^{\frac{1}{2}}\varphi) = -m^2[\varphi - \frac{1}{6}\gamma_0\varphi^3 + \ldots]$ in (1.16), so that γ_0 scales the s-G nonlinearity just as c scales the NLS nonlinearity. However, φ is a real field for s-G and sinh-G, whilst it is complex for the two NLS models.

It is plain too that sinh-G is a continuation of s-G as $\gamma_0 \to -\gamma_0$ (and vice-versa) just as the NLS equations continue in c. We use this fact later. The places $\gamma_0 = 0$, $c = 0$ are exceptional: $\gamma_0 = 0$ in sinh-G or s-G is the linear Klein-Gordon (K-G) model

$$\varphi_{xx} - \varphi_{tt} = m^2\varphi \;; \tag{1.17}$$

c = 0 is the linear Schrödinger (LS) model

$$-i\varphi_t = \varphi_{xx} . \tag{1.18}$$

Evidently both linear models are still Hamiltonian (set $\gamma_o \to 0$ in (1.14)). Both are free-field models in an obvious sense.

There are other integrable models of physical interest: the Landau-Lifshitz model is

$$H[\varphi] = \frac{1}{2} \int_{-\infty}^{\infty} dx(\mathbf{S}^2_{,x} - \mathbf{S}.J.\mathbf{S} + J_3)$$

$$= \frac{1}{2} \int_{-\infty}^{\infty} dx\ (\mathbf{S}^2_{,x} + (J_3 - J_1)S_1^2 + (J_3 - J_2)S_2^2) \tag{1.19}$$

with $J = \mathrm{diag}(J_1,J_2,J_3)$, $0 < J_1 < J_2 < J_3$: $\mathbf{S}(x,t) = (S_1,S_2,S_3)$, $|\mathbf{S}|=1$, and the bracket is

$$\{S_\alpha, S_\beta\} = -\varepsilon_{\alpha\beta\gamma}\ S_\gamma \delta(x - x')\ . \tag{1.20}$$

Hamilton's equations (1.4) are

$$\mathbf{S}_{,t} = \mathbf{S} \times \mathbf{S}_{,xx} + \mathbf{S} \times (J.\mathbf{S})\ . \tag{1.21}$$

This model contains both the s-G and attractive NLS models as special cases [4,5]; the repulsive cases can also be found. In the isotropic case $J_1 = J_2 = J_3$ the model is the integrable Heisenberg ferromagnet [6,7] with equation of motion $\mathbf{S}_{,t} = \mathbf{S} \times \mathbf{S}_{,xx}$.

As for the NLS models the classical s-G models (s-G and sinh-G) have extensions to quantum theories: the commutation relations are

$$[\pi(x,t),\ \varphi(x',t)] = -i\delta(x - x') \tag{1.22}$$

and the equations of motion ((1.10) and (1.11)) and the Hamiltonians ((1.14)) must be placed in normal order. This cannot be done in practice until we define operators of annihilation and creation type from φ. We develop such apparatus later.

One of the more remarkable properties of quantum s-G, evidently a massive boson problem, is its strict equivalence to the massive Thirring model (MTM), a massive fermion problem. "Equivalence" means equivalence of expectation values--especially the eigenspectrum [8]. The MTM is

$$H[\varphi] = \int dx[-i\varphi^\dagger\sigma_3\varphi_x + m_o\varphi^\dagger\ \sigma_1\varphi + 2g_o\varphi_1^\dagger\varphi_2^\dagger\varphi_2\varphi_1]\ ;$$

$$\varphi = \begin{bmatrix} \varphi_1 \\ \varphi_2 \end{bmatrix}\ ,\quad [\varphi,\varphi^\dagger]_+ = I\ \delta(x - x')\ , \tag{1.23}$$

in normal order: σ_3, σ_1 are Pauli matrices. The free field has $g_0 = 0$, and $0 \le g_0 \le \infty$. In connection with the quantum spin-$\frac{1}{2}$ XYZ-model

$$H[\varphi] = -\sum \left(J_1 S_1^n S_1^{n+1} + J_2 S_2^n S_2^{n+1} + J_3 S_3^n S_3^{n+1} \right) , \qquad (1.24)$$

which has spin commutation relations $[S_\alpha^n, S_\beta^m] = i\epsilon_{\alpha\beta\gamma} S_\gamma^m \delta_{mn}$ and reduces to the Landau-Lifshitz model in classical and continuum limit, one defines a coupling constant μ [9] which relates [10] to g_0 as $\mu = \cot^{-1}(-\frac{1}{2}g_0)$ [11]. The spin-$\frac{1}{2}$XYZ model (1.24) maps by Jordan-Wigner transformation [14] and continuum limit to the MTM, and it then proves that for the MTM-quantum s-G equivalence $\frac{1}{8}\gamma_0 = \pi - \mu$. So $g_0 = 0$ is $\gamma_0 = 4\pi$ in the quantum s-G model. Because s-G is covariant one finds a "bare" γ_0 renormalises to

$$\gamma_0'' = \gamma_0[1 - \gamma_0/8\pi]^{-1} . \qquad (1.25)$$

Thus $\gamma_0 = 4\pi$ is $\gamma_0'' = 8\pi$, and it is at this value one finds there are no quantised breather states of quantum s-G [5]. Thus $g_0 = 0$ (free fermions) in the MTM corresponds to no quantised breathers of quantum s-G. Similarly $\gamma_0 = 0$ (free bosons) in the s-G corresponds to $g_0 = \infty$ in quantum MTM. This is the infinitely attractive massive fermion problem and compares with $c = \infty$ in the repulsive NLS, which is the so-called 'impenetrable bose gas' [15].

Finally, for quantum sinh-G $\gamma_0 < 0$, and for $\gamma_0 < 0$ one would expect any related fermion model to be repulsive. In particular for $-4\pi < \gamma_0 < 0$ and $\mu = \cot^{-1}(-\frac{1}{2}g_0)$ we find $0 > g_0 > -\infty$, repulsive, but on a different branch of $\mu (\pi < \mu < 3\pi/2)$ compared with that for the attractive MTM $(\pi/2 < \mu < \pi)$. Korepin's repulsive MTM [13] with his different relation [11] between g_0 and μ has $0 > g_0 > -\pi$ which means $4\pi < \gamma_0 < 8\pi \, (\pi/2 > \mu > 0)$. If however $\mu = \cot^{-1}(-\frac{1}{2}g_0)$ then $4\pi < \gamma_0 < 8\pi$ means $-\infty < g_0 < 0$, repulsive, but on the branch of μ connecting smoothly to the attractive case. Without further investigation it therefore seems that Korepin's repulsive MTM is not equivalent to quantum sinh-G case. We shall not explore the MTM-s-G equivalence in the repulsive regime any further here. We confine attention to quantum sinh-G which is still to be explored as an MTM.

The renormalisation of γ_0 shown in (1.25) arises in connection with the renormalisation of mass m in the (relativistic) quantum s-G model [16]; it appeared first in the work of DHN [17] and necessarily appears in equivalent form in the BA theory of the MTM [10]. Evidently the corresponding result for quantum sinh-G is [18,19]

$$\gamma_o'' = \gamma_o[1 + \gamma_o/8\pi]^{-1}, \quad \gamma_o > 0 ; \tag{1.26}$$

and in this case the only free particle model seems to be where $\gamma_o=0$ (§5). Still we shall see that for quantum sinh-G with $\gamma_o > 0$ there is still a description in terms of bosons and a second, equivalent, description in terms of fermions. Similarly, although quantum repulsive NLS is a boson model with commutation (1.5), it has an equivalent fermion description. It was this fermion description which was used in the work of [1] on the quantum statistical mechanics.

These bose-fermi equivalent descriptions are equivalent. Thus in §5 we shall show how to follow out the work of [1] wholly in a boson description. This equivalence apparently depends on the fact that the classical integrable model are embedded in infinite dimensional Lie algebras. The simplest is an embedding in an "untwisted loop" algebra [20]

$$[L_m^a, L_n^b] = i\,\varepsilon^{abc} L_{m+n}^c \; ; \quad m,n \in \mathbb{Z} \tag{1.27}$$

in which ε^{abc} are structure constants of a finite dimensional Lie algebra g, dimension dim g. All the classical models we have mentioned have the Lie algebra $g = s\ell(2, \mathbb{C})$ which has dim g = 3 and satisfies

$$[L^1, L^2] = L^3 \; ; \quad [L^3, L^1] = 2L^1 \; ; \quad [L^3, L^2] = -2L^2 \, . \tag{1.28}$$

If $\lambda \in \mathbb{C}$ and $L^a \in g$, then the algebra of $\lambda^m L^a$ is isomorphic to (1.27). For dim $g = \infty$ an algebra $g\ell(\infty)$ [21] or a larger algebra [22] replaces (1.27). Embedded in $g\ell(\infty)$ is the integrable Kadomtsev-Petviashvili equations in two space and one time dimensions (2+1 dimensions) [21,22]. In 2+1 dimensions much remains to be understood. But $g\ell(\infty)$ has both "bose" and "fermi" equivalent representations and $g\ell(n)$ carries this down to 1+1 dimensions (one space and one time dimension). Perhaps still more important for present purposes is that the algebras (1.27) determine the "integrability" of the models. The Kostant-Symes-Adler theorem [23] shows that an arbitrary loop algebra (1.27) induces a symplectic structure with infinite sets of quantities commuting under a Poisson bracket; it also yields hierarchies of classical integrable nonlinear evolution equations [23]. Evidently we are concerned with the classical integrability of the classical models and a quantum integrability of the quantum models. We look at the former from a wholly elementary point of view in §2 next. Later (in §4) we introduce the quantum R-matrix first recognized by the Leningrad school in their work on the quantum inverse method [2,16,24,25]. The R-matrix allows us to define quantum integrability and Wadati and Akutsu [26] in their lectures show how it defines the integrability of their IRF models in two-dimensional classical statistical mechanics (2+0 dimensions). The loop algebras (1.27) extend to Kac-Moody-Lie algebras by inclusion of the centre commuting with all the elements of the algebra [20]. The quantum integrable models are embedded in such algebras [20]. De Vega [27] has established some particular connections between these algebras and corresponding algebras associated with the R-matrix(also see [25]).

2. CLASSICAL INTEGRABILITY

We shall exploit both classical and quantum integrability of the integrable models in the statistical mechanics of 1+1 dimensions we develop in these lectures. By classical integrability we shall mean 'complete integrability' in the sense of Liouvile-Arnold [28]. Recall that for a classical Hamiltonian system with a finite number of degrees of freedom N, that is N q-coordinates and N p-coordinates ($\{p_i, q_j\} = \delta_{ij}$), the system is completely integrable when there are N independent first integrals I_k, $k = 1, \ldots, N$ (say) which are in involution, that is commute under the bracket, $\{I_k, I_\ell\} = 0$. Then if the manifold of level lines I_k = const. is compact and connected Arnold [28] proves the motion lies on an N-dimensional torus.

The simplest torus is a 1-dimensional torus, and the simplest integrable system is the 1-dimensional harmonic oscillator with $H = \frac{1}{2}p^2 + \frac{1}{2}\omega^2 q^2$. Since H = E, the energy, the 1-D case is always integrable. The energy E defines the action $\omega^{-1} E \equiv P$ and there is a canonical so-called angle variable $Q = \omega t + \delta$. With $H = \omega P$, $Q_t = \omega$ and $P_t = 0$ are Hamilton's equations. Evidently $p = -(2\omega P)^{\frac{1}{2}} \sin Q$, $q = (2\omega^{-1}P)^{\frac{1}{2}} \cos Q$. Note that the differential 1-forms $dp = -\omega^{\frac{1}{2}}(2P)^{-\frac{1}{2}} dP \sin Q - \omega^{\frac{1}{2}}(2P)^{\frac{1}{2}} \cos Q\, dQ$, and $dq = \omega^{-\frac{1}{2}}(2P)^{-\frac{1}{2}} dP \cos Q - \omega^{-\frac{1}{2}}(2P)^{\frac{1}{2}} \sin Q\, dQ$. Consequently the 2-form $\mathbb{D} = dp \wedge dq = dP \wedge dQ$ and $d\mathbb{D} = 0$. This is just the condition for a canonical transformation [28]. One easily checks that $\{P,Q\} = \left\{\frac{\partial P}{\partial p}\frac{\partial Q}{\partial q} - \frac{\partial P}{\partial q}\frac{\partial Q}{\partial p}\right\} = 1$. This is the other way (invariance of the Poisson brackets) of looking at a canonical transformation. The check follows from $P = \omega^{-1}[\frac{1}{2}p^2 + \frac{1}{2}\omega^2 q^2]$, $Q = \tan^{-1}(p/\omega q)$.

It is easy to extend the theory to any number of oscillators and any corresponding number of degrees of freedom--in particular, by not being too searching about it, to a continuously infinite number of degrees of freedom. This is just the situation for a field $\varphi(x,t)$: there is a q-type coordinate $\varphi(x,t)$ for each label x, and a p-type coordinate $\Pi(x,t)$ likewise; and x is not countable in any segment $-b\leq x\leq d$ in $\mathbb{R}$. Thus, for the linear K-G model (1.17) with $x\in\mathbb{R}$ in particular, we expect an H

$$H = \int_{-\infty}^{\infty}\omega(k)P(k)\,dk \; ; \quad \omega(k) = (m^2+k^2)^{\frac{1}{2}} \tag{2.1}$$

and a $Q(k)$, $0\leq Q(k) < 2\pi$ so that $\{P(k),Q(k')\} = \delta(k-k')$.

In order to show this, note that under the complex Fourier transform $\varphi(x,t)\to\bar{\varphi}(k,t)$ (1.17) becomes

$$\bar{\varphi}_{tt} + (\omega(k))^2\bar{\varphi} = 0 . \tag{2.2}$$

On the other hand, scaling $\bar{\varphi}$, $H \to \int_{-\infty}^{\infty}[\frac{1}{2}\bar{\varphi}_t\bar{\varphi}_t^* + \frac{1}{2}(\omega(k))^2\bar{\varphi}\bar{\varphi}^*]dk$, where $\bar{\varphi}^*(k) = \bar{\varphi}(-k)$, while $\int_{-\infty}^{\infty}(d\Pi\wedge d\varphi)dx = \int_{-\infty}^{\infty}(d\bar{\varphi}_t\wedge d\bar{\varphi}^*)dk$ and the bracket is $\{\bar{\varphi}_t(k),\bar{\varphi}^*(k')\} = \delta(k-k')$. Then $\int_{-\infty}^{\infty}(d\bar{\varphi}_t\wedge d\bar{\varphi}^*)dk = i\int_{-\infty}^{\infty}(da(k)\wedge da^*(k))dk$ $= \int_{-\infty}^{\infty}(dP\wedge dQ)dk$ where $a = (2\omega)^{-\frac{1}{2}}(\bar{\varphi}_t + i\omega\bar{\varphi})$, $a^* = (2\omega)^{-\frac{1}{2}}(\bar{\varphi}_t^* - i\omega\bar{\varphi}^*)$; then $P = |a|^2$, $Q = \arg a$. Thus $(\Pi(x),\varphi(x))\to(P(k),Q(k))$ is canonical, while $\{P(k),Q(k')\} = \delta(k-k')$, $0\leq P(k)<\infty$, $0\leq Q(k) < 2\pi$, and

$$H = \int_{-\infty}^{\infty}\omega(k)|a(k)|^2\,dk = \int_{-\infty}^{\infty}\omega(k)P(k)\,dk \tag{2.3}$$

which is (2.1). Notice that (2.3) is characteristic of a linear problem: the LS model (1.18) yields the same result with $\omega(k) = k^2$. Notice too that the Fourier transform $(\Pi(x),\varphi(x))\to(\bar{\varphi}_t(k),\bar{\varphi}(k))$ is canonical.

That the Fourier transform is canonical extends to the spectral transform which is also canonical [3,5,29,30]. Recall that the spectral (or scattering) transform is used to solve nonlinear models like sinh-G [3]. In light-cone coordinates, $\varphi_{xt} = m^2\sinh\varphi$ has the spectral transform [3,5]

$$i\begin{bmatrix}\partial/\partial x & \frac{1}{2}i\varphi_x\\ \frac{1}{2}i\varphi_x & -\partial/\partial x\end{bmatrix}\begin{bmatrix}v_1\\ v_2\end{bmatrix} = \zeta\begin{bmatrix}v_1\\ v_2\end{bmatrix} . \tag{2.4}$$

This eigen-condition, with eigen- or spectral-parameter ζ, can be written $\hat{L}v = \zeta v$. (The hat on $\hat{L}$ indicates only that it is a (matrix) differential operator). This eigen-condition is the basis of the inverse scattering, or inverse spectral, method for solving sinh-G. Combined with

$$\frac{im^2}{4\zeta}\begin{bmatrix} \cosh\varphi & i\sinh\varphi \\ i\sinh\varphi & -\cosh\varphi \end{bmatrix}\begin{bmatrix} v_1 \\ v_2 \end{bmatrix} = \begin{bmatrix} v_1 \\ v_2 \end{bmatrix}_{,t} , \tag{2.5}$$

or $\hat{A}v = v_{,t}$ one finds [3] that $\hat{L}_{,t} = [\hat{A}, \hat{L}]$ is $\frac{1}{2}\varphi_{xt} = \frac{1}{2}m^2 \sinh\varphi$, together with the isospectral condition $\zeta_{,t} = 0$.

The matrix operator $\hat{L}$ is self-adjoint and has only real eigenvalues. These are dense on the real ζ-axis. The spectral transform $\hat{L}v = \zeta v$ maps $\varphi(x)$ (at fixed t) to spectral data $S = \{a(k), b(k); k \in \mathbb{R}\}$ and $|a|^2 - |b|^2 = 1$. From (2.5) one finds [5]

$$\begin{aligned} a(k,t) &= a(k,0) \\ \arg b(k,t) &= \omega(k)t + \delta(k), \end{aligned} \tag{2.6}$$

in which $\omega(k) = m^2k^{-1}$ is the dispersion relation of the linear K-G in light-cone coordinates. Surprisingly, in view of the nonlinearity of sinh-G, one can now prove [5], that H can be written in the form

$$H = \int_{-\infty}^{\infty} \omega(k)P(k)dk , \tag{2.7}$$

with this $\omega(k)$. One checks [5] that if $P(k) = (\pi k)^{-1} \ln|a(k)|$, constants of the motion, and $Q(k) = \arg b(k,t)$ then

$$\int_{-\infty}^{\infty} \{ d(\tfrac{1}{2}\varphi_x) \wedge d(\tfrac{1}{2}\varphi) \} dx = \int_{-\infty}^{\infty} (dP \wedge dQ)\, dk \tag{2.8}$$

and the transformation is canonical: P(k), Q(k) are action-angle variables, the phase space is $0 \leq P(k) < \infty$, $0 \leq Q(k) < 2\pi$ for each k, and $\{P(k), Q(k')\} = \delta(k - k')$. One includes the coupling constant by $P(k) = (\gamma_o \pi k)^{-1} \ln|a(k)|$ when $\gamma_o \neq 1$.

A second constant of the motion for sinh-G in light cone coordinates then proves to be

$$P = 2\int_{-\infty}^{\infty} k\, P(k)\, dk , \tag{2.9}$$

and covariant combinations are H±P: these are interpreted as

$$H = \int_0^{\infty} \left[\frac{m'^2}{8\xi} + 2\xi \right] P(\xi) d\xi = \int_{-\infty}^{\infty} \omega(k) P'(k)\, dk$$

$$P = \int_0^{\infty} \left[\frac{m'^2}{8\xi} - 2\xi \right] P(\xi) d\xi = \int_{-\infty}^{\infty} k\, P'(k)\, dk \tag{2.10}$$

in laboratory coordinates. We have set $k \equiv (m'^2/8\xi - 2\xi)$, $P(\xi)d\xi = P'(k)dk$, and $m' \equiv 2m$. The mapping $\xi \to k$ is 2:1 since both $-\infty < \xi < 0$ and map to $-\infty < k < \infty$. On the other hand one checks that $\frac{1}{2} \int_{-\infty}^{\infty} (d\Pi(x) \wedge d\varphi(x))dx = \int_{-\infty}^{\infty} \{d(\frac{1}{2}\gamma_0^{-1}\varphi_\xi(\xi)) \wedge d(\frac{1}{2}\varphi(\xi))\}\, d\xi$ when $\xi = x+t$, $\eta = x-t$. Then $4\varphi_{\xi\eta} = m'^2 \sinh\varphi = \varphi_{xx} - \varphi_{tt}$ (using $m' \equiv 2m$). Further one finds γ_0', the coupling constant for the covariant problem is $\gamma_0' \equiv 4\gamma_0$. More detail is given in [5].

Next, using

$$h(\xi) = \tfrac{1}{2}(h'(\xi) + p'(\xi)), \quad p(\xi) = \tfrac{1}{2}(h'(\xi) - p'(\xi)) \tag{2.11}$$

one finds $h'(\xi) = (m'^2 + p'(\xi)^2)^{\frac{1}{2}}$, $p'(\xi) = k$, so that $\omega(k) = (m^2+k^2)^{\frac{1}{2}}$. Dropping the primes we thus have the somewhat remarkable results that

$$H[p] = \int_{-\infty}^{\infty} \omega(k)P(k)dk \quad , \quad P[p] = \int_{-\infty}^{\infty} kP(k)dk \, . \tag{2.12}$$

The notation $H[p]$, etc. indicates that these quantities are expressed in action-angle variables. Under the mapping $\xi \to k$, $Q(\xi) \to Q(k)$, and $\{P(k), Q(k')\} = \delta(k-k')$ with $0 \le P(k) < \infty$, $0 \le Q(k) < 2\pi$. Thus $H[p]$,etc. depend only on the action variables.

In the same way one proves for the repulsive NLS that

$$N[\varphi] \equiv \int_{-\infty}^{\infty} \varphi^*\varphi\, dx \leftrightarrow N[p] = \int_{-\infty}^{\infty} P(k)\, dk$$

$$P[\varphi] \equiv \int_{-\infty}^{\infty} i\varphi_x^*\varphi\, dx \leftrightarrow P[p] = \int_{-\infty}^{\infty} kP(k)\, dk$$

$$H[\varphi] \text{ (eqn.(1.2))} \leftrightarrow H[p] = \int_{-\infty}^{\infty} \omega(k)P(k)dk \tag{2.13}$$

and $\omega(k) = k^2$. The last is the energy of a free non-relativistic particle of mass $m = \frac{1}{2}$, a result consistent with the form of the quantum problem (1.7). These results for repulsive NLS indicate why that model can be thought of as a non-relativistic limit of the covariant s-G model.

However, the more remarkable feature of (2.12) and (2.13) is that these expressions are exactly those one would find respectively for the linear K-G and linear LS models. There is no actual incompatibility here since the results for the linear problems are found (essentially) by Fourier transform, those for the nonlinear problems are found by the spectral transform. They are both canonical transforms with different inverse transforms. Unfortunately the very real difference this represents becomes obscured once we embark on the statistical mechanics. For in the evaluation of the partition function Z for the free energy one expects to average over the action (and angle) variables. We therefore look at this problem next.

3. CLASSICAL STATISTICAL MECHANICS

For a classical 1 D- oscillator

$$Z = \int \exp -\beta H[q](2\pi)^{-1} \, dpdq = \int \exp -\beta H[p](2\pi)^{-1} dPdQ \; ; \qquad (3.1)$$

the notation follows §2 so that

$$H[q] = \tfrac{1}{2}p^2 + \tfrac{1}{2}\,\omega^2 q^2 \; ; \quad H[p] = \omega P \, . \qquad (3.2)$$

Then indeed $Z = (\beta\omega)^{-1}$ from the first expression and $Z = (\beta\omega)^{-1}$ from the second. Implicit is the Jacobian $\partial(p,q)/\partial(P,Q) = 1$ for a canonical transformation. Then the two equal measures $(2\pi)^{-1}dpdq$ and $(2\pi)^{-1}dPdQ$ appeal to the fact that $\hbar = 1$, $h = 2\pi$ so $(2\pi)^{-1}$ is h^{-1}.

In the case of a field $\varphi(x,t)$ with running index x (§2) one evidently needs a continuous infinity of integrals placed in product form in the expressions for Z. This is a functional integral. The first expression for Z in (3.1) applies to any classical system with one degree of freedom and given H[q]. Generalising this to $q \to \varphi(x,t)$, $p \to \Pi(x,t)$ we thus get the functional integral

$$Z = \int \mathbb{D}\Pi \; \mathbb{D}\varphi \, \exp -\beta H[\varphi] \qquad (3.3)$$

and $H[\varphi]$ would be given, for example, by (1.14) in the case of s-G or its continuation in the case of sinh-G. We can then expect to write down the equivalent of the second form for Z in (3.1), namely

$$Z = \int \mathbb{D}\mu \, \exp -\beta H[p] \qquad (3.4)$$

where, now particularly for sinh-G, $H[p]$ is given by (2.12). However, although (3.4) is to follow from (3.3) by canonical transformation we must still find the correct measure $\mathbb{D}\mu$ for the transformed integral. This proves to be [18]

$$\mathbb{D}\mu = \lim_{N\to\infty} (2\pi)^{-(N+1)} \prod_{n=-\frac{1}{2}N}^{+\frac{1}{2}N} dP_n dQ_n \qquad (3.5)$$

and the normalisations on each element $dP_n dQ_n$ are natural. The precise definitions of the P_n and Q_n are given below (4.2), but $P_n \to P(k)$ (actually $P(k)dk$), $Q_n \to Q(k)$ as $N \to \infty$.

It is now easy to evaluate both (3.3) and (3.4) for the linear K-G equation. Some details are given in [5]: one point is that as $\mathbb{D}\Pi\mathbb{D}\varphi$ is discretized to $\prod_n d\Pi_n d\varphi_n$ the natural measure factors $(2\pi)^{-1}$ must again appear with each $d\Pi_n d\varphi_n$. In this way the results for both (3.3) and (3.4) for linear K-G coincide [5] in

$$F_{KG} \equiv -\beta^{-1} \ln Z_{KG} = \beta^{-1} a^{-1}(\ln \beta a^{-1} + \frac{1}{2} ma) \ . \qquad (3.6)$$

The parameter a is a lattice parameter which arises by writing the functional integrals for Z, as their definitions demand, as products of a finite number N+1 of integrals and taking the limit $N \to \infty$. The system is therefore placed on a lattice with a finite number of independent degrees of freedom N and ultimately the limits $N\to\infty$ and $a \to 0$ are to be taken. The lim $a \to 0$ cannot be done in (3.6), but this is because of the physics. The classical statistical mechanics has ultra-violet divergences which can be removed (i.e. renormalised) only in the quantum theories.

4. CLASSICAL STATISTICAL MECHANICS OF THE SINH-G MODEL

Plainly we are now faced with a difficulty. If we use (3.4) for Z with $H[p]$ given by (2.12) we can only regain the classical K-G result (3.6)! In a quantum calculation (§§4 and 5) we would recognize this as a free-field result. On the other hand we can actually use (3.3) directly in the classical cases even though $H[\varphi]$ describes a nonlinear system. The method is the transfer integral method (TIM) and appeals to periodic boundary conditions on φ[5,31]. The calculation has been done for sinh-G [18], and the result of the classical analysis is the low-T asymptotic series

$$\lim_{L\to\infty} FL^{-1} = m\beta^{-1}\left[\frac{1}{4}t - \frac{1}{8}t^2 + \frac{3}{16}t^3 - \frac{53}{128}t^4 \ldots - \frac{478935069186}{2^{23}}t^{12} + \ldots\right] + F_{KG}; \quad (4.1)$$

F_{KG} is the classical free energy of the free K-G given by (3.6), and T = temperature = β^{-1}, $t = (M\beta)^{-1}$ and $M \equiv 8m\gamma_o^{-1}$ ($\gamma_o > 0$), which would also be the mass of the kink or antikink in the s-G model. This result shows that the free energy of the classical sinh-G model is not F_{KG}, and it becomes this only when $\gamma_o = 0$. A corollary is that for linear K-G the periodic b.c.s. of the TIM do not change the result (3.6) for F_{KG}. In §3 periodic b.c.s. were not imposed for F_{KG}.

This linear result is misleading and does not extend to the non-linear models. The error in simply using (2.12) in (3.4) is a failure to define with sufficient care a proper thermodynamic limit. An exactly equivalent problem arises in constructing an applicable form of the quantum inverse method where [24], by defining the Hamiltonian operator $\hat{H}_\mu = \hat{H} - \mu\hat{N}$ on a finite interval $L \in \mathbb{R}$ (μ is a chemical potential and $\hat{N}$ is the number operator of §1 and (4.6) below) the energy separation between the vacuum $|0>$ and the minimum of $\hat{H}_\mu$ is rendered finite. The method is once again to impose periodic b.c.s. of period L. We sketch the argument later in this section and more details can be found in ref.[2,24-26]. There we also show how the quantum inverse method is equivalent to the BA method. Lieb and Liniger [32] were perhaps the first deliberately to solve such a problem by imposing periodic b.c.s., but periodic b.c.s. lie at the heart of the BA method anyway [2]. Notice that this procedure ensures a finite particle density $\bar{n} = NL^{-1}$.

We now see that we can expect to reach a proper classical or quantum thermodynamic limit of finite density by imposing periodic b.c.s. Conveniently the system is first placed on a lattice, spacing a, say, so that N particles in a period L under periodic b.c.s. mean $(N+1)a = L$. A proper thermodynamic limit is then $\lim_{N,L\to\infty} NL^{-1} = \bar{n} > 0$. One such limit is evidently a^{-1}. However for a field $\varphi(x,t)$ we also need $a \to 0$. We shall see how this combination is achieved shortly. First we address the problem of periodic boundary conditions itself.

The action-angle variables of §2 were all found for $x \in \mathbb{R}$ with decaying boundary conditions at $\pm\infty$. Since L is now *a priori* infinite the particle density is zero. Evidently we need action-angle variables under periodic b.c.s. with period $L < \infty$.

For the integrable systems it is natural to start with an integrable lattice with these b.c.s. Izergin and Korepin [33,34] have introduced integrable lattices reducing to s-G and sinh-G as $a \to 0$. We have found action-angle variables for these lattices [18,35,36] valid to $O(L^{-1})$. We sketch some results for sinh-G.

With $L = (N+1)a$ we find

$$H[p] = \sum_{n=-\frac{1}{2}N}^{+\frac{1}{2}N} \omega(\bar{k}_n)P_n + O(L^{-1}). \tag{4.2}$$

We find it convenient to carry both $n = -\frac{1}{2}N$ and $n = +\frac{1}{2}N$ in (4.2) and choose N even: the period is still $(N+1)a = L$. The P_n in (4.2) relate to the $P(k)$, as $L \to \infty$, through $P_n = P(\bar{k}_n)(2\pi L^{-1} + O(L^{-1})) \to P(\bar{k})d\bar{k}$ as $2\pi nL^{-1} = k_n \to k$ as $L \to \infty$. Thus as $L \to \infty$, $(4.2) \to \int_{-\infty}^{\infty}\omega(k)P(k)dk$ providing the modes $\bar{k}_n$ properly fill the k-space as $L \to \infty$.

The modes $\bar{k}_n$ are the allowed modes. The periodic b.c.s. restrict the allowed modes to those satisfying the integral equation [18,35]

$$\bar{k}_n = k_n - \sum_{m=-\frac{1}{2}N,\, m\neq n}^{\frac{1}{2}N} \Delta_c(\bar{k}_n,\bar{k}_m)P_m \,. \tag{4.3}$$

The analysis determines the phase shifts Δ_c and for classical sinh-G and repulsive NLS these are [18,35]

$$\Delta_c(k,k') = \frac{-2m^2(\frac{1}{8}\gamma_o)}{k\omega(k') - k'\omega(k)} \tag{4.4}$$

(with $\gamma_o > 0$ and $\omega(k) = (m^2+k^2)^{\frac{1}{2}})$ and

$$\Delta_c(k,k') = -2c(k - k')^{-1} \tag{4.5}$$

respectively. For details we refer the reader to [35] where they are found by using classical Floquet theory on the integrable sinh-G and NLS lattices. Notice a does not appear, i.e. lim $a \to 0$ is already taken. However, for $a > 0$ $\omega(k)$ in (4.4) actually involves a and this will show itself in the free field contribution F_{KG} to the free energy as (3.6) has already shown.

To present the classical Floquet theory [5,18,35,36] here would divert from our main theme. Instead we attempt to motivate the all important result (4.3) by describing the corresponding approach in the quantum inverse method. This allows us to bring together the quantum inverse and BA methods and this way we can relate both of these to our own work [5,18,36-38].

The quantum repulsive NLS model has played a fundamental role in the development of the quantum inverse method [2,24,25]. The sketch of the argument we give here is for that model and largely follows Korepin [39]. The Hamiltonian of the model is the normally ordered operator (1.6). The commutation relation is (1.5). Evidently

$$\hat{N} = \int \varphi^{\dagger}\varphi dx \ , \qquad \hat{P} = i \int \varphi_x^{\dagger}\varphi dx \tag{4.6}$$

commute with $\hat{H}$. The corresponding BA problem is (1.7) solved in [2,32] and used in [1].

First we need a remark about the classical NLS under periodic b.c.s. This is solved by introducing the monodromy matrix

$$T(\lambda) = \begin{bmatrix} A(\lambda) & B(\lambda) \\ C(\lambda) & D(\lambda) \end{bmatrix} , \quad \lambda \in C \tag{4.7}$$

(say). For an integrable lattice $\varphi = \varphi_n$ on distinct sites n spaced with lattice spacing a the spectral operator $\hat{L}$, such that $\hat{L}v = \zeta v$, of §2 is replaced by $L(n|\lambda)$ such that $v_{n+1} = L(n|\lambda)v_n$ and a Lax pair description can be found [24,25,35,39]. Under periodic b.c.s. $\varphi_{n+N} = \varphi_n$ so

$$T(\lambda) = \prod_{n=\frac{1}{2}N-1}^{-\frac{1}{2}N} L(n|\lambda) \tag{4.8}$$

which is ordered from $n = \frac{1}{2}N-1$ to the right (and we again assume N even). For present purposes it is sufficient to assume that the relevant NLS lattice is integrable when $a \to 0$. Then to $0(a^2)$

$$L(n|\lambda) = \begin{bmatrix} 1-\frac{1}{2}i\lambda a & -i\sqrt{c}\ \varphi_n^* \\ i\sqrt{c}\ \varphi_n & 1+\frac{1}{2}i\lambda a \end{bmatrix} \tag{4.9}$$

while

$$\varphi_n = a^{-1} \int_{x_n-1}^{x_n} \varphi(x)\ dx \tag{4.10}$$

so that

$$\{\varphi_n, \varphi_m^*\} = i\ \delta_{mn} \ . \tag{4.11}$$

It is now plain that the bracket condition (4.11) on the φ_n induces bracket relations on the elements of $T(\lambda)$. The Leningrad school introduces the classical r-matrix $r(\lambda,\mu)$(a 4 × 4 matrix for NLS) such that

$$\{T(\lambda) \overset{\otimes}{,} T(\mu)\} = [T(\lambda)\otimes T(\mu), r(\lambda,\mu)]. \tag{4.12}$$

As usual ⊗ is Kronecker product: so $T(\lambda) \otimes T(\mu)$ is 4 × 4 and the right side is the usual matrix commutator. The left side is simply the corresponding matrix of Poisson brackets. For the repulsive NLS

$$r(\lambda,\mu) = \frac{c}{\lambda-\mu}\begin{bmatrix} 1 & 0 & 0 & 0 \\ 0 & 0 & 1 & 0 \\ 0 & 1 & 0 & 0 \\ 0 & 0 & 0 & 1 \end{bmatrix} \equiv \frac{c}{\lambda-\mu}\,\Pi . \tag{4.13}$$

There are natural 8 × 8 matrices $r_{12}(\lambda,\mu) \equiv r(\lambda,\mu) \otimes I$ and $r_{23}(\lambda,\mu) = I \otimes r(\lambda,\mu)$ ($I = I_2$ is the 2 × 2 unit matrix). The 12 and 23 notation describes two of the three ways of embedding a 2-particle configuration space in a 3-particle space. Evidently there is $r_{13}(\lambda,\mu)$ also and we find

$$[r_{13}(\lambda,\nu), r_{23}(\mu,\nu)] + [r_{12}(\lambda,\mu), r_{13}(\lambda,\nu)] + [r_{12}(\lambda,\mu), r_{23}(\mu,\nu)]=0. \tag{4.14}$$

(4.14) is a statement on the scattering of three particles-namely that this breaks down to a product of two-particle scatterings. It is in fact the semi-classical limit of the Yang-Banter condition also introduced by Wadati and Akutsu [26] in their contribution.

The trace of the right side of (4.12) vanishes. So

$$\{\Delta(\lambda), \Delta(\mu)\} = 0 ; \quad \Delta(\lambda) \equiv \mathrm{Tr}T(\lambda) . \tag{4.15}$$

Consequently $\Delta(\lambda)$ is a generator of classical Hamiltonians H with a continuous infinity of commuting constants $\{H,\Delta(\lambda)\} = 0$ (the use of $\Delta(\lambda) \equiv \mathrm{Tr}T(\lambda)$ here should cause no confusion with the use of Δ for phase shift in what follows). The existence of the r-matrix satisfying (4.14) guarantees the existence of a Lax representation for the lattice if it applies exactly for that lattice, it guarantees a Lax representation for the field $\varphi(x,t)$ derived from any lattice as $a \to 0$, and it also guarantees the integrability of both in the spectral transform sense, and their complete integrability in the Liouville-Arnold sense. Thus (4.14) is a classical integrability condition.

Korepin's integrable NLS lattice is chosen to preserve the same r-matrix for both the lattice and its continuum limit. The lattice has

$$L(n|\lambda) = \begin{bmatrix} 1-\frac{1}{2}i\lambda a+\frac{1}{2}c\psi_n^* \psi_n a^2 & -i\sqrt{c}\ \psi_n^*\rho_n a \\ +i\sqrt{c}\ \rho_n\psi_n a & 1+\frac{1}{2}i\lambda a+\frac{1}{2}c\psi_n^*\psi_n a^2 \end{bmatrix} \qquad (4.16)$$

with $\rho_n \equiv [1+\frac{1}{4}c\psi_n^*\psi_n a^2]^{\frac{1}{2}}$, $\{\psi_n^*, \psi_m\} = -ia^{-1}\delta_{nm}$ and $\{\psi_n,\psi_m\} = \{\psi_m^*,\psi_m^*\} = 0$. The transfer matrix $L(n|\lambda)$ satisfies

$$\{L(n|\lambda) \otimes L(m|\mu)\} = \delta_{nm}[L(n|\lambda) \otimes L(m|\mu),\ r(\lambda,\mu)] \ , \qquad (4.17)$$

$r(\lambda,\mu) = c(\lambda-\mu)^{-1}\Pi$ (i.e. (4.13)), and (4.17) implies (4.12) for $T(\lambda)$ defined by (4.8).

This summarises the r-matrix theory of the classical integrable models epitomized by repulsive NLS. The generalisation to the quantum case and the quantum inverse method is now relatively straightforward: φ_n^*, φ_n become operators $\varphi_n^\dagger$, φ_n and

$$[\varphi_n,\ \varphi_m^\dagger] = \delta_{nm}a^{-1} \ . \qquad (4.18)$$

Then $\hat{L}(n|\lambda)$ is the operator

$$\hat{L}(n|\lambda) = \begin{bmatrix} 1-\frac{1}{2}ia\lambda & -i\sqrt{c}a\varphi_n^\dagger \\ i\sqrt{c}a\varphi_n & 1+\frac{1}{2}ia\lambda \end{bmatrix} + 0(a^2) \ . \qquad (4.19)$$

Define the ordered operator

$$\hat{T}(m,n|\lambda) = \hat{L}(m|\lambda) \ldots \hat{L}(n|\lambda) \ . \qquad (4.20)$$

The periodic b.c.s. mean $\hat{T}(m+N,n+N|\lambda) = \hat{T}(m,n|\lambda)$ so focus on

$$\hat{T}(\tfrac{1}{2}N-1,\ -\tfrac{1}{2}N|\lambda) \equiv \hat{T}(\lambda) \qquad (4.21)$$

for which

$$\hat{\Delta}(\lambda) \equiv \mathrm{Tr}\hat{T}(\lambda) \qquad (4.22)$$

is an operator. It can be shown that

$$R(\lambda,\mu)\hat{T}(\lambda)\otimes \hat{T}(\mu) = \hat{T}(\mu)\otimes \hat{T}(\lambda)R(\lambda,\mu) \ . \qquad (4.23)$$

The matrix $R(\lambda,\mu)$ is a 4 × 4 c-number matrix

$$R(\lambda,\mu) = \Pi\!\Pi - \frac{ic}{\lambda-\mu} I_4 , \tag{4.24}$$

where I_4 is the 4 × 4 unit matrix and $\Pi\!\Pi$ is defined in (4.13). $R(\lambda,\mu)$ is the R-matrix and (4.23) are the commutation relations for the operator elements $\hat{A}(\lambda)$, $\hat{B}(\lambda)$, etc. of the monodromy matrix operator $\hat{T}(\lambda)$. Evidently $\hat{T}(\lambda)$ is the quantum form of (4.7) and (4.23) that of (4.12). The matrix R satisfies the Yang-Baxter relation

$$(I \otimes R(\lambda,\mu))\ (R(\lambda,\nu) \otimes I)\ (I \otimes R(\mu,\nu)) =$$

$$(R(\mu,\nu) \otimes I)(I \otimes R(\lambda,\nu))(R(\lambda,\mu) \otimes I) . \tag{4.25}$$

A solution of this relation (namely (4.24)) determines the commutation relations (4.23). From

$$R\hat{T}(\lambda) \otimes T(\mu)R^{-1} = \hat{T}(\mu) \otimes \hat{T}(\lambda) \tag{4.26}$$

and invariance of the trace of 4 × 4 matrices

$$\hat{\Delta}(\lambda)\ \hat{\Delta}(\mu) = \hat{\Delta}(\mu)\hat{\Delta}(\lambda) . \tag{4.27}$$

Thus $\hat{\Delta}(\lambda)$ is a generator of Hamiltonian operators $\hat{H}$ and

$$[\hat{H},\hat{\Delta}(\lambda)] = 0 . \tag{4.28}$$

The existence of R guarantees an (operator) Lax representation. Thus the Yang-Baxter relation (4.25) is a quantum integrability condition.

For the present purposes the following results, which can be derived in the standard way(see for example ref.[2]),are what we need. First of all for the repulsive NLS model there are states$|\{\lambda_j\};N>$, $j = 1,2,\ldots, N$ such that

$$\hat{N}|\{\lambda_j\};N> = N|\{\lambda_j\};N> \tag{4.29a}$$

$$\hat{P}|\{\lambda_j\};N> = \left[\sum_{j=1}^{N} \lambda_j\right] |\{\lambda_j\};N> \tag{4.29b}$$

$$\hat{H}|\{\lambda_j\};N> = \left[\sum_{j=1}^{N} \lambda_j^2\right]|\{\lambda_j\};N> \tag{4.29c}$$

and these results correspond to (2.12). The operator $\hat{B}(\nu)$ is a raising operator and $\hat{B}(\nu)|\{\lambda_j\};N> = |\nu,\{\lambda_j\};\ N+1>$, so $\hat{B}(\nu)$ creates the eigenstates of (4.29). The eigenvalues in (4.29) are evidently the free field eigenvalues. However, the second result we need is that under the periodic b.c. only certain λ_j are allowed, namely those satisfying

$$\lambda_n = 2\pi n\ L^{-1} + L^{-1} \sum_{m=\frac{1}{2}N-1,\, m\neq n}^{-\frac{1}{2}N} \Delta(\lambda_m - \lambda_n) \ ,$$

$$\Delta(\lambda_m - \lambda_n) = -2\ \tan^{-1}[c/(\lambda_m - \lambda_n)] \ . \tag{4.30}$$

We can associate λ_n and $2\pi nL^{-1}$ simply to label λ_n by n. However, for c = 0, $\Delta = 0$, so we can identify λ_n with allowed wave vectors $\tilde{k}_n$. Then

$$\tilde{k}_n = k_n - L^{-1} \sum_{m=\frac{1}{2}N-1,\, m\neq n}^{-\frac{1}{2}N} \Delta(\tilde{k}_n, \tilde{k}_m) \tag{4.31}$$

with

$$\Delta(k,k') = -\ 2\ \tan^{-1}[c/(k-k')] \tag{4.32}$$

(We note in passing a problem in maintaining the limits $-\frac{1}{2}N \leq n \leq \frac{1}{2}N-1$ if the $\tilde{k}_n$ are allowed wave vectors(compare with(4.39) below)).

The results (4.29) and (4.31) with (4.32) are exactly those found by the BA method also [2]. The BA method starts from (1.8), makes an ansatz for ψ which solves it exactly when (4.31) with (4.32) applies, and thus arrives at (4.29) for the eigenvalues. Thus the results (4.29) and (4.31) bring together these two otherwise different calculational procedures.

One can show that the result corresponding to (4.31) for sinh-G is (after writing exp λ for λ so that the new λ is a rapidity)

$$m_1 \sinh 2\lambda_n = 2\pi nL^{-1} - L^{-1} \sum_{m=\frac{1}{2}N-1,\, m\neq n}^{-\frac{1}{2}N} \Delta(\lambda_n, \lambda_m) \tag{4.33}$$

where,

$$\Delta(\lambda,\lambda') = -i\ \ln \frac{\sinh(\lambda-\lambda'+i\mu)}{\sinh(\lambda-\lambda'-i\mu)} \ , \tag{4.34}$$

m_1 is a partially renormalised mass [16,40], and μ is the coupling constant (§1) related to γ_o by $\mu = \pi(1 + (\gamma_o/8\pi))$. However, since sinh-G is covariant it is necessary to fill a Dirac sea with particles: this dresses the "bare" particles and the result is finally (4.31) again [18,40] with

$$\Delta(k,k') = -2 \tan^{-1}\left[\frac{m^2 \sin\left[\frac{\gamma_o''}{8}\right]}{k\omega(k') - k'\omega(k)}\right], \tag{4.35}$$

where $k = m \sinh\lambda$, $\omega(k) = m\cosh\lambda$ and γ_o'' is given by (1.26). In this expression m is the fully renormalised mass [16,40] and can be identified with the physical m for sinh-G.

The form (4.31) is therefore rather general and apparently applies whenever the classical H can be expressed in the form (4.2), that is in the forms of (2.12) and (2.13). However, we must still specify the branches of the $\tan^{-1}$ expressions for the various Δ. We choose these in the first place to be the continuous branch running between $\Delta(k,k') = -2\pi$ for $k \to -\infty$ to $\Delta(k,k') = 0$ for $k \to +\infty$ at fixed k'. When $k = k'$, $\Delta = -\pi$. We then indicate this particular choice of branch by writing Δ_f for Δ. However, it is evident that Δ_f is a form of the 2-body S-matrix phase shift of the chosen quantum model (refer to reference [2] for quantum NLS). The true S-matrix phase shift should vanish for $k \to +\infty$ and for $k \to -\infty$. We therefore define a phase shift

$$\Delta_b(k,k') = \Delta_f(k,k') + 2\pi\Theta(k'-k) \tag{4.36}$$

with this property: $\Theta(k)=1$, $k>0$, $\Theta(k)=0$, $k<0$. Evidently Δ_b is singular at $k = k'$. However, for $c \to 0$, $\gamma_o \to 0$ Δ_b then has the semi-classical forms for NLS and sinh-G

$$\Delta_c = -2c/(k-k'), \tag{4.37}$$

$$\Delta_c = -2m^2(\tfrac{1}{8}\gamma_o)/(k\omega(k') - k'\omega(k)), \tag{4.38}$$

respectively. These are precisely the expressions (4.5) and (4.4) for Δ_c found in the classical Floquet theoretical analysis [18,35].

The classical condition (4.3) for the allowed modes $\tilde{k}_n$ found by classical Floquet theory, and the result (4.31) found by the quantum inverse method as sketched, or by the BA method as given in e.g. [2], now all become consistent if Δ in (4.3) is Δ_f and

$$P_n = 1, \quad \tilde{k}_n \text{ allowed}; \quad P_n = 0, \quad \tilde{k}_n \text{ not allowed}. \tag{4.39}$$

It is clear from (2.3) in §2 that $P_n \to P(\bar{k}_n)(2\pi L^{-1}+0(L^{-1})) \to P(\bar{k})d\bar{k}$ can be interpreted as a particle number in the mode $\bar{k}_n$. Indeed comparison of (2.13), (4.6) and (4.29) confirms this view. Thus (4.39) is a fermion description. Thacker [2] shows by BA that choosing the smooth branch for Δ is equivalent to making a fermion description despite the bose character of the model. We can then intuit that Δ_b is a description in terms of bosons so that

$$P_n = m_n \; ; \; m_n = 0, 1, 2, \ldots . \tag{4.40}$$

Then $P_n \leftrightarrow P(\bar{k})d\bar{k}$ is the natural extension from this description in terms of bosons to a description in terms of classical Maxwell-Boltzmann (MB) particles. Accordingly, by using appropriate phase shifts Δ_f, Δ_b or Δ_c, (4.3) generalises the BA or quantum inverse method condition (4.31) to descriptions in terms of fermions or bosons in the quantum case and then extends it to the classical case as well.

5. QUANTUM STATISTICAL MECHANICS

A fundamental object to consider is the quantum partition function Z: like the classical partition function (3.3) this will be a functional integral. We choose

$$Z = \int \mathbb{D}\Pi \, \mathbb{D}\varphi \, \exp S[\varphi] , \tag{5.1}$$

$$S[\varphi] = \int_0^\beta d\tau \left[\int \Pi(x,\tau)\varphi(x,\tau)_{,\tau} dx - H[\varphi]\right] . \tag{5.2}$$

Note that $H[\varphi]$, Π and φ are classical quantities: $\{\Pi,\varphi\} = \delta(x-x')$. This form can be reached [5,41] by starting from the quantum mechanical Feynman propagator expressed in Hamiltonian form and performing the Wick rotation $t \to -i\tau$ followed by integration on 'time' τ from 0 to β. Since the trace is required for Z, $\varphi(x,0)$ runs to $\varphi(x,0)$ as $\tau \to \beta$. Thus (5.1) is periodic in τ of period β. But $\varphi(x,\tau)$ is also periodic in x, of period L, because of the thermodynamic limit. Thus (5.1) is defined on the space-time torus $-\frac{1}{2}L \le x < \frac{1}{2}L$; $0 \le \tau \le \beta$ (and τ includes $\tau=\beta$).

If one notes that $\int_0^\beta d\tau$ in (5.2) is really $\hbar^{-1} \int_0^{\beta h} d\tau$ and lets $\hbar \to 0$, one regains (3.3) [5] from (5.1) and (5.2). Thus for models like repulsive NLS and sinh-G which have no soliton solutions and have action-angle variables $P(k)$, $Q(k)$, $k \in \mathbb{R}$ but no others, the quantum form of (3.4) will be

$$Z = \int \mathbb{D}\mu \ \exp S[\rlap{/}p] \tag{5.3}$$

with the measure $\mathbb{D}\mu$ given by (3.5):

$$S[\rlap{/}p] = \int_0^\beta d\tau \left[i \int P(k)Q(k)_{,\tau} dk - H[p] \right] , \tag{5.4}$$

and H[p] is given by forms like (2.12). We should stress again that all of these quantities are classical quantities. Moreover P(k), Q(k) and H[p] are all real.

Now we can see that $F = -\beta^{-1} \ln Z$ is a real free energy when (and more-or-less only when)

$$\int P_n Q_{n,\tau} d\tau = \oint P_n dQ_n = 2\pi m_n \tag{5.5}$$

in which m_n is an integer. Since $\hbar = 1$, $h = 2\pi$ as noted (§3), so (5.5) is Bohr semi-classical quantisation. If we set $m_n = 0,1,2, \ldots$ (bosons) or $m_n = 0,1$ (fermions), then $P_n = 0,1,2, \ldots$ (bosons) or $P_n = 0,1$ (fermions). This is consistent with the results of the last section. Then from (5.3) and (5.4)

$$Z = \int \mathbb{D}\mu \ \exp - \beta H[p] , \tag{5.6}$$

together with the quantisations $P_n = 0,1,2, \ldots$ or $P_n = 0,1$ and the allowed modes

$$\bar{k}_n = k_n - \sum_{m \neq n} \Delta(\bar{k}_n, \bar{k}_m)_{b \text{ or } f} P_m \tag{5.7}$$

in the two cases. Equation (4.3) for classical MB particles now fits smoothly into this: (4.3) is (5.7) with Δ_c replacing Δ_b or Δ_f and P_m is the classical action variable. Notice that (5.6) for the quantum cases is formally equivalent to the classical (3.4). Note too that in (5.7) it is now convenient to let m,n run from $-\frac{1}{2}N$ to $+\frac{1}{2}N$ rather than to $\frac{1}{2}N-1$ (refer by (4.2)).

The functional integral (5.6) must keep the same value evaluated in terms of bosons or in terms of fermions. Before demonstrating this by an actual calculation, we first evaluate Z by the very different route pioneered by Yang and Yang [1]. Here a fermion description was used, so we shall differ from that work by making the equivalent analysis in terms of bosons. This calculation is reported [18,42] and Wadati [43] has done a similar calculation, but beyond this nothing seems to have been done in boson description in either the BA or quantum inverse methods. In particular the status of (5.7) is that it derives from the commutation relations (4.23) and the Yang-Baxter relations but then emerges in fermion description. We assume it generally true and proceed. The classical form (4.3) is proved rigorously in the classical case [5,18,35,36,44,45]. We handle both quantum repulsive NLS and quantum sinh-G together.

Since P_n is a particle number in $\tilde{k}_n$, $P_n L^{-1}$ is a number density (of bosons) in $\tilde{k}_n$. Then $P_n L^{-1} \to P(\tilde{k}_n) L^{-1} (2\pi L^{-1} + O(L^{-1})) \to \rho(k)dk$ (say), as $L \to \infty$; and $\rho(k) \equiv P(k)L^{-1}$. From (2.13), and guided by (4.29) for NLS or from (2.12) for sinh-G, we can then define energy, momentum, and number densities through

$$\lim_{L\to\infty} EL^{-1} = \int_{-\infty}^{\infty} \omega(k)\rho(k)\, dk \; ; \quad \lim_{L\to\infty} PL^{-1} = \int_{-\infty}^{\infty} k\rho(k)dk \; ;$$

$$\lim_{L\to\infty} NL^{-1} \equiv \bar{n} = \int_{-\infty}^{\infty} \rho(k)dk \; . \qquad (5.8)$$

Evidently $\omega(k) = k^2$ for NLS, while $\omega(k) = (m^2+k^2)^{\frac{1}{2}}$ for sinh-G upto ultraviolet divergence corrections (§3). Notice that the independent variable is $\tilde{k}_n$ which is assumed properly to fill the k-space with the vectors $\tilde{k}$ as $L \to \infty$. We can therefore set $k = h(\tilde{k})$ and can define a density of allowed states through $dh/d\tilde{k} \equiv 2\pi f(\tilde{k})$. Then, dropping the tilda again as in (5.8), (5.7) becomes

$$2\pi f(k) = 1 + \int \frac{d\Delta_b(k,k')}{dk} \rho(k')dk' \; . \qquad (5.9)$$

We then reach a second relation between $f(k)$ and $\rho(k)$ by defining a free energy F and minimising this: $F = E - \beta^{-1}S$ and we need an entropy S.

Following Yang and Yang [1], but referring to bosons, the number of possible states in dk consistent with ρ and f is

$$(L(\rho+f)dk)!/(L\rho dk)!\,(Lfdk)! \; . \qquad (5.10)$$

So after using Stirling's formula one finds an entropy per unit length

$$SL^{-1} = \int [(f+\rho)\ln(f+\rho) - f\ln f - \rho\ln\rho]\, dk \; . \qquad (5.11)$$

We now minimise $FL^{-1} = (E - \beta^{-1}S)L^{-1}$: SL^{-1} is given by (5.11) and EL^{-1} by (5.8). A condition for minimum FL^{-1} is that the functional derivative $\delta(FL^{-1})/\delta\rho = 0$. After a little manipulation one finds

$$\delta(SL^{-1})/\delta\rho = \ln\{(f+\rho)\rho^{-1}\} - \frac{1}{2\pi}\int\frac{d}{dk}\Delta_b(k,k')\ln\{(f(k')+\rho(k'))/f(k')\}dk' \tag{5.12}$$

since f relates to ρ through (5.9). Consequently the condition $\delta(FL^{-1})/\delta\rho = 0$ becomes

$$\ln\{(f+\rho)\rho^{-1}\} - \beta\omega(k) - \frac{1}{2\pi}\int_{-\infty}^{\infty}\frac{d}{dk}\Delta_b(k,k')\ln\{(f+\rho)f^{-1}\}dk' = 0. \tag{5.13}$$

It is therefore convenient to set

$$(f+\rho)\rho^{-1} = f\rho^{-1} + 1 \equiv \exp\beta\,\varepsilon(k) \tag{5.14}$$

in which $\varepsilon(k)$ are energies. These energies are allowed energies if, and only if, they satisfy (5.13), and this integral equation becomes

$$\varepsilon(k) = \omega(k) + \frac{1}{2\pi\beta}\int_{-\infty}^{\infty}\frac{d\Delta_b(k,k')}{dk}\ln(1 - e^{-\beta\varepsilon(k')})dk' \tag{5.15a}$$

while $\varepsilon(k) = \varepsilon(-k)$. From this result, and by using (5.9), we go on to show that the free energy FL^{-1} per unit length is given by

$$\lim_{L\to\infty} FL^{-1} = (2\pi\beta)^{-1}\int_{-\infty}^{\infty}dk\,\ln(1 - e^{-\beta\varepsilon(k)}), \tag{5.15b}$$

where the $\varepsilon(k)$ are determined by (5.15a). The free-field results are (c=0 for NLS, $\gamma_o = 0$ for sinh-G) then just (5.15b) with $\varepsilon(k)=\omega(k)$.

It is now plain that these results (5.15) in boson form have the classical limit

$$\varepsilon(k) = \omega(k) + (2\pi\beta)^{-1}\int_{-\infty}^{\infty}\frac{d}{dk}\Delta_c(k,k')\ln(\beta\varepsilon(k'))dk' \tag{5.16a}$$

$$\lim_{L\to\infty} FL^{-1} = (2\pi\beta)^{-1}\int_{-\infty}^{\infty}dk\,\ln(\beta\varepsilon(k)). \tag{5.16b}$$

Yang and Yang [1] prove that the fermion form of (5.15) (equations (5.19) below) iterates to yield an iterated expansion for FL^{-1} for the quantum NLS model. There is a problem in iterating the classical form (5.16) for the NLS model [45], but no such problem arises for the classical sinh-G model. Iteration of (5.16) in this case yields a sequence of integrals all of which can be done and the result is exactly the TIM result (4.1) [18]. Moreover, rather than take the classical limit of (5.15), one can derive (5.16) for classical MB particles. Hence the low temperature asymptotic result (4.1) confirms the consistency of all the calculations.

We now convert (5.15) to its fermion form. Introduce a chemical potential μ at (5.12) by minimising the negative pressure $-p = FL^{-1} - \mu NL^{-1} = (E - \beta^{-1}S - \mu N)L^{-1}$; $NL^{-1} = \bar{n}$ is given by (5.8) so that $\omega(k) \to \omega(k) - \mu$ in (5.15a) and $\mu\bar{n}$ adds to the right side of (5.15b). From (4.57), in which Δ_f is smooth,

$$\frac{d}{dk}\Delta_b(k,k') = \frac{d}{dk}\Delta_f(k,k') - 2\pi\delta(k-k') \ . \tag{5.17}$$

If one defines $\bar{\varepsilon}(k)$ through

$$\ln(1+e^{-\beta\bar{\varepsilon}(k)}) = -\ln(1 - e^{-\beta\varepsilon(k)}) \ , \tag{5.18}$$

one then finds

$$\bar{\varepsilon}(k) = \omega(k) - \mu - (2\pi\beta)^{-1}\int_{-\infty}^{\infty}\frac{d}{dk}\Delta_f(k,k')\ln(1+e^{-\beta\bar{\varepsilon}(k')})dk' \tag{5.19a}$$

and

$$\lim_{L\to\infty} FL^{-1} = \mu\bar{n} - (2\pi\beta)^{-1}\int_{-\infty}^{\infty}\ln(1 + e^{-\beta\bar{\varepsilon}(k')})dk' \ . \tag{5.19b}$$

Note that it is now Δ_f which appears under the integral sign. For the NLS $\Delta(k,k') = \Delta(k-k')$ and $d\Delta/dk = -d\Delta/dk'$. With this choice equations (5.19) are exactly the result derived in [1] by fermion description. It is now clear that the bose and fermi descriptions yield the same values for FL^{-1}.

However, for NLS in particular there are two interesting free particle limits. Since

$$d\Delta_f/dk' = 2c[(k-k')^2 + c^2]^{-1} \ , \tag{5.20}$$

and this $\to 0$ for $c \to \infty$ and $-2\pi\delta(k-k')$ for $c \to 0$, $c \to \infty$ yields $\varepsilon(k) = k^2 - \mu$, a gas of free fermions (and this is the 'impenetrable bose gas' [1,15]), while for $c \to 0$, $\bar{\varepsilon}(k) = \omega(k) - \mu - \beta^{-1}\ln(1+e^{-\beta\bar{\varepsilon}(k)})$ so $\varepsilon(k) = \omega(k) - \mu$, for a gas of free bosons. The free boson case follows directly from (5.15a) of course (once μ is included).

In the case of sinh-G one sees from (4.35) that, when $\gamma_o \to 0$, one again obtains a gas of free bosons. However since $\sin(\gamma_o''/8) \to$ 0, 1, 0 as $\gamma_o \to 0, 8\pi/15, 8\pi/7$ there seems to be no free fermion limit.

From (5.19) for both repulsive NLS and sinh-G one can examine the excitations above the quantum ground state by defining $k = k_F$ where $\bar{\varepsilon}(\pm k_F) = 0$ (note that $\bar{\varepsilon}(-k) = \bar{\varepsilon}(k)$). Then $\mu = \mu_F$ is fixed by this condition also and $\bar{\varepsilon}(k) > 0$, $|k| > k_F$, while $\bar{\varepsilon}(k) < 0$, $|k| < k_F$. Thus for $\beta^{-1} \to 0$

$$\bar{\varepsilon}(k) \sim \omega(k) - \mu_F - (2\pi\beta)^{-1} \int_{-k_F}^{+k_F} \frac{d\Delta_f}{dk} \ln(e^{-\beta\bar{\varepsilon}(k')})dk'$$

$$\sim \omega(k) - \mu_F + (2\pi)^{-1} \int_{-k_F}^{+k_F} \frac{d\Delta_f}{dk} \bar{\varepsilon}(k')\, dk' \ . \tag{5.21}$$

The excitation energies $E - E_o \equiv E_1$ above the ground state energy E_o correspond to lifting a particle to the state k_p above k_F so creating a hole at k_h, $|k_h| < k_F$: then [2] $E_1 = \bar{\varepsilon}(k_p) - \bar{\varepsilon}(k_h)$. Moreover, the ground state energy E_o can be calculated from $E_o L^{-1} = \int_{-k_F}^{+k_F} \omega(k)\rho(k)dk$ + constant by solving (5.22). For at $\beta^{-1} = 0$, (5.14) in the form $f\rho^{-1} = \exp \beta\bar{\varepsilon}$ means that $\rho(k) = 0$, $|k| > k_F$ and (5.9) then means that ρ satisfies

$$2\pi\rho(k) = 1 + \int_{-k_F}^{+k_F} \frac{d\Delta_f(k,k')}{dk} \rho(k')dk' \tag{5.22}$$

for $|k| < k_F$. We should note however that this integral equation is for the bose particle density $\rho(k)$: we discuss elsewhere how it relates to the fermion density used by Yang and Yang [1]. Otherwise the finite temperature results (5.19) evidently contain all of the quantum theory of the models at $\beta^{-1} = T = 0$ as we should expect.

6. EQUIVALENT RESULTS FOR THE s-G MODEL

The TIM result (4.1) for the low temperature expansion of FL^{-1} for classical sinh-G has the continuation in γ_o to $-\gamma_o$ which is (4.1) with $\gamma_o \to -\gamma_o$ i.e. $t \to -t$. The series in (4.1) becomes negative in every term, but F_{KG}, independent of γ_o, is unchanged. However in addition (4.1) now gains terms in $e^{-1/t}$, $e^{-2/t}$, ... [36] which apparently come from the kinks and antikinks: the term in $e^{-1/t}$ is [36]

$$- \beta^{-1}\left[\frac{8}{t\pi}\right]^{\frac{1}{2}} e^{-1/t}\left[1 - \frac{7}{8}t - \frac{59}{128}t^2 - \frac{897}{1024}t^3 - \frac{75005}{32768}t^4 - \ldots\right]. \tag{6.1}$$

It would be natural to associate the series for s-G as arising from the classical breathers but sinh-G has no breathers and the series is the continuation from the series in (4.1).

On the other hand, the whole of the classical MB analysis which leads to (5.16) for sinh-G has been generalised to classical s-G. The argument uses an H[p] which includes kink, anti-kink and K-G contributions but not breather contributions [36].

However, it is known that in the quantum s-G [16] (or MTM [10]) filling the Dirac sea (using the modes $\bar{k}_n$) dresses the system to quantum breather-like terms plus kink-antikink terms (cf. e.g. [19,37,38,46,47]). It has been shown that the classical limit of these quantum statistical mechanical integral equations for quantum s-G yields the classical continuation $\gamma_0 \to -\gamma_0$ of (4.1) [37,38,47]. The method of §5 which leads to (5.15) and (5.19) for quantum sinh-G and (5.16) for classical sinh-G we call [18,36] a 'generalised BA method'. The same method yields the quantum statistical mechanics of the s-G model [48]. So indeed does the functional integral method generalising (5.6) with (5.7). We return to calculate the functional integral (5.6) for quantum sinh-G in the next section.

7. EVALUATION OF THE FUNCTIONAL INTEGRALS FOR CLASSICAL AND QUANTUM Z

The classical Z for sinh-G or repulsive NLS is (3.4): it is important to notice that

$$H[p] = \sum_{-\frac{1}{2}N}^{\frac{1}{2}N} \omega(\bar{k}_n)P_n \tag{7.1}$$

in which $\omega(\bar{k}_n)$ depends on $\bar{k}_n$, while

$$\bar{k}_n = k_n - \sum_{m \neq n} \Delta_c(\bar{k}_n, \bar{k}_m)P_m . \tag{7.2}$$

The functional integral (3.4) can be evaluated by iterating (7.2) in (7.1) and the result for F can be put in the form [18]

$$FL^{-1} = (2\pi\beta)^{-1} \int_{-\infty}^{\infty} dk \ln(\beta\omega(k)) - (2\pi\beta)^{-2} \int_{-\infty}^{\infty} dq[\omega(q)]^{-1} \int_{-\infty}^{\infty} dk$$

$$\times \Delta_c(k,q)d(\ln\omega(k))/dk + (2\pi\beta)^{-3} \int_{-\infty}^{\infty} dp[\omega(p)]^{-1} \int_{-\infty}^{\infty} dq$$

$$\times \Delta_c(q,p)\frac{\partial}{\partial q} \left[[\omega(q)]^{-1} \int_{-\infty}^{\infty} dk\ \Delta_c(k,q)d(\ln\omega(k))/dk\right]$$

$$- \frac{1}{2}(2\pi\beta)^{-3} \int_{-\infty}^{\infty} dq[\omega(q)]^{-2} \left[\int_{-\infty}^{\infty} dk\ \Delta_c(k,q)d(\ln\omega(k))/dk\right]^2$$

$$+ \ . \ . \ . \ \ . \qquad (7.3)$$

This is exactly the (formal) iteration of the classical free energy (5.16b) with energies $\varepsilon(k)$ given by the classical integral equation (5.16a)!

For sinh-G the same iteration (or the iteration of the system (5.16)) yields the asymptotic series expression (4.1) for FL^{-1}; F_{KG} is given by the first integral in (7.3). Similar work for the classical repulsive NLS is not possible beyond a formal iteration since (5.16a) does not iterate then [45]. Otherwise the classical functional integral and generalised BA methods come together.

It is now plain that in the corresponding quantum analysis we use (5.6) with (5.7) and the quantisation $P_n = 0,1$ (fermion form) or $P_n = 0,1,2,\ldots$ (boson form). One readily finds that iteration of (5.7) through (5.6) with either of these quantisation conditions yields (5.19) (fermions) or (5.15) (bosons). Thus the functional integral methods derived from (5.3) and (5.4) in terms of action-angle variables reach, in either of these fashions, exactly the same results as the quantum forms of the generalised BA method described in §5.

It is worth stressing that the functional integral method follows Feynman's original approach by using the classical action throughout: but the action-angle variables are used after canonical transformation, and a measure problem arises and is solved (for sinh-G, NLS [18] (and also for s-G [36])): the semiclassical quantisations which follow discretize the action variables and this is more-or-less necessary for a real quantum free energy F. All of the quantum mechanics created by normal ordering then rests in the periodicity conditions (5.7), namely in the 2-body S-matrix phase shifts Δ_b or Δ_f. As noted in the theory presented in §5, Δ_f comes essentially from the commutation relations (4.23) of the quantum inverse method, therefore from the R-matrix, and so from the Yang-Baxter relations. We shall show how this can be done directly elsewhere [48]. The corresponding analysis for bosons has not been worked out yet (apparently).

Now how different the final form of the functional integral is from Feynman's original form [49]. The classical action is used, but this is in Hamiltonian form, the trivial (Bohr) quantisation conditions are imposed, so no time discretization is needed to evaluate the quantum functional integrals and the periodicity conditions carry the deeper quantum mechanics.

It is now evident that this functional integral method extends to quantum and classical integrable models other than sinh-G and repulsive NLS, namely to all of the integrable models. Bose-fermi equivalence plays a fundamental role even in the classical cases. We cite the case of classical s-G [36] as a non-trivial example of this extension: the results of the functional integral method coincide with those found by the generalised BA method described in §6. Results for the quantum s-G model seem to show a wholly similar measure of agreement [48].

Corresponding results for other integrable models (e.g. Landau-Lifshitz, §1) are to be reported [50].

REFERENCES

[1] C. N. Yang, C. P. Yang, J. Math. Phys. **10** (1969) 1115.
[2] H. B. Thacker, Rev. Mod. Phys. **53** (1981) 253.
[3] R. K. Bullough, P. J. Caudrey, in **Solitons**, ed. R. K. Bullough and P. J. Caudrey (Springer, Heidelberg, 1980) and the other chapters there.
[4] E. K. Sklyanin, LOMI preprint E-3-1979 (1979).
[5] R. K. Bullough, in **Nonlinear Phenomena in Physics**, ed. F. Claro (Springer, Heidelberg, 1985), pp. 70-102; R. K. Bullough, D. J. Pilling, J. Timonen, in **Nonlinear Phenomena in Physics**, pp. 103-128.
[6] M. Lakshmanan, Phys. Lett. **61A** (1977) 53.
[7] L. A. Takhtadzhyan, Phys. Lett. **64A** (1977) 235.
[8] S. Coleman, Phys. Rev. **D11** (1975) 2088.
[9] M. Takahashi, M. Suzuki, Prog. Theor. Phys. **48** (1972) 2187.
[10] H. Bergnoff, H. B. Thacker, Phys. Rev. **D19** (1979) 3666.
[11] Korepin [12,13] in effect uses the different relation between

μ and g_o that $\mu = \frac{1}{2}(\pi+g_o)$. In [12] he treats the attractive MTM with $0 < g_o < \pi$ ($\pi/2<\mu<\pi$ or, since $\pi-\mu = \frac{1}{8}\gamma_o$ (see below), $4\pi > \gamma_o > 0$) and in [13] he treats a repulsive MTM with $-\pi < g_o < 0$ ($0 < \mu < \pi/2$ or $8\pi > \gamma_o > 4\pi$).

[12] V. E. Korepin, TMP (USSR) **41** (1979) 169.
[13] V. E. Korepin, Comm. Math. Phys. **76** (1980) 165.
[14] A. Luther, Chap. 12 in **Solitons** Ref. [3], pp. 355-372.
[15] M. Jimbo, T. Miwa, Y. Mori, M. Sato, Physica **1D** (1980) 80 and references.
[16] E. K. Sklyanin, L. A. Takhtadzhyan, L. D. Faddeev, Theor. Mat. **40** (1979) 194.
[17] R. F. Dashen, B. Hasslacher, A. Neveu, Phys. Rev. **D11** (1975)3424.
[18] R. K. Bullough, D. J. Pilling, J. Timonen, J. Phys. A: Math. Gen. **19** (1986) L955.
[19] R. K. Bullough, D. J. Pilling, J. Timonen, in **Physics of Many-Particle Systems**, ed. A. S. Davydov (Ukrainian Academy of Sciences of the USSR, Kiev, 1986).
[20] P. Goddard, D. Olive, Int. J. Mod. Phys. **A1**, No. 2 (1986) 303.
[21] M. Jimbo, T. Miwa, in **Vertex Operators in Mathematical Physics**, ed. J. Lepowsky, S. Mandelstam and I. M. Singer (Springer, Heidelberg, 1984), pp. 275-290 and other papers there.
[22] V. Kac, **Infinite Dimensional Lie Algebras--An Introduction**, 2d ed (Cambridge University Press, Cambridge, 1985).
[23] For example S. Olafsson, R. K. Bullough, to be published.
[24] P. P. Kulish, E. K. Sklyanin, "Quantum Spectral Transform Method: Recent Developments", in **Proc. of the Tvärminne Symposium, Finland, 1981**, ed. J. Hietarinta and C. Montonen (Springer, Heidelberg, 1982).
[25] L. D. Faddeev, in **Proc. Ecole d'Eté de Physique Theorique, Les Houches 1982**, ed. R. Stora and J. B. Zuber (North-Holland, Amsterdam, 1983).
[26] M. Wadati and Y. Akutsu, Exactly Solvable Models in Statistical Mechanics, in this volume.
[27] H. J. de Vega, in **Proc. Symp. on Topological and Geometrical Methods in Field Theory** (World Scientific, Singapore, 1986), in press.
[28] V. I. Arnold, **Mathematical Methods of Classical Mechanics**(Springer, Heidelberg, 1978).
[29] R. K. Dodd, R. K. Bullough, Physica Scr. **20** (1979) 514.
[30] L. D. Faddeev, Chap. 11 in **Solitons** Ref. [3], pp. 339-354.
[31] D. J. Scalapino, M. Sears and R. S. Ferrell, Phys. Rev. **B6** (1972) 3409.
[32] E. H. Lieb and W. Liniger, Phys. Rev. **130** (1963) 1605.
[33] A. G. Izergin and V. E. Korepin, Lett. Math. Phys. **5** (1981) 199.
[34] A. G. Izergin and V. E. Korepin, in **Problems in Quantum Field Theory and Statistical Physics 3**, ed. P. P. Kulish and V. N. Popov, LOMI Vol. 120 "Nauka" Leningrad.
[35] Yi Cheng, Ph.D. Thesis, University of Manchester (1987).
[36] J. Timonen, M. Stirland, D. J. Pilling, Yi Cheng and R. K. Bullough, Phys. Rev. Lett. **56** (1986) 2233.
[37] J. T. Timonen, R. K. Bullough and D. J. Pilling, Phys. Rev. **B34** (1986) 6525.
[38] J. T. Timonen, R. K. Bullough and D. J. Pilling, Classical Limit of Bethe Ansatz Statistical Mechanics for the Massive Thirring Model, to be published.
[39] V. E. Korepin, in **Completely Solvable Systems in Field Theory**, Lectture Notes at SERC-LMS Symposium, University of Durham (1986).
[40] D. J. Pilling, R. K. Bullough and J. Timonen, to be published.
[41] C. Itzykson and J. B. Zuber, **Quantum Field Theory** (McGraw-Hill, New York, 1980)

[42] J. Timonen, D. J. Pilling and R. K. Bullough, in **Coherence, Cooperation and Fluctuations**, ed. F. Haake, L. M. Narducci and D. F. Walls (CUP, Cambridge, 1986), pp. 18-34. Unfortunately, in this paper a copying error replacing $k = h(\bar{k})$ by $\bar{k} = h(k)$ introduced sign errors in the equivalents of our equations (5.9) and (5.22).
[43] M. Wadati, J. Phys. Soc. Japan **54** (1985) 3727.
[44] R. K. Bullough, D. J. Pilling and J. Timonen, in **Magnetic Excitations and Fluctuations**, ed. S. W. Lovesey, U. Balucani, F. Borsa, V. Tognetti (Springer, Berlin, 1984), pp. 80-85.
[45] R. K. Bullough, D. J. Pilling and J. Timonen, in **Dynamical Problems in Soliton Systems**, ed. S. Takeno (Springer, Heidelberg, 1985), pp. 105-114.
[46] S. G. Chung, Y. C. Chang, Phys. Rev. Lett. **50** (1983) 791.
[47] N. N. Chen, M. D. Johnson and M. Fowler, Phys. Rev. Lett. **56** (1986) 907; **56** (1986) 1427 (Erratum).
[48] Yu- zhong Chen, D. J. Pilling, R. K. Bullough and J. Timonen, to be published.
[49] R. P. Feynman and A. R. Hibbs, **Quantum Mechanics and Path Integrals** (McGraw-Hill Book Co., New York, 1965).
[50] Yu-zhong Chen, D. J. Pilling, R. K. Bullough and J. Timonen, to be published.
[51] M. Karowski, H. J. Thun, J. T. Truong and P. H. Weiss, Phys. Lett. **67B** (1977) 321.

Exactly Solvable Models in Statistical Mechanics

M. Wadati[1] *and Y. Akutsu*[2]

[1]Institute of Physics, College of Arts and Sciences,
University of Tokyo, Meguro-ku, Tokyo 153, Japan
[2]Institute of Physics, Kanagawa University, Rokkakubashi,
Kanagawa-ku, Yokohama 221, Japan

Recent studies on exactly solvable models in statistical mechanics are reviewed. A brief summary of the quantum inverse scattering method is given to emphasize the soliton theoretic aspect of the theory. Introducing a class of lattice models called the IRF models, it is shown that there exists an infinite number of exactly solvable models in 2-dimensional statistical mechanics. Significances both in physics and mathematics are discussed.

1. INTRODUCTION

Twenty years have passed since the discovery of soliton [1]. Advances during the last two decades may be summarized as follows.

(1) Ubiquity of soliton equations. Canonical soliton equations, such as the Korteweg-de Vries equation, the nonlinear Schrödinger equation and the sine-Gordon equation, appear in almost all branches of physics.
(2) Systematic methods to solve soliton equations. The soliton equations can be solved by analytical methods such as the inverse scattering method, the Bäcklund transformation and Hirota's method.
(3) Universality of soliton picture. Soliton system as a completely integrable system includes solvable models in quantum field theory, quantum spin systems and statistical mechanics.

In this lecture note we review our recent contribution to the theory of exactly solvable models in statistical mechanics [2-11]. By doing this as elementary as possible, we like to convey our finding that the extension of the soliton theory implied in (3) is surprisingly wide and deep. Before we proceed to the main theme, we briefly summarize the quantum inverse scattering method [12]. This way of writing may be helpful for readers to understand the fundamental strategy of the theory.

An operator version of the auxiliary linear problem is

$$\left(\frac{d}{dx} + Q(x,\lambda)\right)G_L(x,\lambda) = 0, \quad |x| \le L. \tag{1.1}$$

Here $Q(x, \lambda)$ and $G_L(x, \lambda)$ are M x M matrix operators, and λ is the spectral parameter. We assume the boundary conditions:

$$G_L(-L,\lambda) = I,$$

$$G_L(L, \lambda) = T_L(\lambda), \tag{1.2}$$

where I is an identity operator and $T_L(\lambda)$ is called the transition matrix.

For later discussions, it is more transparent to formulate the quantum inverse scattering method (QISM hereafter) on lattice. The system size 2L is divided into 2N intervals each of which has a length $\Delta = 2L/2N$. The local transition matrix $L_n(\lambda)$

$$L_n(\lambda) = I - \int_{x_n-\Delta}^{x_n} Q(x,\lambda)dx \tag{1.3}$$

describes the change of the Jost function matrix $G_L(x,\lambda)$ over the n-th interval. In terms of the operator $L_n(\lambda)$, the transition matrix $T_L(\lambda)$ is expressed as

$$T_L(\lambda) = \lim_{\Delta\to 0} L_N(\lambda)\, L_{N-1}(\lambda) \;.\;.\;.\; L_{-N+1}(\lambda). \tag{1.4}$$

For a quantum integrable system we can associate $L_n(\lambda)$ such that

$$R(\lambda,\mu)\cdot[L_n(\lambda) \otimes L_n(\mu)] = [L_n(\mu) \otimes L_n(\lambda)]\cdot R(\lambda,\mu), \tag{1.5}$$

where symbol $\otimes$ denotes the direct product of the matrices. The c-number matrix $R(\lambda,\mu)$, of course, depends on the model. The relation (1.5) is referred as the (local) Yang-Baxter relation. If $L_n(\lambda)$'s with different n commute, we further have

$$R(\lambda,\mu)\cdot[T_L(\lambda) \otimes T_L(\mu)] = [T_L(\mu) \otimes T_L(\lambda)]\cdot R(\lambda,\mu). \tag{1.6}$$

This is the (global) Yang-Baxter relation.

The QISM provides not only a powerful method to study completely integrable systems, but also a unified viewpoint on the structure of solvable models in (1+1)-dimensional quantum field theory and 2-dimensional statistical mechanics. We list the consequences of the Yang-Baxter relation.

(1) The relation (1.6) indicates the existence of an infinite number of conserved quantities as seen from the relation

$$[T(\lambda), T(\mu)] = 0, \quad T(\lambda) = \mathrm{Tr}\, T_L(\lambda). \tag{1.7}$$

Further, off-diagonal elements of (1.6) yield an algebraic formulation [12-14] of the Bethe ansatz method.

(2) When we regard the elements of $L_n(\lambda)$ and $R(\lambda,\mu)$ as those of scattering matrix (S-matrix), the relation (1.5) is nothing but the factorization equation for the S-matrix [15].

(3) When we consider $L_n(\lambda)$ as vertices of a vertex model in statistical mechanics, (1.5) is a condition that the transfer matrices $T(\lambda) = \mathrm{Tr}\, T_L(\lambda)$ with different λ's commute, which implies the model is exactly solvable.

It is to be remarked that we have already suggested a common feature of exactly solvable models, that is, the commutability of the "transfer matrices." We shall explain this in detail in Section 3.

2. IRF MODELS

To be concrete, we shall mainly consider a type of statistical mechanical models in two dimensions, called IRF (interaction round a face) models [16]. Let us introduce an IRF model. Spins σ_i (σ_i = 0, 1, . . . , k-1) are located on the lattice points (sites) of a square lattice and the Boltzmann weight is assigned on each face (or plaquette) depending on the spin configurations round the face. By $\varepsilon(a,b,c,d)$, we denote the energy of a plaquette with spin configuration (a,b,c,d). The corresponding Boltzmann weight is

$$w(a,b,c,d) = \exp[-\varepsilon(a,b,c,d)/k_B T], \tag{2.1}$$

where k_B is the Boltzmann constant and T is the temperature (Fig. 1).

Let N be the particle number. The partition function Z_N, the free energy per particle f and the spin density $\langle\sigma_1\rangle$ (the average of a certain spin, say σ_1) are defined by

$W(a,b,c,d) =$ 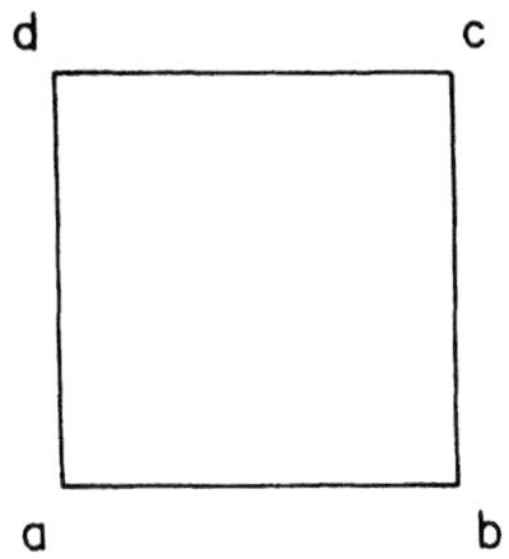

Fig. 1 Boltzmann weight w(a,b,c,d)

$$Z_N = \sum_{\sigma_1} \cdots \sum_{\sigma_N} \prod_{(i,j,k,l)} w(\sigma_i, \sigma_j, \sigma_k, \sigma_l), \tag{2.2}$$

$$f = - k_B T \lim_{N\to\infty} N^{-1} \log Z_N, \tag{2.3}$$

$$\langle\sigma_1\rangle = \frac{1}{Z_N} \sum_{\sigma_1} \cdots \sum_{\sigma_N} \sigma_1 \prod_{(i,j,k,l)} w(\sigma_i, \sigma_j, \sigma_k, \sigma_l). \tag{2.4}$$

Here, the product is over all faces of the lattice and the sum is over all values of all spins.

Exact solvability of a model in statistical mechanics means that we can evaluate physical quantities such as the free energy and the spin density without any approximation. It is known that the Ising model, the next nearest neighbor Ising model, the 6-vertex model, the 8-vertex model and the 3-spin model are exactly solvable [16]. Note that the IRF model is not special but very general. Most of the exactly solvable models are expressed in the form of IRF models. For instance, the Ising model is defined by

$$\varepsilon(a,b,c,d)$$

$$= - \frac{J}{2}[(2a-1)(2b-1)+(2b-1)(2c-1)+(2c-1)(2d-1)+(2d-1)(2a-1)], \tag{2.5}$$

$$a, b, c, d, = 0, 1,$$

and the 8-vertex model is defined by

$$\varepsilon(a, b, c, d)$$

$$= -J(2a-1)(2c-1) - J'(2b-1)(2d-1) \quad J_4(2a-1)(2b-1)(2c-1)(2d-1),$$

$$a, b, c, d, = 0, 1. \tag{2.6}$$

In general, the model is called k-state model when spin variables can take k values, $\sigma_i = 0, 1, \ldots, k-1$. The Ising model and the 8-vertex model are, as seen from (2.5) and (2.6), two-state models.

3. TRANSFER MATRICES

Exactly solvable models in statistical mechanics are used to be counted on the fingers. This situation has been drastically changed by the following two discoveries:

(1) Concept of the commuting transfer matrices,

(2) Evaluation of physical quantities via the corner transfer matrices.

The row-to-row transfer matrix V has elements (Fig. 2)

$$V_{\sigma\sigma'} = \prod_{j=1}^{n} w(\sigma_j, \sigma_{j+1}, \sigma'_{j+1}, \sigma'_j), \tag{3.1}$$

where $\sigma = \{\sigma_1, \ldots, \sigma_n\}$, $\sigma' = \{\sigma'_1, \ldots, \sigma'_n\}$, $\sigma_{n+1} = \sigma_1$ and $\sigma'_{n+1} = \sigma'_1$.

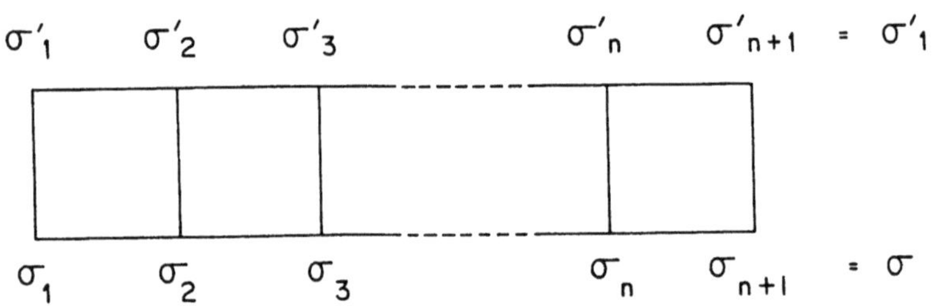

Fig. 2 Row-to-row transfer matrix $V_{\sigma\sigma'}$

Similarly, we define V' with w replaced by w'

$$V'_{\sigma\sigma'} = \prod_{j=1}^{n} w'(\sigma_j, \sigma_{j+1}, \sigma'_{j+1}, \sigma'_j). \tag{3.2}$$

Then, the elements of the matrix product VV' are

$$(VV')_{\sigma\sigma'} = \sum_{\sigma''} V_{\sigma\sigma''} V'_{\sigma''\sigma'}$$

$$= \sum_{\sigma''_1 \cdots \sigma''_n} \prod_{j=1}^{n} X(\sigma_j, \sigma''_j, \sigma'_j | \sigma_{j+1}, \sigma''_{j+1}, \sigma'_{j+1}), \tag{3.3}$$

where

$$X(a,b,c \mid a',b',c') = w(a,a',b',b)w'(b,b',c',c). \tag{3.4}$$

We regard $X(a,c \mid a',c')$ as the $k \times k$ matrix with element $X(a,b,c \mid a',b',c')$ in row b and column b' (Fig. 3). Then, (3.3) can be written as

$$(VV')_{\sigma\sigma'} = \mathrm{Tr}\, X(\sigma_1, \sigma_1' | \sigma_2, \sigma_2') X(\sigma_2, \sigma_2' | \sigma_3, \sigma_3') \ldots X(\sigma_n, \sigma_n' | \sigma_1, \sigma_1'). \tag{3.5}$$

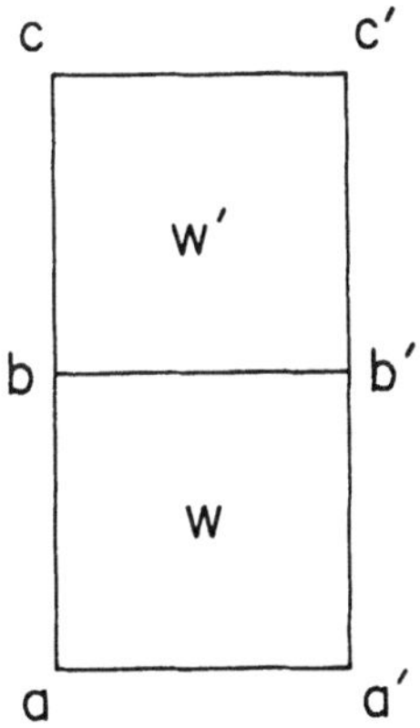

Fig. 3 $X(a,b,c \mid a',b',c')$ in (3.4)

Similarly, we define X' with w and w' interchanged in (3.4). Then we have

$$(V'V)_{\sigma\sigma'} = \mathrm{Tr}\, X'(\sigma_1, \sigma_1' | \sigma_2, \sigma_2') X'(\sigma_2, \sigma_2' | \sigma_3, \sigma_3') \ldots X'(\sigma_n, \sigma_n' | \sigma_1, \sigma_1'). \tag{3.6}$$

From (3.5) and (3.6), we see that V and V' commute if there exist $k \times k$ matrices $M(a,a')$ such that

$$X(a,a'|b,b') = M(a,a')X'(a,a'|b,b')[M(b,b')]^{-1}. \tag{3.7}$$

Multiplying $M(b,b')$ from the right and writing the element (c,d) of $M(a,a')$ as $w''(c,a,d,a')$, we obtain

$$\sum_c w(b,d,c,a)w'(a,c,f,g)w''(c,d,e,f)$$

$$= \sum_c w''(a,b,c,g)w'(b,d,e,c)w(c,e,f,g). \tag{3.8}$$

Equation (3.8) is a sufficient condition for the commuting transfer matrices and will be referred as star-triangle equation (Fig. 4).

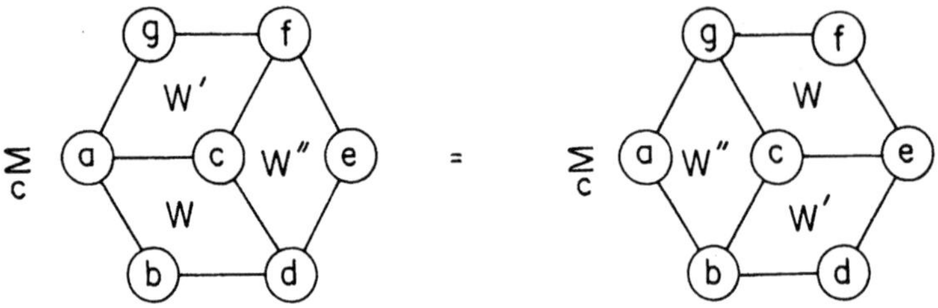

Fig. 4 Schematic explanation of the star-triangle equation (3.8)

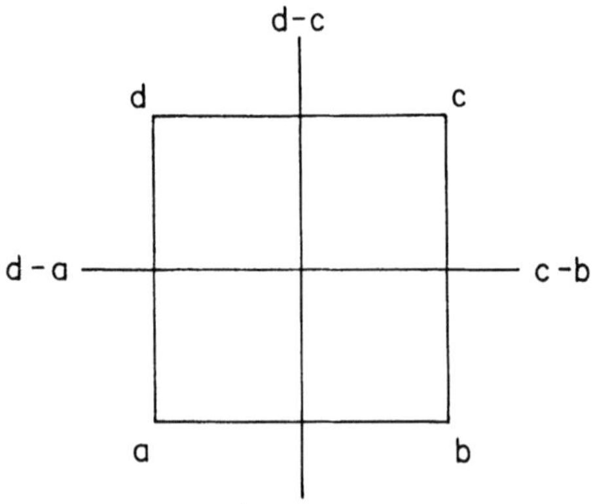

Fig. 5 Wu-Kadanoff-Wegner transformation

Commutability of the transfer matrices is a common feature of exactly solvable models. By Wu-Kadanoff-Wegner transformation (Fig. 5), the star-triangle equation for vertex models becomes

$$\sum_{\gamma\mu''\nu''} w(\mu,\alpha|\gamma,\mu'')w'(\nu,\gamma|\beta,\nu'')w''(\nu'',\mu''|\nu',\mu')$$

$$= \sum_{\gamma\mu''\nu''} w''(\nu,\mu|\nu'',\mu'')w'(\mu'',\alpha|\gamma,\mu')w(\nu'',\gamma|\beta,\nu') . \qquad (3.9)$$

The relation (3.9) is schematically shown in Fig. 6. Furthermore, when we set an arrow on each bond of vertex and regard the vertex as trajectories of two particles (Fig. 7), (3.9) is a condition imposed on the 3-body scattering from initial state (ν,μ,α) to final state (μ',ν',β):

$$\sum_{\gamma\mu''\nu''} S^{\alpha\gamma}_{\mu\mu''} S'^{\gamma\beta}_{\nu\nu''} S''^{\mu'\nu'}_{\nu''\mu'}$$

$$= \sum_{\gamma\mu''\nu''} S''^{\mu\nu''}_{\nu\mu''} S'^{\alpha\gamma}_{\mu''\mu'} S^{\gamma\beta}_{\nu''\nu'} . \qquad (3.10)$$

This relation is called factorization equation for the S-matrix.

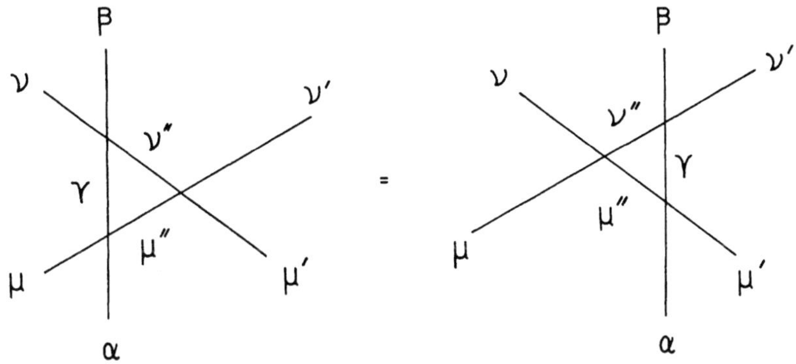

Fig.6 The star-triangle equation for vertex model

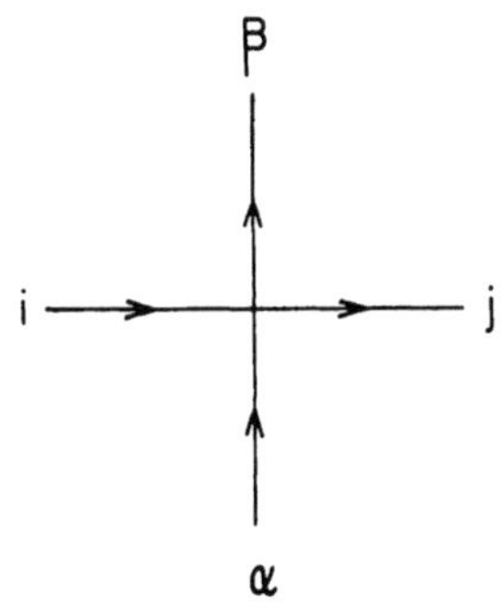

Fig. 7 2-body scattering $S_{ij}^{\alpha\beta}$

We have shown that (3.8), (3.9) and (3.10) are equivalent. Baxter noticed the importance of (3.9) in the study of 8-vertex model [17] and Yang [18] introduced (3.10) as a consistency condition of the Bethe ansatz method [19]. After them, these equations in general are often called Yang-Baxter relation as already used in Section 1.

We consider corner transfer matrices. The lattice is divided into four quadrants as indicated in Fig. 8. The boundary spins are fixed to their ground state values. For spin configurations (Fig. 9) in the lower-right quadrant, we define corner transfer matrix A:

$$A_{\sigma\sigma'} = \delta(\sigma_1, \sigma_1')\Sigma\Pi\, w(\sigma_i, \sigma_j, \sigma_k, \sigma_l). \tag{3.11}$$

Note that the spins $\sigma = \{\sigma_1, \ldots, \sigma_m\}$ and $\sigma' = \{\sigma_1', \ldots, \sigma_m'\}$ are not summed over. Similarly, we introduce corner transfer matrices B, C and D in the upper-right, upper-left and lower-left quadrants. The partition function Z_N and the spin density $\langle\sigma_1\rangle$ are expressed as

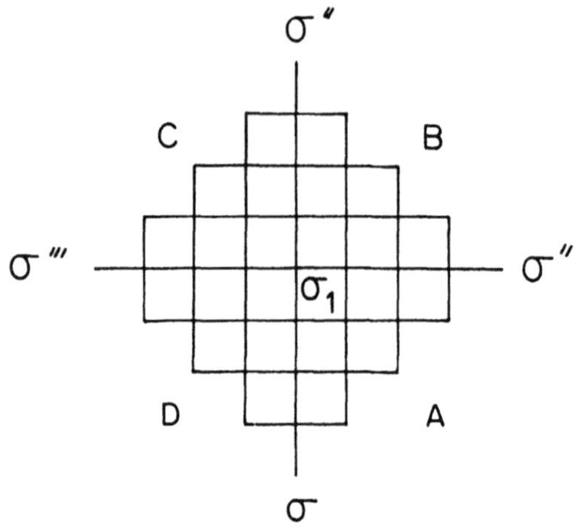

Fig. 8

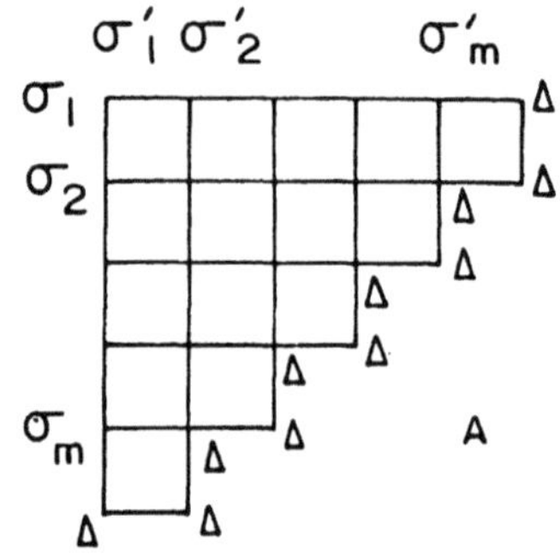

Fig. 9

Fig. 8 The lattice is divided into four quadrants. The center spin is σ_1.

Fig. 9 Corner transfer matrix $A_{\sigma\sigma'}$. Boundary spins denoted by Δ are fixed to their ground state values.

$$Z_N = \mathrm{Tr}(A\ B\ C\ D), \tag{3.12}$$

$$\langle\sigma_1\rangle = \lim_{m\to\infty} \mathrm{Tr}(\sigma_1\ A\ B\ C\ D)/\mathrm{Tr}(A\ B\ C\ D). \tag{3.13}$$

Corner transfer matrices can be diagonalized in the thermodynamic limit $m \to \infty$ using the star-triangle equation and the analytic and periodic properties of the Boltzmann weights [16]. In contrast with the row-to-row transfer matrix the eigenvalues of the corner transfer matrices remain discrete. Recently, Thacker [20] found a link to the field theory that corner transfer matrix may be interpreted as the boost operator, [21].

4. ROGERS-RAMANUJAN IDENTITY AND PARTITION THEORY

One of the fascinating aspects which we encounter in the study of IRF models is a close relation to the theory of partitions in number theory. There exists a direct correspondence of the conditions on the spins and the constraints on the partition of natural numbers. Appearance of the Rogers-Ramanujan type identities in the evaluation of the spin density through the corner transfer matrix is a consequence of this correspondence.

Celebrated Rogers-Ramanujan identities [22] are

$$G(q) = 1 + \sum_{n=1}^{\infty} \frac{q^{n^2}}{(1-q)(1-q^2)\dots(1-q^n)}$$

$$= \prod_{n=0}^{\infty} \frac{1}{(1-q^{5n+1})(1-q^{5n+4})}, \tag{4.1a}$$

$$H(q) = 1 + \sum_{n=1}^{\infty} \frac{q^{n^2+n}}{(1-q)(1-q^2)\dots(1-q^n)}$$

$$= \prod_{n=0}^{\infty} \frac{1}{(1-q^{5n+2})(1-q^{5n+3})} . \tag{4.1b}$$

It is a famous story that in 1917 S. Ramanujan who had known (4.1) incidentally found a paper by Rogers in the Proceedings of London Mathematical Society. We shall give a short introduction to the theory of partitions [23], which leads to the proofs of (4.1) and its extension.

A partition of a positive integer n is a finite nonincreasing sequence of positive integers whose sum is n. The number of partitions of n is denoted by P_n. For instance, we have

$$\begin{aligned} 3 &= 1 + 1 + 1 \\ &= 2 + 1 \\ &= 3 \\ P_3 &= 3. \end{aligned}$$

It is useful to introduce a notation that makes explicit the number of times that a particular integer occurs as a part:

$$\begin{aligned} n &= \sigma_1 . 1 + \sigma_2 . 2 + \sigma_3 . 3 + \sigma_4 . 4 + \dots \\ &= (1^{\sigma_1} 2^{\sigma_2} 3^{\sigma_3} 4^{\sigma_4} \dots). \end{aligned} \tag{4.2}$$

We often deal with problems that the number of partitions is enumerated under restriction (restricted partition problems). We denote the number of partitions of n under restriction R by $P_n(R)$. It is known [23] that

$$P_n(\text{all partitions with distinct parts})$$

$$= P_n(\text{all partitions with odd parts}). \qquad (4.3)$$

For instance, we have

$$6 = (1\ 2\ 3) = (1\ 5) = (2\ 4) = (6),$$

$$6 = (1^6) = (1^3\ 3) = (1\ 5) = (3^2),$$

which confirms (4.3) for n = 6.

The partition problem is called equipartition if for two restrictions R and R' it holds that

$$P_n(R) = P_n(R') \quad \text{for all } n. \qquad (4.4)$$

Define the generating function Z(q,R) by

$$Z(q,R) = \sum_{n=0}^{\infty} q^n P_n(R). \qquad (4.5)$$

In terms of the generating functions, equipartition condition (4.4) is expressed as

$$Z(q,R) = Z(q,R'). \qquad (4.6)$$

The following identity is useful for later discussions:

$$\sum_{n=0}^{\infty} P_n(R) q^n = \prod_{n \in R} (1 - q^n)^{-1}. \qquad (4.7)$$

A simple example of (4.7) is

$$\sum_{n=0}^{\infty} P_n\,(\text{all partitions with odd parts}).q^n$$

$$= \prod_{n=1}^{\infty} (1 - q^{2n-1})^{-1}. \qquad (4.8)$$

We are in a position to prove Rogers-Ramanujan identity (4.1). Define a function $F(\sigma_o;\ q)$ by

$$F(\sigma_0;\ q) = \Sigma' q^{\sigma_1 + 2\sigma_2 + 3\sigma_3 + \ldots} \qquad (4.9)$$

where the sum Σ' is over $\sigma_1, \sigma_2, \ldots$ under the constraints

$$R: \sigma_i = 0, 1 \text{ and } 0 \le \sigma_i + \sigma_{i+1} \le 1, \ i \ge 0. \tag{4.10}$$

By a calculation, we see that $F(0;q)$ and $F(1:q)$ coincide with the infinite series of (4.1) respectively. We can prove [23] that the partition with the restriction R in (4.10) is equipartition to that with restriction R' defined by

$$R': \sigma_i = 0 \quad \text{for } i = 0, \pm(2 - \sigma_0) \bmod 5$$

$$\sigma_i = 0, 1, 2, \ldots, \text{ otherwise.} \tag{4.11}$$

Applying the identity (4.7) to (4.10) which is equivalent to (4.11), we obtain the infinite products of (4.1). Thus, the Rogers-Ramanujan identities (4.1) are expressed as

$$F(\sigma_0; q) = \prod_{n \neq 0, \pm(2-\sigma_0) \bmod 5} (1 - q^n)^{-1}. \tag{4.12}$$

In 1961, B. Gordon [24] discovered a generalization of the Rogers-Ramanujan identities. Define

$$F_k(\sigma_0; q) = \Sigma' \, q^{\sigma_1 + 2\sigma_2 + 3\sigma_3 + \cdots} \tag{4.13}$$

where the sum Σ' is over $\sigma_1, \sigma_2, \ldots$ under restriction R given by

$$R: \sigma_i = 0, 1, 2, \ldots, (k-1)$$

$$\text{and } \ 0 \le \sigma_i + \sigma_{i+1} \le k - 1, \ i \ge 0. \tag{4.14}$$

Equipartition holds for the restriction R and the following restriction R'

$$R': \sigma_i = 0 \quad \text{for } i = 0, \pm (k - \sigma_0) \bmod (2k+1),$$

$$\sigma_i = 0, 1, 2, \ldots, \text{ otherwise.} \tag{4.15}$$

Then, we arrive at the identity

$$F_k(\sigma_0; q) - \prod_{\substack{n=1 \\ n \neq 0, \pm(k-\sigma_0) \bmod (2k+1)}}^{\infty} (1 - q^n)^{-1}. \tag{4.16}$$

This is called Gordon's generalization of the Rogers-Ramanujan identities. The case k = 2 corresponds to (4.12).

Ramanujan further discovered a list of identities including

$$G(x^6)H(x) - x\, G(x)H(x^6) = \frac{P(x)}{P(x^3)}, \qquad (4.17)$$

$$P(x) = \prod_{n=1}^{\infty} (1 - x^{2n-1}).$$

5. GORDON'S GENERALIZATION HIERARCHY

We go back to the IRF models. As shown in Section 3, if the Boltzmann weights satisfy the star-triangle equation (3.8), the transfer matrices commute and then the model is solvable. This fact offers an extremely powerful method to construct exactly solvable models. Finding of solvable models consists of two steps.

(1) Introduce a model with appropriate physical requirements (symmetries, number of spin states, etc).
(2) Solve the star-triangle equation (STE, for short) for the model.

In general, k-state IRF model has k^4 independent Boltzmann weights and the number of the STE to be satisfied is k^6. As k increases, the system of equations is extremely overdetermined and does not have interesting solutions. Therefore, the setting of physical requirements is crucial in this game.

We introduce the spectral parameter

$$w = w(u), \quad w' = w(u + v), \quad w'' = w(v). \qquad (5.1)$$

Then, the STE (3.8) reads as

$$\sum_g w(a,b,g,f;u)\, w(f,g,d,e;u+v)\, w(g,b,c,d;v)$$

$$= \sum_g w(g,c,d,e;\, u)\, w(a,b,c,g;\, u+v)\, w(f,a,g,e;\, v). \qquad (5.2)$$

We assume diagonal exchange symmetry among the Boltzmann weights

$$w(a,b,c,d;\, u) = w(c,b,a,d;\, u)$$

$$= w(a,\, d,\, c,\, b;\, u). \qquad (5.3)$$

In 1980, Baxter [25] found that the hard hexagon model is exactly solvable. In the hard hexagon model, spin variable can take either of 0 or 1, and the Boltzmann weight is non-zero only if there are no neighbouring pairs of 1-spins. Then, the physical requirements on the model are

$$\sigma_i = 0, 1, \tag{5.4a}$$

$$0 \leq \sigma_i + \sigma_j \leq 1 \quad \text{for neighbouring spins } \sigma_i \text{ and } \sigma_j. \tag{5.4b}$$

With the hard-core condition (5.4b), spin-1 particle looks "hexagon shape molecule" on the lattice. The model has been successfully applied to the adsorption problem of He^4 monolayer on carbon surface [16].

Due to the hard-core condition (5.4b) and the symmetry (5.3), the hard hexagon model has 5-independent Boltzmann weights:

$$\omega_1 = w(0, 0, 0, 0; u),$$

$$\omega_2 = w(0, 0, 0, 1; u) = w(0, 1, 0, 0; u),$$

$$\omega_3 = w(0, 0, 1, 0; u) = w(1, 0, 0, 0; u),$$

$$\omega_4 = w(0, 1, 0, 1; u),$$

$$\omega_5 = w(1, 0, 1, 0; u). \tag{5.5}$$

The STE to be satisfied by these Boltzmann weights are

$$\omega_1\omega_2'\omega_1'' + \omega_3\omega_4'\omega_3'' = \omega_2\omega_1'\omega_2'',$$

$$\omega_3\omega_1'\omega_1'' + \omega_5\omega_2'\omega_3'' = \omega_1\omega_3'\omega_2'',$$

$$\omega_1\omega_2'\omega_3'' + \omega_3\omega_4'\omega_5'' = \omega_4\omega_3'\omega_2'',$$

$$\omega_3\omega_1'\omega_3'' + \omega_5\omega_2'\omega_5'' = \omega_2\omega_5'\omega_2'',$$

$$\omega_3\omega_2'\omega_3'' + \omega_5\omega_4'\omega_5'' = \omega_4\omega_5'\omega_4'', \tag{5.6}$$

where the arguments of ω_i, ω_i' and ω_i'' are u, u+v and v, respectively. The solution of these functional equations is expressed in terms of elliptic theta function. It is convenient to employ the definition

$$H(u;\ p^2) = \sin u \prod_{n=1}^{\infty} (1 - 2p^{2n} \cos 2u + p^{4n})(1 - p^{2n}). \qquad (5.7)$$

We hereafter write H(u) for $H(u;p^2)$. The solution is

$$\omega_1 = \frac{H(3\lambda - u)}{H(3\lambda)}, \quad \omega_2 = \frac{H(\lambda - u)}{H(\lambda)},$$

$$\omega_3 = \frac{H(u)}{[H(\lambda)H(2\lambda)]^{\frac{1}{2}}}, \quad \omega_4 = \frac{H(4\lambda - u)}{H(4\lambda)},$$

$$\omega_5 = \frac{H(2\lambda - u)}{H(2\lambda)}, \qquad (5.8)$$

where

$$\lambda = \pi/5. \qquad (5.9)$$

Using an addition formula

$$H(a+x)H(a-x)H(b+y)H(b-y)$$
$$= H(a+y)H(a-y)H(b+x)H(b-x) + H(a+b)H(a-b)H(x+y)H(x-y), \qquad (5.10)$$

we can prove that the solution (5.8) satisfies the STE (5.6).

The hard hexagon model has four regimes; I(the vacuum), II(the triangular ordering), III(the vacuum) and IV(the square ordering). The parameter u and p^2 measure the anisotropy of coupling constants and the deviation from critical temperature. In regime I, we use x and w defined by

$$p^2 = -e^{-\varepsilon}, \quad x = -e^{\pi^2/5\varepsilon}, \quad w = e^{2\pi u/\varepsilon}. \qquad (5.11)$$

By using the analytical properties of corner transfer matrices and the explicit forms of the Boltzmann weights, we can show [16] that in regime I the spin density is given by

$$\langle\sigma_1\rangle = \frac{r_0^2\, F(1;\ s^2)}{F(0;s^2) + r_0^2 F(1;\ s^2)}, \qquad (5.12a)$$

$$s^2 = x^6, \quad r_0^2 = -\, xG(x)/H(x). \qquad (5.12b)$$

Here, $F(\sigma_0;q)$, G(x) and H(x) have been introduced in (4.9), (4.1a) and (4.1b). Using (4.12) and (4.17) in (5.12), we arrive at

$$\langle\sigma_1\rangle = -\frac{xG(x)H(x^6)P(x^3)}{P(x)}. \tag{5.13}$$

We can analyze the critical behavior exactly. For instance, critical indices across the I-II regime boundary are $\alpha = 1/3$ and $\beta = 1/9$.

The hard-hexagon model is a 2-state model where spin variable σ_i takes 0 or 1. We introduce a k-state model

$$\sigma_i = 0, 1, \ldots, (k-1) \tag{5.14a}$$

subject to the constraint

$$0 \le \sigma_i + \sigma_j \le k-1 \quad \text{for adjacent spins } \sigma_i \text{ and } \sigma_j. \tag{5.14b}$$

It should be remarked that the constraint (5.14) corresponds to the restriction (4.14) which leads to the Gordon's generalization of the Rogers-Ramanujan identities. We call the k-state model defined by (5.14) GG(k) model. The number of independent Boltzmann weights is $k(k+1)(k+2)(k+3)/24$. We succeeded in finding the solution of the STE for the GG(3) [2], GG(4) [5,7] and GG(5) [8,11] models by solving 37, 158 and 510 functional equations. We also presented a method to construct the Boltzmann weights for the GG(k) model [11]. A beautiful mathematical structure of the Boltzmann weights was found in the construction.

As an example, we write down the solution for the GG(3) model which are obtained by solving 37 independent STE's.

$$w(0,0,0,0;u) = \frac{H(5\lambda-u)H(3\lambda-u)}{H(5\lambda)H(3\lambda)} - \frac{H(2\lambda)}{H(\lambda)H(3\lambda)^2}\, H(u)H(\lambda-u),$$

$$\begin{aligned} w(1,0,0,0;u) &= w(0,0,1,0;u) \\ &= \varepsilon_1\left[\frac{H(2\lambda)}{H(3\lambda)}\right]^{\frac{1}{2}} \frac{H(u)H(4\lambda-u)}{H(\lambda)H(4\lambda)}, \end{aligned}$$

$$\begin{aligned} w(2,0,0,0;u) &= w(0,0,2,0;u) \\ &= \varepsilon_2\, \frac{H(u)H(3\lambda-u)}{[H(\lambda)H(3\lambda)]^{\frac{1}{2}}H(2\lambda)}, \end{aligned}$$

$$w(0,0,0,1;u) = w(0,1,0,0;u) = \frac{H(\lambda-u)H(4\lambda-u)}{H(\lambda)H(4\lambda)},$$

$$w(0,0,0,2;u) = w(0,2,0,0;u) = \frac{H(\lambda-u)H(5\lambda-u)}{H(\lambda)H(5\lambda)},$$

$$w(1,0,1,0;u) = \frac{H(3\lambda-u)H(4\lambda-u)}{H(3\lambda)H(4\lambda)},$$

$$w(2,0,2,0;u) = \frac{H(2\lambda-u)H(3\lambda-u)}{H(2\lambda)H(3\lambda)},$$

$$w(0,1,0,1;u) = \frac{H(5\lambda-u)H(4\lambda-u)}{H(5\lambda)H(4\lambda)},$$

$$w(0,2,0,2;u) = \frac{H(6\lambda-u)H(5\lambda-u)}{H(6\lambda)H(5\lambda)},$$

$$w(1,0,0,1;u) = w(1,1,0,0;u) = w(0,1,1,0;u)$$

$$= w(0,0,1,1;u) = \varepsilon_2 \frac{H(u)H(\lambda-u)}{[H(\lambda)H(2\lambda)]^{\frac{1}{2}}H(3\lambda)},$$

$$w(1,0,2,0;u) = w(2,0,1,0;u) = \varepsilon_1\varepsilon_2 \frac{H(u)H(6\lambda-u)}{[H(\lambda)H(2\lambda)]^{\frac{1}{2}}H(3\lambda)},$$

$$w(0,2,0,1;u) = w(0,1,0,2;u) = \frac{H(\lambda-u)H(2\lambda-u)}{H(\lambda)\ H(2\lambda)},$$

$$w(0,1,1,1;u) = w(1,1,0,1;u) = \varepsilon_1 \frac{H(u)H(5\lambda-u)}{[H(2\lambda)H(3\lambda)]^{\frac{1}{2}}H(\lambda)},$$

$$w(1,0,1,1;u) = w(1,1,1,0;u) = \frac{H(\lambda-u)H(3\lambda-u)}{H(\lambda)\ H(3\lambda)},$$

$$w(1,1,1,1;u) = \frac{H(3\lambda-u)H(5\lambda-u)}{H(3\lambda)\ H(5\lambda)}, \qquad (5.15)$$

where $\varepsilon_i = \pm 1$ (i=1,2,3) and the parameter λ (the crossing point) is

$$\lambda = \pi/7. \qquad (5.16)$$

Independently, this solution has been found by Baxter and Andrews [26]. For the general k, the Boltzmann weights are expressed in polynomials of products of (k-1) elliptic theta functions and λ is given by [11]

$$\lambda = \pi/(2k+1). \qquad (5.17)$$

Thus, there exists a series of exactly solvable models, that is, the GG(k) model for k = 2 (hard hexagon model), 3, 4, 5, ... We call it Gordon's Generalization (GG) hierarchy. The finding of the GG hierarchy is a result of our belief that exactly solvable IRF models are closely related to the theory of partitions.

6. GRAND HIERARCHY OF EXACTLY SOLVABLE MODELS

Besides the GG hierarchy explained in the previous section, we know a hierarchy of solvable models which we call ABF hierarchy [27]. The ABF hierarchy is also closely related to the theory of partitions. Originally, those models (restricted 8V SOS models) are defined as the solid-on-solid (SOS) models where the height variables $\{h_i\}$ are assigned on the lattice sites. In terms of the spin variables $\{\sigma_i\}$, the model is defined by the restrictions

$$\sigma_i = 0, 1, 2, \ldots, (k-1), \qquad (6.1a)$$

$$k-2 \leq \sigma_i + \sigma_j \leq k-1 \quad \text{for adjacent spins } \sigma_i \text{ and } \sigma_j. \qquad (6.1b)$$

Recently, we discovered a new hierarchy (KAW hierarchy hereafter) where the model is defined by [6]

$$\sigma_i = 0, 1, 2, \ldots, (k-1), \qquad (6.2a)$$

$$k-2 \leq \sigma_i + \sigma_j \leq k \quad \text{for adjacent spins } \sigma_i \text{ and } \sigma_j. \qquad (6.2b)$$

Furthermore, we found a hierarchy (the special S_2 hierarchy) where the constraints are the same as (6.1) but the highest spin $\sigma = (k-1)$ is replaced by a doublet [4].

Those hierarchies are classified by the following constraints:

(1) number of states:

$$\sigma_i = 0,\ 1,\ 2,\ \ldots,\ (k-1), \tag{6.3a}$$

(2) hard-core conditions for adjacent spins σ_i and σ_j:

$$L \leq \sigma_i + \sigma_j \leq L + f, \tag{6.3b}$$

(3) additional constraints on adjacent spins σ_i and σ_j for $f \geq 3$. (6.3c)

We consider an (L,f) plane, $L \geq 0$, $f \geq 1$ (see Fig. 10). The ABF and KAW hierarchies correspond to $f = 1$ and $f = 2$ line, respectively. The $L = 0$ line is the GG hierarchy. Based on this observation, we have predicted [6] that there exists an exactly solvable model at every lattice point in the (L,f) plane. A set of those models which contains $\infty \times \infty$ models will be referred to as Grand Hierarchy.

The existence of the Grand Hierarchy was confirmed as follows [9]. We start from the eight vertex solid-on-solid model (8VSOS model, for short) which is also an IRF model (Fig. 11). Each height variable h_i takes integer values under constraints

$$|h_i - h_j| = 1 \quad \text{for adjacent heights } h_i \text{ and } h_j. \tag{6.4}$$

From the homogeneous model, we can construct the Z-invariant 8VSOS model (the term "Z-invariant" by origin means that the partition function Z_N remains invariant under the shift of lines forming the irregular lattice [28]). We construct the Z-invariant model on the square lattice

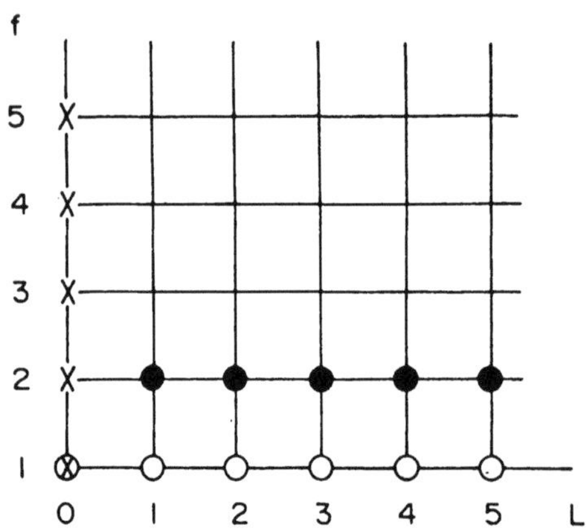

Fig. 10 The (L,f) plane showing the Grand Hierarchy. The symbols (x), (O) and (●) correspond to the GG, ABF and KAW hierarchies, respectively.

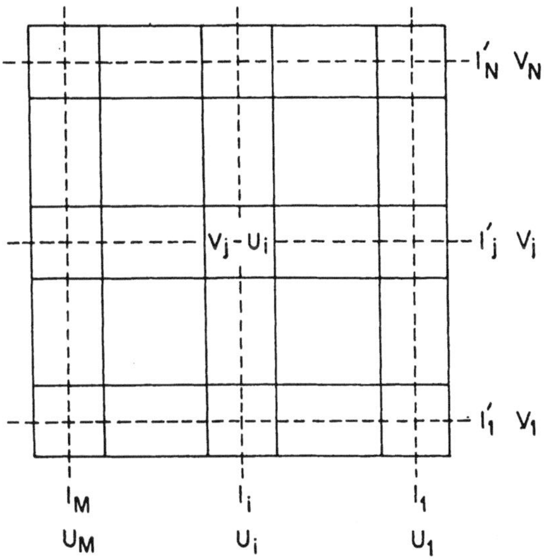

Fig. 11

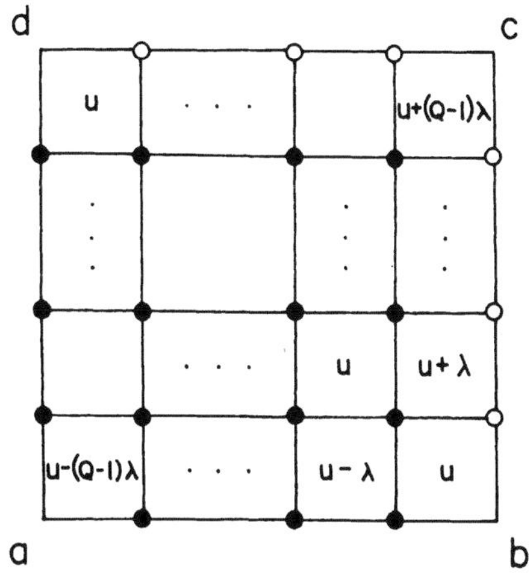

Fig. 12

Fig. 11 A Z-invariant 8VSOS model. Height variables are assigned on the sites of the lattice formed by the full lines. To each dashed vertical (horizontal) line ℓ_i (ℓ'_j), spectral parameter u_i (v_j) is assigned. The spectral parameter of the IRF plaquette which contains the intersection of ℓ_i and ℓ'_j is $v_j - u_i$.

Fig. 12 Boltzmann weight $w_{QQ}(a,b,c,d;u)$ of the IRF model. The symbol ● denotes the summation over the "inner" height. The weight does not depend on the edge heights denoted by O.

in the following manner. On a plane we place M vertical lines $\{\ell_j\}$ and N horizontal lines $\{\ell'_j\}$ forming a square lattice. To each line $\ell_i(\ell'_i)$ we assign a spectral parameter u_i (v_i). An inhomogeneous 8VSOS model is introduced on the dual square lattice. The spectral parameter of an IRF-plaquette which contains an intersection point of ℓ_i and ℓ'_j is assigned to be $v_j - u_i$. We call the resultant 8VSOS model the Z-invariant 8VSOS model. The homogeneous model is recovered by setting $v_j = u$ (all j) and $u_i = 0$ (all i).

We consider $Q \times Q$ periodic reduction for the Z-invariant 8VSOS model, which means that we set (see Fig. 12)

$$u_i/\lambda = i - 1 \bmod Q,$$

$$(v_j-u)/\lambda = j - 1 \bmod Q. \tag{6.5}$$

We also set (r = positive integer, r $\geq$ 5)

$$\lambda = \pi/r, \tag{6.6}$$

and regard Q as f in (6.3). From the edge-height independence (Fig.12), resultant Q $\times$ Q blocks can be considered as plaquettes of an IRF model in the (L,f) plane. Hence the partition function of the IRF model is the same as the Q $\times$ Q periodic Z-inavriant 8VSOS model. Thus,the solvability of the Grand Hierarchy is a consequence of the periodic reduction of the Z-invariant 8VSOS model [9].

It is an extremely interesting fact that the special S_2 hierarchy [4] is not included in the Grand Hierarchy. It may be possible that we extend the special S_2 hierarchy into the Z-invariant model and again by the periodic reduction we construct a new set of solvable hierarchies.

7. SUMMARY AND FURTHER DEVELOPMENTS

(1) We have shown that an infinite number of solvable models in statistical mechanics, at least $\infty \times \infty$ number of models, can be constructed by solving the star-triangle equation. Thus, the existence of the Grand Hierarchy has been established [6,9]. Significance of this achievement in statistical mechanics cannot be overemphasized. Physical applications, for instance, to the interfacial problems are very interesting.

(2) Let us relate our discussion to the conformal theory [29,30]. The conformal theory enables us to predict the possible values of critical index η in (1+1)-field theory or 2-dimensional statistical mechanics. Since any analytic function of the coordinate z = x + iy is the conformal transformation, conformal group in two dimensions is infinite dimensional. Generators L_n, n = 0, ±1, ±2, ..., of the transformations satisfy the Virasoro algebra [31]:

$$[L_n, L_m] = (n-m)L_{n+m} + \frac{c}{12}(n^3 - n)\delta_{n+m,0}. \tag{7.1}$$

From the representation theory of the Virasoro algebra [32] and the physical requirements such as unitarity [30], the value of central charge c is restricted to

(1) $c \geq 1$

(2) $c = 1 - \dfrac{6}{m(m+1)}$, $m = 2, 3, 4, \ldots$. (7.2)

Further, in the case of (2), the possible value of critical index η is given by [30]

$$\eta = 2(h + \bar{h}),$$

$$h, \bar{h} = \frac{[(m+1)p - mq]^2 - 1}{4m(m+1)}, \quad m = 2, 3, 4, \ldots$$

$$1 \leq p \leq m+1, \quad 1 \leq q \leq p. \qquad (7.3)$$

It is quite interesting that the critical index η calculated from the exact values of α and β for the ABF hierarchy through a scaling relation $\eta = 4\beta/(2-\alpha)$ agrees with the one predicted by (7.3) with $p = q$ [33]. In other words, the ABF hierarchy is a partial realization of the conformal theory. The detailed analysis of the Grand Hierarchy is under progress and at the moment we predict that there exists an exactly solvable model in each universality class.

We note that while the conformal theory deals with the phenomena only at the criticality the exactly solvable models can offer information both on- and off-critical temperatures.

(3) As we have seen in Section 5, the Rogers-Ramanujan type identities appear in the spin density calculation by the corner transfer matrix method. Conversely, we expect that there exists a solvable model corresponding to any Rogers-Ramanujan type partition identity [2,34]. We remark that elliptic functions (more generally, modular functions) play an important role again in the exact theory of statistical mechanics as we experienced in the soliton theory. It is interesting that the modular functions are closely related to the partition theory and the representation theory of Kac-Moody algebra [35].

(4) Very recently, we have found that the study of exactly solvable models in statistical mechanics has deep connections with the von Neumann algebra theory and the knot theory [10]. In 1984, V. Jones [36] discovered new polynomial invariant (Jones polynomial) which is more powerful than the Alexander polynomial to classify knots and links. A central role is played by the algebra $A_{q,n}$ defined by

$$e_i^2 = e_i,$$

$$e_i e_j = e_j e_i, \quad |j - i| \geq 2, \tag{7.4}$$

$$e_i e_{i\pm1} e_i = \tau e_i.$$

A certain trace is associated with the algebra $A_{q,n}$ making it the von Neumann algebra. Due to the existence condition of the trace, the value of τ in (7.4) is restricted to

(1) $\tau^{-1} \geq 4$

(2) $\tau^{-1} = 4\cos^2(\pi/k), \quad k = 3, 4, \ldots .$ (7.5)

On the other hand, the star-triangle equation for the ABF hierarchy (restricted 8-vertex SOS model with the height variable $h_j = 1,2,\ldots, r-1$) at criticality yields the Temperley-Lieb algebra $\{U_j\}$ [37]:

$$U_j^2 = q^{1/2} U_j,$$

$$U_i U_j = U_j U_i, \quad |j-i| \geq 2, \tag{7.6}$$

$$U_i U_{i\pm1} U_i = U_i,$$

with q given by

$$q = 4\cos^2(\pi/r). \tag{7.7}$$

We see from (7.4) and (7.6) that both algebras are the same when we set $U_j = q^{1/2} e_j$ and

$$\tau^{-1} = q. \tag{7.8}$$

Moreover, it is known [33] that the (r-1)-state 8VSOS model exhibits the generic (r-2)-fold multicriticality whose exponent realizes the discrete series (7.3) of the Virasoro algebra through the correspondence

$$m = r - 1. \tag{7.9}$$

Combining the above results we arrive [10] at a novel relation between the discrete series of the Virasoro algebra and the indices τ^{-1} for subfactors of II_1 factors [38]:

$$m + 1 = r = k. \tag{7.10}$$

Note that the critical 8VSOS hierarchy intermediates the completely different objects in mathematics.

More recently, we have found that new invariant polynomials for knots and links can be constructed from solvable (vertex and IRF) models describing critical statistical systems [39].

In conclusion we summarize this note by emphasizing that the soliton theory including solvable models in statistical mechanics has much more wide and deep extensions in physics and mathematics than we expected a few years ago.

References

[1] N. J. Zabusky and M. D. Kruskal, Phys. Rev. Lett. **15** (1965) 240.
[2] A. Kuniba, Y. Akutsu and M. Wadati, J. Phys. Soc. Jpn. **55**(1986)1092.
[3] Y. Akutsu, A. Kuniba and M. Wadati, J. Phys. Soc. Jpn. **55**(1986)1466.
[4] Y. Akutsu, A. Kuniba and M. Wadati, J. Phys. Soc. Jpn. **55**(1986)1880.
[5] A. Kuniba, Y. Akutsu and M. Wadati, J. Phys. Soc. Jpn. **55**(1986)2166.
[6] A. Kuniba, Y. Akutsu and M. Wadati, J. Phys. Soc. Jpn. **55**(1986)2605.
[7] A. Kuniba, Y. Akutsu and M. Wadati, Phys. Lett. **116A** (1986) 382.
[8] A. Kuniba, Y. Akutsu and M. Wadati, Phys. Lett. **117A** (1986) 358.
[9] Y. Akutsu, A. Kuniba and M. Wadati, J. Phys. Soc. Jpn. **55**(1986) 2907.
[10] A. Kuniba, Y. Akutsu and M. Wadati, J. Phys. Soc. Jpn. **55**(1986)3285.
[11] A. Kuniba, Y. Akutsu and M. Wadati, J. Phys. Soc. Jpn. **55**(1986)3338.
[12] L. D. Faddeev, Sov. Sci. Rev. Math. Phys. **C1** (1981) 107;
H. B. Thacker, Rev. Mod. Phys. **53** (1981) 253;
P. P. Kulish and E. K. Sklyanin, **Lecture Notes in Physics** (Springer,Berlin, New York, 1982), Vol. 151, p. 61.
M. Wadati, in **Dynamical Problems in Soliton Systems**, ed S. Takeno (Springer,Berlin, 1985), p. 68.
[13] L. A. Takhatajan and L. D. Faddeev, Russian Math. Survey **34**(1979)11.
[14] H. J. de Vega, in **Dynamical Problems in Soliton Systems**, ed S. Takeno (Springer,Berlin, 1985), p. 81.
[15] M. Karowski, H. J. Thun, T. T. Truong and P. H. Weisz, Phys. Lett. **67B** (1977) 321;
A. B. Zamolodchikov and A. B. Zamolodchikov, Ann. of Phys. (N.Y.) **120** (1979) 253;
K. Sogo, M. Uchinami, A. Nakamura and M. Wadati, Prog. Theor. Phys. **66** (1981) 1284;
K. Sogo, M. Uchinami, Y. Akutsu and M. Wadati, Prog. Theor. Phys. **68** (1982) 508.
[16] R. J. Baxter, **Exactly Solved Models in Statistical Mechanics** (Academic Press, London, 1982).
[17] R. J. Baxter, Ann. of Phys. (N.Y.) **70** (1972) 193.
[18] C. N. Yang, Phys. Rev. Lett. **19** (1967) 1312.
[19] E. Lieb and W. Liniger, Phys. Rev. **130** (1963) 1605;
B. Sutherland, Phys. Rev. Lett. **20** (1968) 98.
[20] H. B. Thacker, Physica **18D** (1986) 348.
[21] K. Sogo and M. Wadati, Prog. Theor. Phys. **69** (1983) 431.
[22] L. J. Rogers, Proc. London Math. Soc. **25** (1894) 318;
S. Ramanujan, Proc. Camb. Phil. Soc. **19** (1919) 214.
[23] G. E. Andrews, **The Theory of Partitions** (Addison,London, 1976).

[24] B. Gordon, Amer. J. Math. **83** (1961) 393.
[25] R. J. Baxter, J. Phys. **A13** (1980) L61.
[26] R. J. Baxter and G. E. Andrews, J. Stat. Phys. **44** (1986) 249.
[27] G. E. Andrews, R. J. Baxter and P. J. Forrester, J. Stat. Phys. **35** (1984) 193.
[28] R. J. Baxter, Philos. Trans. R. Soc. London **A289** (1978) 315; A. B. Zamolodchikov, Sov. Sci. Rev. **A2** (1980) 1; R. J. Baxter, Proc. Roy. Soc. London **A404** (1986) 1; J. H. H. Perk and F. Y. Wu, Physica **138A** (1986) 100.
[29] A. A. Belavin, A. B. Zamolodchikov and A. M. Polyakov, J. Stat. Phys. **34** (1984) 763; A. A. Belavin, A. M. Polyakov and A. B. Zamolodchikov, Nucl. Phys. **B241** (1984) 333.
[30] D. Friedan, Z. Qiu and S. Shenker, Phys. Rev. Lett. **52** (1984) 1575.
[31] M. Virasoro, Phys. Rev. **D1** (1970) 2933.
[32] V. G. Kac, **Lecture Notes in Physics** (Springer,Berlin, 1979), Vol. 94, p. 441; B. L. Feigin and D. B. Fuchs, Funct. Anal. Appl. **16** (1982) 114.
[33] D. A. Huse, Phys. Rev. **B30** (1984) 3908.
[34] C. A. Tracy, Preprint (1986).
[35] J. Lepowsky and R. L. Wilson, Proc. Nat. Acad. Sci. USA **78**(1981)7254.
[36] V. Jones, Bull. Amer. Math. Soc. **12** (1985) 103.
[37] H. N. V. Temperley and E. H. Lieb, Proc. Roy. Soc. London **A322** (1971) 251.
[38] O. Bratteli and D. W. Robinson, **Operator Algebras and Quantum Statistical mechanics I, II** (Springer,Berlin,New York , 1979,1981).
[39] Y. Akutsu and M. Wadati, "Knot Invariants and the Critical Statistical Systems," J. Phys. Soc. Jpn. **56** (1987) 839.

Part V

Applications: Physics and Biology

Solitons and Some Other Special Solutions in Field Theory

K. Babu Joseph

Department of Physics, Cochin University of Science and Technology, Cochin 682022, India

Solitons in gauge theories such as monopoles and vortices are discussed. The applications of the Hirota bilinear operator and group analysis methods for the SU(2) Higgs model are described.

1. INTRODUCTION

In field theory a soliton characterises a stable and extended solution of a nonlinear field equation, having finite energy or action, whose existence may sometimes be related to the topological properties of the fields at spatial infinity. Though solitons arise in certain gauge as well as scalar field theories we shall confine ourselves with gauge theory solitons. Magnetic monopoles are the most widely studied solitons of this category. In contrast to the singular point monopole discovered by Dirac [1] in abelian gauge theory, the 't Hooft-Polyakov non-abelian monopole [2] is a finite energy static field configuration carrying magnetic charge. It is a solution of the SU(2) Yang-Mills-Higgs model, where the gauge group is SU(2), which is spontaneously broken to U(1) by a triplet of Higgs scalar fields. Solitons carrying both magnetic and electric charges, called dyons, have also been constructed [3]. However, monopole and dyon solutions can be expressed in closed analytic form only in a special limit.

Vortices or strings form another class of gauge theory solitons. These objects contain quantized units of magnetic flux. The abelian Higgs model possesses a finite energy vortex solution [4]. A scale-dependent vortex solution has been obtained recently [5], and it is the first solution of its kind to be derived. It is important to note that a pure gauge theory, abelian or non-abelian, has no finite energy solutions.

Finite action solutions of four-dimensional non-abelian gauge theories in Euclidean space-time are called instantons [6]. Since instantons are localized in time as well as space, they do not qualify as particles. In the corresponding quantized theory instantons tunnel between different gauge vacua.

All of the foregoing types of soliton have associated topological conservation laws. But in some models solitons arise from the conservation of an ordinary charge, and these are called non-topological solitons [7]. Instantons and non-topological solitons will not be treated in these lectures. The topics we discuss here include application of the Hirota bilinear operator [8] and group analysis [9] methods to the construction of soliton and some time-dependent solutions of the SU(2) Higgs model. Among these are the celebrated Prasad-Sommerfield (PS) monopole [10] and related solutions, which were discovered by mere guess-work. No systematic derivation of these solutions has so far been given in the literature. This explains the importance of the approaches herein presented. There is a short section on vortices also. A general familiarity with gauge theories at the classical level is assumed on the part of the reader.

2. TOPOLOGICAL CRITERION FOR SOLITON SOLUTIONS

The existence and stability of a topological solution are guaranteed by the conservation of a topological quantity called the winding number or Pontryagin index. Two maps from a topological space to another are said to be homotopic, if one can be continuously deformed into the other. A homotopy class is an equivalence class of maps which are homotopic to one another. Each such class is labelled by a winding number which is conserved, implying that maps belonging to two distinct homotopy classes are not deformable into each other.

To illustrate the winding number idea, let us consider the SU(2) Higgs model mentioned in the introduction. It is described by the Lagrangian

$$\mathcal{L} = -\frac{1}{4} F^a_{\mu\nu} F^{\mu\nu a} + \frac{1}{2} D_\mu\varphi^a - V(\varphi) \tag{2.1}$$

where the $F^a_{\mu\nu}$ are the three components of the SU(2) field strength tensor

$$F^a_{\mu\nu} = \partial_\mu A^a_\nu - \partial_\nu A^a_\mu + g\varepsilon_{abc} A^b_\mu A^c_\nu , \tag{2.2}$$

D_μ is the gauge covariant derivative operator, defined by

$$D_\mu\varphi^a = \partial_\mu\varphi^a + g\varepsilon_{abc} A^b_\mu \varphi^c, \tag{2.3}$$

and $V(\varphi)$ is the Higgs potential,

$$V(\varphi) = \frac{\lambda}{4}(\varphi^2 - \frac{m^2}{\lambda})^2 \qquad (2.4)$$

with $\varphi^2 = \varphi^a\varphi^a$.

For $m^2 > 0$, the minimum of $V(\varphi)$ corresponds to $\varphi^2 = m^2/\lambda$. By convention, the classical vacuum is identified to be at spatial infinity, and thus

$$\varphi^2 \to m^2/\lambda \quad \text{as } r \to \infty. \qquad (2.5)$$

This condition defines a map of a sphere at spatial infinity to a sphere of radius $m/\sqrt{\lambda}$ in the field space spanned by the Higgs fields φ^a. We denote this map by $S^2 \to S^2$. Now, the homotopy classes of the map $S^2 \to S^2$ form a group called the second homotopy group, $\Pi_2(S^2)$, and $\Pi_2(S^2) = Z$, the group of integers. Each element of $\Pi_2(S^2)$ defines a homotopy class and represents a winding number. Hence the winding numbers of the present model are integers.

The total energy of any solution of the above model is

$$E = \int d^3x[\tfrac{1}{4}F^a_{ij}F^a_{ij} + \tfrac{1}{2}D_i\varphi^a D_i\varphi^a + V(\varphi) + \tfrac{1}{2}F^a_{oi}F^a_{oi} + \tfrac{1}{2}D_o\varphi^a D_o\varphi^a]. \qquad (2.6)$$

For a static solution of the Yang-Mills fields the energy will be finite if and only if $V(\varphi)$ vanishes or $\varphi^2 \to m^2/\lambda$ as $r \to \infty$. Hence, a way is open to classify the solitons according to their winding number. The n=1 soliton is a magnetic monopole which generates (asymptotically) the magnetic fields,

$$B_i = \frac{1}{g}\frac{\hat{r}_i}{r^2}, \qquad (2.7)$$

where $\hat{r}$ is a unit vector. The stability of a monopole stems from the topological conservation law, $\Delta n = 0$. The conservation of magnetic charge is a consequence of the Higgs field topology.

3. HIROTA'S BILINEAR OPERATOR METHOD: CONSTRUCTION OF PS MONOPOLE SOLUTION

The limit, $m^2 \to 0$, $\lambda \to 0$, keeping m^2/λ finite, is called the PS limit in which an explicit, static monopole solution was obtained [10] by ingenious guess-work. In this section we discuss an application of Hirota's method [8] to a systematic derivation of the PS monopole and other related solutions [11].

The bilinear operator method of solving a nonlinear differential equation consists in expressing the dependent variable h as the ratio of two functions, $h = g/f$. When this ratio is substituted in the original equation, an equation with two dependent variables, g and f, is obtained. The derivatives of functions can always be expressed in terms of Hirota's bilinear derivatives, defined by

$$D_x f(x,t).g(x,t) = \left(\frac{\partial}{\partial x} - \frac{\partial}{\partial x'}\right) f(x,t) g(x',t)\Big|_{x=x'} . \qquad (3.1)$$

The nonlinear equation for g and f is split into two coupled equations, and the functions g and f are expanded as power series in a parameter ε. The individual functions in the power series are evaluated by successively integrating the differential equations that follow from equating the coefficients of equal powers of ε on either side of each of the split equations. Solutions can be obtained either by terminating the series by some technique, or by actual summation.

Some useful identities that follow from (3.1) are listed below:

$$\frac{\partial^n}{\partial x^n} g = D_x^n\, g.1 \qquad (3.2)$$

$$\frac{\partial}{\partial x}(g/f) = \frac{D_x g.f}{f^2} \qquad (3.3)$$

$$\frac{\partial^2}{\partial x^2}(g/f) = \frac{D_x^2 g.f}{f^2} - \frac{g}{f}\,\frac{D_x^2 f.f}{f^2} \qquad (3.4)$$

$$D_x^n(g.f) = D_x^{n-1}(g_x.f - f_x.g) . \qquad (3.5)$$

Turning to the monopole problem, the equations of motion derived from (2.1) reduce to the following form, when a suitable ansatz [2] which is spherically symmetric and static, with winding number 1, is made use of:

$$r^2 K'' = K(K^2 + H^2 - 1), \qquad (3.6a)$$

$$r^2 H'' = H(2K^2 + \frac{\lambda}{g^2} H^2 - m^2 r^2), \qquad (3.6b)$$

where K and H are both functions of r, and a prime denotes differentiation with respect to the argument. The energy integral is

$$E = \frac{4\pi}{g^2}\int_0^\infty dr\,\{(K')^2 + \frac{(rH'-H)^2}{2r^2} + \frac{(K^2-1)^2}{2r^2} + \frac{K^2H^2}{r^2} + \frac{\lambda r^2}{4g^2}(\frac{H^2}{r^2} - \frac{g^2m^2}{\lambda})^2\}. \tag{3.7}$$

For finiteness of this integral, the necessary boundary conditions are

$$H \to 0, \quad K \to 1, \text{ as } r \to 0 \tag{3.8a}$$

$$H \to \frac{gm}{\sqrt{\lambda}}\,r, \quad K \to 0, \text{ as } r \to \infty\,. \tag{3.8b}$$

In the PS limit eqs. (3.6) become

$$r^2K'' = K(K^2 + H^2 - 1) \tag{3.9a}$$

$$r^2H'' = 2HK^2. \tag{3.9b}$$

One makes the dependent variables transformation:

$$K(r) = A(r)/B(r), \quad H(r) = C(r)/B(r). \tag{3.10}$$

Using (3.4), eqs. (3.9) are rewritten in the form

$$r^2(BD^2A.B - AD^2B.B) = A(A^2 - B^2 + C^2) \tag{3.11a}$$

$$r^2(BD^2C.B - CD^2B.B) = 2CA^2 \tag{3.11b}$$

where $D^2 = D_r^2$, a second order bilinear operator.

In the next step, (3.11b) is split into two coupled equations with the help of a function G(r), which will be determined later:

$$r^2D^2B.B + GB^2 = -2A^2 \tag{3.12a}$$

$$r^2D^2C.B + GCB = 0. \tag{3.12b}$$

One readily verifies that solutions to (3.12) are solutions of the original set (3.11). Although other splitting patterns may also be effective, the present one is advantageous in reducing the degree of nonlinearity from three to two. (3.11a) now becomes

$$r^2BD^2A.B = A[C^2 - (G+1)B^2 - A^2]. \tag{3.12c}$$

The functions A, B and C are expanded as power series in a small parameter ε:

$$A(r) = \varepsilon A_1(r) + \varepsilon^2 A_2(r) + \ldots \quad (3.13a)$$

$$B(r) = 1 + \varepsilon B_1(r) + \varepsilon^2 B_2(r) + \ldots \quad (3.13b)$$

$$C(r) = 1 + \varepsilon C_1(r) + \varepsilon^2 C_2(r) + \ldots \quad (3.13c)$$

Inserting (3.13) into (3.12) and comparing the zeroth power of ε on both sides, one finds that $G(r) = 0$ for consistency. Extending the same procedure to higher powers of ε yields equations of the following type:

$$\varepsilon \Rightarrow B_1'' = 0, \quad C_1'' = 0, \quad A_1'' = 0 \quad (3.14a)$$

$$\varepsilon^\alpha \Rightarrow r^2(2D^2 B_2.1 + D^2 B_1.B_1) = -2A^2$$

$$D^2 C_2.1 + D^2 C_1.B_1 + D^2 B_2.1 = 0$$

$$r^2(D^2 A_2.1 + D^2 A_1.B_1 + B_1 D^2 A_1.1) = 2A_1(C_1 - B_1). \quad (3.14b)$$

A consistent set of solutions of (3.14) is the following:

$$A_1 = ar, \quad A_2 = acr^2, \quad A_3 = -dacr^2 + ac^2 r^3/2! \quad (3.15a)$$

$$B_1 = br+d, \quad B_2 = N^2 r^2/2!, \quad B_3 = -dN^2 r^2/2! + (N^2 b - 2a^2 c) r^3/3! \quad (3.15b)$$

$$C_1 = cr+d, \quad C_2 = (2bc - N^2) r^2/2!,$$

$$C_3 = -d(2bc - N^2) r^2/2! + [(c-2b)N^2 + 2c(a^2+b^2)] r^3/3! \quad (3.15c)$$

where $N = (b^2 - a^2)^{\frac{1}{2}}$.

A solution of (3.9) by the above procedure is labelled by five parameters a, b, c, d and ε. However, by suitable choice of these parameters, one can make summable series out of (3.15). There are two trivial solutions too. For, setting $a = b = c = d = 0$ gives

$$K = 0, \quad H = 1. \quad (3.16)$$

Similarly, the choice $a = b$, $c = 0$ yields

$$K = \eta r/(1+\eta r), \quad H = 1/(1+\eta r) \quad (3.17)$$

where $\eta = a\varepsilon/(1+d)$, an arbitrary constant. Both of these solutions represent point monopoles of infinite energy.

For $a \neq b$, $c = 0$, and $|d\varepsilon| < 1$,

$$A(r) = \varepsilon a r, \tag{3.18a}$$

$$B(r) = \frac{\varepsilon}{\beta}(b \sinh\beta r + N \cosh\beta r), \tag{3.18b}$$

where $\beta = N\varepsilon/(1+d\varepsilon)$. Due to the large arbitrariness of d and ε, one can take β to be independent of a and b. After a straightforward calculation, the series for C(r) becomes

$$C(r) = \frac{\varepsilon}{\beta}[(N - b\beta r)\cosh\beta r + (b - N\beta r)\sinh\beta r]. \tag{3.18c}$$

Transforming back to the original pair of dependent variables, one finds

$$K(r) = \alpha\beta r\, e^{\beta r}/(\alpha^2 e^{2\beta r} - 1) \tag{3.19a}$$

$$H(r) = -\beta r(\alpha^2 e^{2\beta r} + 1)/(\alpha^2 e^{2\beta r} - 1) + 1, \tag{3.19b}$$

with $\alpha = a/(b - \sqrt{b^2 - a^2})$, which is arbitrary. This coincides with the general point monopole solution obtained by Ju [12]. Setting $\alpha = e^{\gamma}$ (say), one recovers the Protogenev solution [13]. The most interesting solution emerges with $\alpha = 1$, which is the PS monopole solution. It is, however, not possible to sum the series for B(r) and C(r) for non-zero c.

The Hirota method provides a unified scheme for deriving the PS and some other solutions, which were originally obtained by trial and error. Nevertheless, the procedure fails to yield closed expressions in the general case away from the PS limit.

4. GROUP ANALYSIS OF SU(2) MONOPOLE EQUATIONS

The group analysis of partial differential equations (PDE) is a standard procedure [9] that exposes their invariances and often, leads to solutions. For a second order PDE,

$$\mathcal{H}(x,t,u,u_x,u_t,u_{xx},u_{xt},u_{tt}) = 0 \tag{4.1}$$

a similarity transformation $x \to x^*$, $t \to t^*$, $u \to u^*$, is one which leaves the equation invariant. In infinitesimal form one writes

$$x^* = x + \varepsilon X(x,t,u) + 0(\varepsilon^2) \tag{4.2a}$$

$$t^* = t + \varepsilon T(x,t,u) + 0(\varepsilon^2) \tag{4.2b}$$

$$u^* = u + \varepsilon U(x,t,u) + 0(\varepsilon^2), \tag{4.2c}$$

where X, T, U denote the infinitesimals of the transformation. From these equations it follows that

$$u(x+\varepsilon X+0(\varepsilon^2);\ t+\varepsilon T+0(\varepsilon^2)) = u + \varepsilon U + 0(\varepsilon^2). \tag{4.3}$$

Expanding, and equating the $0(\varepsilon)$ terms on either side,

$$X \frac{\partial u}{\partial x} + T \frac{\partial u}{\partial t} = U. \tag{4.4}$$

This is called the invariant surface condition and yields the characteristic equations:

$$dx/X = dt/T = du/U \tag{4.5}$$

which imply,

$$\frac{dx}{dt} = f(x,t,u) \tag{4.6a}$$

$$\frac{du}{dt} = g(x,t,u). \tag{4.6b}$$

When (4.6) is independent of u, one deduces the solution,

$$x = x(t,\ k_1,\ k_2) \tag{4.7a}$$

$$u = u(t,\ k_1;\ k_2) \tag{4.7b}$$

where k_1 and k_2 are arbitrary constants. One of these constants, say k_1, plays the role of a new independent variable, henceforth denoted by ρ; then k_2 becomes the dependent or similarity variable; thus,

$$u(x,t) = F(\rho). \tag{4.8}$$

On inserting this into (4.1) there results an ODE for $F(\rho)$.

From a knowledge of the infinitesimals X, T, U, one can construct a set of infinitesimal operators which generate the infinitesimal similarity transformations and the Lie algebra corresponding to them. It characterizes the group of invariances **G** of the system of equations, and the procedure herein described, is called group analysis. Its usefulness in field theory is clear from the possibility of generating time-dependent solutions of the SU(2) Higgs model [14]. The time-dependent equations of this model in the PS limit are

$$r^2(K_{rr} - K_{tt}) = K(K^2 + H^2 - 1) \tag{4.9a}$$

$$r^2(H_{rr} - H_{tt}) = 2HK^2. \tag{4.9b}$$

One defines a generic dependent variable u^α ($\alpha = 1,2$) such that $u^1 = K$ and $u^2 = H$, and considers a one-parameter family of infinitesimal transformations in r, t, and u^α with associated infinitesimals R, T and U^α, respectively.

To ensure the invariance of the system (4.9), the second derivatives of u^α must transform according to

$$u^{\alpha *}_{r^*r^*} = u^\alpha_{rr} + \varepsilon[U^\alpha_{rr}] + 0(\varepsilon^2), \tag{4.10a}$$

$$u^{\alpha *}_{t^*t^*} = u^\alpha_{tt} + \varepsilon[U^\alpha_{tt}] + 0(\varepsilon^2), \tag{4.10b}$$

where $[U^\alpha_{xx}]$ denotes a second extension [9].

When the transformed system corresponding to (4.9) is written in view of (4.10), and coefficients of terms of order ε are equated to zero, one finds

$$r^2[U^1_{rr}] - r^2[U^1_{tt}] + 2rR(K_{rr} - K_{tt}) + U^1(1-3K^2-H^2)-2U^2KH) = 0 \tag{4.11a}$$

$$r^2[U^2_{rr}]-r^2[U^2_{tt}] + 2rR(H_{rr}-H_{tt}) - U^2(2K^2+ 4HK) = 0. \tag{4.11b}$$

From these equations one picks up the coefficients of different orders of derivatives of K and H, and sets them equal to zero. This leads to a large number of determining equations which are consistently solved to yield the infinitesimals:

$$R = 2\lambda rt + \varkappa r, \tag{4.12a}$$

$$T = \lambda(r^2 + t^2) + \varkappa t + \sigma, \tag{4.12b}$$

$$U^\alpha = 0, \tag{4.12c}$$

where λ, $\varkappa$, and σ are constants. The occurrence of three independent parameters in these equations, permits one to define the three generators G_a as follows:

$$G_1 = 2rt\frac{\partial}{\partial r} + (r^2 + t^2)\frac{\partial}{\partial t}, \tag{4.13a}$$

$$G_2 = r\frac{\partial}{\partial r} + t\frac{\partial}{\partial t}, \tag{4.13b}$$

$$G_3 = \frac{\partial}{\partial t}, \tag{4.13c}$$

which satisfy the Lie algebra:

$$[G_1,G_2] = -G_1, \quad [G_2,G_3] = -G_3, \quad [G_3,G_1] = 2G_2. \tag{4.14}$$

The group analysis approach yields a set of time-dependent solutions. To this end, one considers a subgroup $G_1 \subset G$ as well as itself, defines a similarity variable in each case, and solves the corresponding similarity-reduced equation.

Under the full group G, with $\lambda \neq 0$, $\varkappa \neq 0$ and $\sigma = \varkappa^2/4\lambda$ the similarity variable is

$$\rho = r/(t^2 - r^2 + \varkappa t/\lambda + \varkappa^2/4\lambda). \tag{4.15}$$

The corresponding similarity-reduced equations have the same form as that of the static SU(2) Higgs model in the PS limit, (3.9), with r being replaced by ρ. This directly gives the solution

$$K(\rho) = C\rho/\sinh(C\rho), \tag{4.16a}$$

$$H(\rho) = C\rho/\coth(C\rho), \tag{4.16b}$$

which coincides with that reported in ref. (15). One, however, obtains a new solution also:

$$K(\rho) = \rho/(A + \rho), \tag{4.17a}$$

$$H(\rho) = A/(A + \rho), \tag{4.17b}$$

where A is a nonzero arbitrary constant. Both $K(\rho)$ and $H(\rho)$ are singular on the surface $A + \rho = 0$.

When a subgroup $G_1 \subset G$, defined by setting $\varkappa = \sigma = 0$, is considered, the similarity variable is

$$\eta = r/(t^2 - r^2), \tag{4.18}$$

and the reduced system is again of the PS form, (3.9), with $\rho \to \eta$. In addition to a PS-like solution, which is reported in ref. (15), one obtains a new solution:

$$K(\eta) = \eta/(A + \eta), \tag{4.19a}$$

$$H(\eta) = A/(A + \eta), \tag{4.19b}$$

with $A \neq 0$.

5. VORTICES

We shall discuss here some aspects of a scale-dependent vortex solution recently obtained in the SU(2) Higgs system [5]. Using a suitable axially symmetric ansatz, that satisfies the boundary condition,

$$K(r) \to 1, \quad H(r) \to 0 \text{ as } r \to 0 \tag{5.1}$$

the equations of motion in the PS limit are obtained:

$$r^2 H_{rr} - rH_r = H(H^2 + K^2 - 1), \tag{5.2a}$$

$$r^2 K_{rr} - rK_r = 2K\,H^2. \tag{5.2b}$$

With the substitution

$$K(r) = 1 - H(r), \tag{5.3}$$

which is in agreement with (5.1), (5.2) are converted into a single nonlinear equation

$$r^2 H_{rr} - rH_r = 2H^2(H - 1). \tag{5.4}$$

So far no analytic solutions of this equation have been found. The authors of ref. (5) have resorted to phase plane analysis, and found that there exists a regular solution (5.4) with the behaviour

$$H(0) = 0, \quad H(\infty) = 1. \tag{5.5}$$

Since (5.4) is invariant under a scale transformation, $r \to \sqrt{\alpha}\, r$, the solution as well as its energy, will be scale-dependent. Thanks to the axial symmetry, this soliton is rightly designated as a vortex or string.

The energy per unit length of a vortex is called its tension. In the present case, the tension is scale-dependent and is given by the integral

$$E = -2\pi \int_0^\infty r^2 dr, \tag{5.6}$$

or

$$E = \frac{\pi}{e^2} \int_0^\infty \frac{dr}{r^3}[3r^2(H_r)^2 - 4rHH_r + 4H^2 - 4H^3 + 3H^4], \tag{5.7}$$

which has been evaluated only numerically [5] for different α values. The scale-dependent vortex, as this solution may be called, is stable because, even though α may be continuously varied, the value $\alpha = 0$ is forbidden. The reason for this is that at this point, the boundary behaviour at $r = \infty$ (5.5) is altered.

The magnetic flux contained in the scale-dependent vortex, is calculated from the formula,

$$\Phi = \int_0^\infty 2\pi r \, dr \, F^3_{12}, \tag{5.8}$$

which gives

$$\Phi = -2\pi/e, \tag{5.9}$$

as is the case with the Nielsen-Olesen string [4].

6. CONCLUSION

In these lectures we have discussed the concept of finite energy solitons in gauge theories at the classical level and the application of two well-known mathematical methods--Hirota's method and group analysis--to the construction of some of these solutions in closed form. Much work has been done in recent years to find multi-soliton solutions in field theory. For instance, multivortex solutions where n vortices are superimposed in an axially symmetric configuration have been discovered [16]. However, there exists no formula for a general multivortex solution. On the other hand, static multi-monopole solutions in the PS limit have been obtained by methods of algebraic geometry [17] and using Bäcklund transformations [18]. The soliton concept has permeated into supergravity and Kaluza-Klein theories and has even been suggested as a model of baryons.

References

[1] P. A. M. Dirac, Proc. Roy. Soc. **A133** (1931) 60; Phys. Rev. **74** (1948) 817.
[2] G. 't Hooft, Nucl. Phys. **B79** (1974) 276; A. M. Polyakov, JETP Lett. **20** (1974) 194.
[3] B. Julia and A. Zee, Phys. Rev. **D11** (1975) 2227.
[4] H. B. Nielsen and P. Olesen, Nucl. Phys. **B61** (1973) 45.
[5] Prakash Mathews, C. M. Ajithkumar and K. Babu Joseph, Phys. Lett. **171B** (1986) 64.
[6] A. A. Belavin, A. M. Polyakov, A. S. Schwartz and Yu. S. Tyupkin, Phys. Lett. **59B** (1975) 85.
[7] R. Friedberg, T. D. Lee and A. Sirlin, Phys. Rev. **D13** (1976) 2739.
[8] R. Hirota in **Solitons** ed R.K. Bullough and P. J. Caudrey (Heidelberg: Springer-Verlag, 1980).
[9] G. W. Bluman and J. D. Cole, **Similarity Methods for Differential Equations** (New York: Springer-Verlag, 1974).
[10] M. K. Prasad and C. M. Sommerfield, Phys. Rev. Lett. **35** (1975)760.
[11] C. M. Ajithkumar, Ph.D. Thesis, Cochin University of Science and Technology (1986) (unpublished).
[12] I. Ju, Phys. Rev. **D17** (1978) 1637.
[13] A. P. Protogenov, Phys. Lett. **67B** (1977) 62.
[14] K. Babu Joseph and B. V. Baby, J. Math. Phys. **26** (1985) 2746.
[15] W. Mecklenburg and D. P. O'Brein, Phys. Rev. **D18** (1978) 1327.
[16] H. J. de Vega and F. A. Schaposnik, Phys. Rev. **D14** (1976) 1100.
[17] R. S. Ward, Comm. Math. Phys. **79** (1981) 317.
[18] P. Forgacs, Z. Horvath and L. Palla, Phys. Lett. **99B** (1981) 232; Phys. Lett. **102B** (1981) 131.

Solitary Waves of the "2-Dimensional Ferromagnet" *

R. Rajaraman

Centre for Theoretical Studies, Indian Institute of Science,
Bangalore 560012, India

Finite energy solitary wave solutions obtained by Polyakov and Belavin for the isotropic 2-dimensional ferromagnet are briefly reviewed. The ideas of Bogomolnyi connecting the winding number and potential energy of the system play an important role in the discussion.

I will derive here some exact solitary waves solutions of the so-called nonlinear σ-model in particle-physics parlance. The system consists of three coupled fields $\varphi^a(\mathbf{x},t)$, a = 1,2,3, in 2+1 (two-space, one-time) dimensions. This system can also be viewed, as I hope to show you, as a simple model of an isotropic 2-dimensional ferromagnet. These solutions were obtained in the late seventies by Polyakov and Belavin [1] using very elegant methods. The ideas of Bogomolnyi [2] will also play an important role in our discussion. These solutions have been known now for several years and have also been covered in reviews [3]. Nevertheless I decided to discuss them in this school for the following reason. Most participants here are from either the applied mathematics or plasma physics wings of the soliton community. Most of you are very familiar, in great detail, with 1+1 dimensional systems, such as the KdV, the sine-Gordon, the nonlinear Schrödinger equation, and with their remarkable exact soliton solutions. But if we relax the requirement of exact solitons (i.e., preservation of shape and velocity even after collisons), and settle for just solitary waves, then there are many interesting solutions in higher dimensions as well. In particular, particle physicists have found several such solutions in 2, 3 and 4 dimensions, by using some elegant analytic methods. These solutions also possess an interesting topological index. The solitary waves of the 2-dimensional ferromagnet, which I shall discuss here, are perhaps the simplest examples of this type. Other examples are the 't Hooft-Polyakov monopole, the Skyrmion, and the Yang Mills instanton[3].

*Notes taken and prepared by M. Daniel.

The equation we are going to solve is

$$\Box\varphi^a - (\varphi^b \Box \varphi^b)\, \varphi^a = 0, \qquad (1)$$

where

$$\varphi^a = \varphi^a(x_1, x_2, t), \quad a = 1,2,3.$$

It is understood that the repeated indices are summed over. Thus, b is summed and a is free. This equation results from the following simple Lagrangian,

$$L = \sum_a \int [\tfrac{1}{2}(\partial_t \varphi^a)^2 - \tfrac{1}{2}(\nabla\varphi^a)^2] d^2x \qquad (2)$$

subject, however, to the constraint

$$\sum_a \varphi^a(\mathbf{x},t)\, \varphi^a(\mathbf{x},t) = 1 \quad \text{at each } \mathbf{x},\ t.$$

This is the Lagrangian of a massless Klein Gordon field in 2-dimensions, under the condition $\sum_a \varphi^a \varphi^a = 1$, that is, the field is not allowed to vary freely but varies only subject to the above condition. Let us show that the field equation (1) is the Euler-Lagrangian equation for Lagrangian (2), in the presence of the constraint $\sum_a \varphi^a \varphi^a = 1$. The constraint can be incorporated using the method of Lagrangian multipliers. Let us write a new Lagrangian by adding the Lagrange multiplier term to the old one,

$$\tilde{L} = L + \int \lambda(x,t)(\varphi^a\varphi^a - 1)\, d^2x\,. \qquad (3)$$

The Euler-Lagrange equation for the Lagrangian (3) is

$$\Box\varphi^a - \lambda\varphi^a = 0. \qquad (4)$$

This may appear like a linear equation, but remember that we must adjust the Lagrangian multiplier $\lambda(\mathbf{x},t)$ in such a way that the constraint is satisfied. Rewriting (4), we have

$$\Box\varphi^a = \lambda\varphi^a\,. \qquad (5)$$

Multiply both sides by φ^a, and sum over a.

$$\begin{aligned} \varphi^a \Box \varphi^a &= \lambda\varphi^a\varphi^a \qquad (6) \\ &= \lambda. \end{aligned}$$

Using equation (6) in (5), we get

$$\varphi^a - (\varphi^b \Box \varphi^b)\varphi^a = 0, \qquad (7)$$

which is just the original equation (1).

So far nothing has been said about ferromagnets. Now one can see the following connection with the ferromagnets. Consider first a one-dimensional ferromagnet, that is, a lattice with a magnetic moment φ_i^a at each point. Here i lebels the lattice site and a are the components of the magnetic moment. Also, we have the constraint $\varphi_i^a \varphi_i^a = 1$, since the individual magnets can only rotate, and not change their magnitudes.

φ_i^a

. . . . ♂ ♂ . . . .

i i+1

The dynamics of these magnets depends on what they energtically favour. Energy is least when they are parallel and most when anti-parallel. We assume that each magnet interacts only with its nearest neighbour and that this interaction is characterized by the following potential energy.

$$V = -\alpha \sum_i (\varphi_i^a \varphi_{i+1}^a - 1) \cdot \qquad (8)$$

Rewriting,

$$V = + \frac{\alpha}{2}(\sum_i (\varphi_{i+1}^a - \varphi_i^a)(\varphi_{i+1}^a - \varphi_i^a). \qquad (9)$$

When the lattice size is small we can take the continuum limit, by taking $\varphi_i^a \longrightarrow \varphi^a(x)$, $\sum_i = \int d\mathbf{x}$. Then the potential energy for this one-dimensional ferromagnet becomes, after absorbing constants,

$$V = \frac{1}{2} \int dx \left(\frac{\partial \varphi^a}{\partial x}\right)^2. \qquad (10)$$

In two dimensions the same arguments can be carried out and one will get

$$V = \frac{1}{2} \int d^2\mathbf{x} (\nabla \varphi^a)^2. \qquad (11)$$

This is the potential energy appearing in the Lagrangian of the system (2). Adding to this the kinetic energy we will get the full Lagrangian (2). The corresponding equation of motion is given in eq. (1).

Equation (1) is our starting point and the task ahead is to solve this equation. To make life simpler we will look for **time-independent** solutions: $\varphi^a(x_1,x_2)$. This will be sufficient for obtaining a class of solitary waves. These time-independent solutions will satisfy the following equations.

$$\nabla^2\varphi^a - (\varphi^b \nabla^2\varphi^b)\varphi^a = 0. \tag{12}$$

Once the stationary solution $\varphi^a(x_1,x_2)$ is found the solution in the moving frame can be constructed because the equation of motion is co-variant under (2+1)-dimensional Lorentz transformations. Hence, by Lorentz transforming, we get the function

$$\varphi^a\left(\frac{x_1 - ut}{(1-u^2)^{\frac{1}{2}}}, x_2\right)$$

which is also a time-dependent solution of (1).

Note: However that such transformations will generate some but not all time-dependent solutions.

Equation (12) was solved by Polakov and Belavin using the following clever trick. The lowest energy trivial solution for the equation is called vacuum solution which is of the form

$$\varphi^a(\mathbf{x}) = u^a \tag{13}$$

where

$$\|u\|^2 = u^a u^a = 1.$$

Thus, we have an infinite number of trivial vacuum solutions, one corresponding to each possible direction the unit vector u^a could take. Next consider non-trivial **x**-dependent solitary wave solutions. The energy of these solutions (entirely potential, since they are static solutions) is

$$\begin{aligned} V &= \frac{1}{2}\int d^2x \ \frac{1}{2}(\nabla\varphi^a)^2 \\ &= \frac{1}{2}\int_0^\infty dr\, r \int_0^{2\pi} d\theta \left\{\frac{1}{2}\left(\frac{\partial\varphi^a}{\partial r}\right)^2 + \frac{1}{2r^2}\left(\frac{\partial\varphi^a}{\partial\theta}\right)^2\right\}. \end{aligned} \tag{14}$$

In order that this energy be finite, the energy density (the integrand in (14)) should go to zero as $r \to \infty$. This yields the boundary conditions

$$\frac{\partial\varphi^a}{\partial r} \to 0 \quad \text{and} \quad \frac{\partial\varphi^a}{\partial\theta} \to 0 \quad \text{as } r \to \infty. \tag{15}$$

In other words, as $r \to \infty$ the field $\varphi^a(\mathbf{x})$ must approach some fixed value, independent of direction. All points on the circle at infinity must have the same value of φ^a. Such a function $\varphi^a(\mathbf{x})$, since it identifies (that is, it has the same value at) the boundary at infinity, compactifies physical space R_2 into a spherical surface S_2. In the interior of this space, $\varphi^a(\mathbf{x})$ could be any continuous function of $\mathbf{x}$, with unit modulus ($\varphi^a\varphi^a = 1$). The allowed values of φ^a at each given $\mathbf{x}$, also forms a sphere S_2 since the modulus of φ^a has to be unity. Thus, any finite energy configuration $\varphi^a(\mathbf{x})$ represents a mapping of one sphere S_2 into another sphere S_2.

We know from mathematics that all such mappings can be divided into homotopy classes, where mappings within each class can be continuously deformed into one another. Each class is characterised by a "winding number." The expression for the winding number n for these mappings is given by

$$n = \frac{1}{8\pi} \int (\varphi^a \partial_\mu \varphi^b).(\partial_\nu \varphi^c) \varepsilon_{abc}\varepsilon_{\mu\nu} . \qquad (16)$$

A useful relation connecting the winding number and potential energy of the system was written down by Bogomolnyi. We have

$$\int d^2\mathbf{x}(\partial_\mu\boldsymbol{\varphi}\pm\varepsilon_{\mu\nu}\boldsymbol{\varphi}\times\partial_\nu\boldsymbol{\varphi})(\partial_\mu\boldsymbol{\varphi}\pm\varepsilon_{\mu\sigma}\boldsymbol{\varphi}\times\partial_\sigma\boldsymbol{\varphi}) \geq 0, \qquad (17)$$

that is,

$$\int d^2\mathbf{x}(\partial_\mu\boldsymbol{\varphi}\cdot\partial_\mu\boldsymbol{\varphi}) + \varepsilon_{\mu\nu}\varepsilon_{\mu\sigma}(\boldsymbol{\varphi}\times\partial_\nu\boldsymbol{\varphi})(\boldsymbol{\varphi}\times\partial_\sigma\boldsymbol{\varphi}) \geq \mp 2\int\varepsilon_{\mu\nu}\partial_\mu\boldsymbol{\varphi}\cdot(\boldsymbol{\varphi}\times\partial_\nu\boldsymbol{\varphi}). \qquad (18)$$

It is easy to check that the first two terms are equal. Then using eq. (16), one can write

$$2\int d^2x(\partial_\mu\boldsymbol{\varphi}\cdot\partial_\mu\boldsymbol{\varphi}) \geq \mp 16\pi n. \qquad (19)$$

Given (1), this is equivalent to the following:

$$4V \geq \mp 16\pi n,$$

that is,

$$V \geq 4\pi|n|. \qquad (20)$$

Equation (20) is the Bogomolńyi inequality. It says that the energy of any configuration is greater than or equal to 4π times its winding number. Now, recall the important fact that any classical static solution extremises the energy functional. In particular, if the **equality** in eq. (20) is satisfied, then the energy is minimised in any given topological sector. Hence, a field $\varphi^a(\mathbf{x})$ for which the equality in eqs. (17-20) holds, is automatically going to solve the field equation (12). This in turn requires, because of (17) that

$$\partial_\mu \varphi^a = \pm \varepsilon_{\mu\nu} \varepsilon^{abc} \varphi_b (\partial_\nu \varphi_c). \tag{21}$$

Thus, instead of solving the second order differential equation (12), we need only to solve the first order equation (21). Any solution of the latter will also be a solution of the former. This can be verified by applying ∂_μ on equation (21). To solve eq. (21), we make the following stereographic projection.

$$\omega_1 = \frac{2\varphi_1}{1-\varphi_3}, \qquad \omega_2 = \frac{2\varphi_2}{1-\varphi_3}. \tag{22}$$

Let $\omega \equiv \omega_1 + i\omega_2 = \frac{2\Phi}{1-\varphi_3}$ where $\Phi = \varphi_1 + i\varphi_2$.

In terms of this, the Bogomolnyi equation (21) now becomes

$$\partial_1 \Phi = \mp i\Phi \overleftrightarrow{\partial}_2 \varphi_3 \tag{23a}$$

$$\partial_2 \Phi = \pm i\Phi \overleftrightarrow{\partial}_1 \varphi_3 . \tag{23b}$$

Using the stereographic projection (22), eq. (23) takes the form

$$\partial_1 \omega = \pm \frac{2}{(1-\varphi_3)^2} [\partial_1 \Phi + \Phi \overleftrightarrow{\partial}_1 \varphi_3]. \tag{24}$$

On making use of (23), eq. (24) becomes

$$\begin{aligned} \partial_1 \omega &= \pm \frac{2}{(1-\varphi_3)^2} [-i\Phi \overleftrightarrow{\partial}_2 \varphi_3 - i\partial_2 \Phi] \\ &= \mp i\partial_2 \omega. \end{aligned} \tag{25}$$

Equation (25) is the Cauchy-Riemann equation for which any analytic function $\omega(z)$ $(z = x_1 + ix_2)$ or $\omega(z^*)$ is a solution. Thus our original equation is solved. In terms of the variables $\omega = 2(\varphi_1 + i\varphi_2)/(1-\varphi_3)$, and $z = x_1 + ix_2$ any analytic function $\omega(z)$ or $\omega(z^*)$ is an exact solution of (21), and also of the original field equation (12). The solution can be rewritten in terms of the original field variables φ^a, by inverting the stereographic projection (22). These solutions will, for reasons given earlier, have finite energy, equal to 4 times its winding number n. In terms of the variable $\omega(t)$, eq. (16) can be written as

$$n = \frac{V}{4\pi} = \frac{1}{4\pi} \int d^2z \frac{\left|\frac{d\omega}{dz}\right|^2}{\left(1+\frac{|\omega|^2}{4}\right)^2} \tag{26}$$

As an example, consider $\omega = [(z-z_o)/\lambda]^{n_o}$, where z_o and λ are constants and n_o is a positive integer. This is an analytic function and hence a solution of the original field equation (12). Upon inserting this function into (26), one can check that

$$n = \frac{1}{4\pi} \int d^2z \frac{n_o^2 |z-z_o|^{2n_o-2} \lambda^{2n_o}}{\left(\lambda^{2n_o} + \frac{|z-z_o|^{2n_o}}{4}\right)^2}$$

$$= n_o = \frac{V}{4\pi} . \tag{27}$$

The winding number is n_o. This is also evident from the fact that in the function $\omega = [(z-z_o)/\lambda]^{n_o}$, the z plane is clearly mapped n_o times into the ω-plane. The constants z_o and λ represent respectively the location and the size of the soliton. Notice that thanks to the translational and scale covariance of the field equation (12), the energy (27) does not depend either on z_o or on λ .

We had taken n_o to be a positive integer. Choosing n_o to be negative yields an equally good solution, with the same energy but opposite winding. That there is a pole at $z = z_o$ when n_o is a negative integer is no cause for concern. The divergence of ω as $z \to z_o$ only means that $\varphi_3 \to 1$.

In this fashion, any meromorphic function of z or z* is a solitary wave solution of the system (12).

REFERENCES

[1] A. A. Belavin and A. M. Polyakov, JETP Letters, **22** (1975) 245.
[2] E. B. Bogomolnyi, Soviet J. of Nuc. Phys. **24** (1976) 449.
[3] R. Rajaraman, **Solitons and Instantons** (Amsterdam: North-Holland Co., 1982).

Soliton Propagation in Optical Fibres

A. Kumar

Department of Physics, Indian Institute of Technology,
Hauz Khas, New Delhi 110016, India

Pulse propagation in optical fibres, including the effects of nonlinear change in the refractive index, group velocity dispersion and losses, is reviewed. An analysis of the effect of fifth-order nonlinearity on the soliton propagation in the model based on the damped nonlinear Schrödinger equation is made. The experimental observation of bright solitons and soliton laser are discussed in short.

INTRODUCTION

Nonlinear pulse propagation in optical fibres has been a subject of intensive research [1-40] primarily because of the possibility of transmitting undistorted pulses of high peak powers which can be used in various fields such as communication, power transmission, medicine and industrial processing [3,4], etc. The main activity in this connection is centred around a suggestion given by Hasegawa and Tappert [1,2] that the nonlinear response of the fibre, which tends to self-confine the pulse, can be exploited to get rid of the pulse distortions, caused by dispersion, frequency chirp, etc., which put a critical limitation in realizing the full bandwidth capability of the linear communication systems using fibres. The distortionless pulse propagation based on this mechanism is usually referred to as soliton propagation. According to this mechanism when the frequency shift due to the nonlinear change in the refractive index of the fibre is balanced with that due to group dispersion in the negative dispersion region of the fibre the optical pulse should form an envelope soliton and a stationary pulse transmission should take place. However, due to the losses in fibre the soliton pulse continuously loses its energy, as a result of which its width increases with the distance of propagation and it tends to vanish [13,18-21,27]. Therefore, for long distance propagation periodic amplification and reshaping of the pulse should be done [14-16,22].

The prediction of Hasegawa and Tappert was successfully verified by Mollenauer, Stolen and Gordon [28]. After this demonstration of the feasibility of the soliton concept a large number of research papers (both theoretical and experimental) dealing with further possible applications of optical solitons appeared in literature. All these investigations led to the creation of what is called a soliton laser, which is a valuable achievement.

The organisation of the lecture is as follows. Section I deals with the basic nonlinear evolution equation for the complex envelope amplitude and the soliton solutions to them in the lossless case. In Section II we see the effect of fibre losses on soliton propagation while in Section III we study the amplification and reshaping of solitons and their effect on the stability of the pulse. Section IV describes the effect of fifth-order nonlinearity on the propagation of a gaussian input pulse taking into account the frequency chirp. Section V deals with the modulational instability in optical fibres and the possibility of generating a train of solitons using induced modulational instability. In the last two sections, i.e., in the sixth and seventh we, in short, talk about the experimental observations of bright soliton pulses and the soliton laser.

I. BASIC NONLINEAR EQUATION: BRIGHT AND DARK SOLITONS

Consider the propagation of a pulse envelope $\varphi(t,x)$ (where x is the longitudinal coordinate along the axis of the fibre) through an optical fibre including the effects of group velocity dispersion, nonlinear change in the refractive index of the fibre and the fibre losses. It is well known [1,13,18,28,37] that the evolution of the envelope is governed by the following differential equation:

$$i\left(K_1 \frac{\partial\varphi}{\partial t} + \gamma\varphi + \frac{\partial\varphi}{\partial x}\right) + \tfrac{1}{2}K_2 \frac{\partial^2\varphi}{\partial t^2} - \frac{K_o}{n_o}\, n_{NL}(|\varphi|^2)\varphi = 0\,, \tag{1.1}$$

where γ is the exponent of amplitude decay in linear regime, $K_1 = \partial K/\partial\omega$ $K_2 = \partial^2 K/\partial\omega^2$ and $n_{NL}(|\varphi|^2)$ represents the intensity dependent nonlinear part of the refractive index:

$$n(\omega,|\mathbf{E}|^2) = n_o(\omega) + n_{NL}(|\mathbf{E}|^2). \tag{1.2}$$

In the case of cubic nonlinearity $n_{NL} = n_2|\mathbf{E}|^2$ and self-focusing medium $n_2 > 0$, using the transformations

$$S = \tau^{-1}(t-K_1x); \quad \xi = |K_2|\tau^{-2}x$$

$$\psi = \tau(\alpha/|K_2|)^{1/2}\varphi; \quad \alpha = \frac{1}{2}\frac{n_2K_o}{n_o}, \tag{1.2a}$$

equation (1.1) can be reduced to the following dimensionless form

$$i\frac{\partial\psi}{\partial\xi}+i\Gamma\psi+\frac{1}{2}\mathrm{Sgn}K_2\frac{\partial^2\psi}{\partial S^2} - |\psi|^2\psi = 0, \tag{1.3}$$

where $\Gamma = \gamma\tau^2/|K_2|$. Note that in the above transformations τ is an arbitrary time scale which allows a pulse of standard duration in the dimensionless retarded time variable s to correspond to a pulse of any desired duration in t [28].

In the absence of losses $\gamma = 0$ and Eq. (1.3) reduces to well-known nonlinear Schrödinger equation (NLS). This equation can be solved for any input structure $\psi(\xi,S)$ at $\xi = 0$ using Zakharov-Shabat scheme based on inverse scattering method. The general response of $\psi(\xi,S)$ can be described by N solitons and continuous modes which vanish at $\xi \to \infty$. The number of solitons N is determined by the area

$$A = \int_{-\infty}^{+\infty} |\psi(0,S)|dS.$$

The soliton solutions obtained by Hasegawa and Tappert [1,2] for optical fibres are one soliton solutions to Eq. (1.3) which are usually called bright or dark solitons depending on whether they exist in the anomalous dispersion region or normal dispersion region respectively.

The bright soliton solutions to Eq. (1.3) in the anomalous dispersion region when $K_2 < 0$ are given by

$$\psi_{BS}(\xi,S) = E_{BS}\,\mathrm{sech}(S/S_o)\,\exp[i\lambda_B\xi], \tag{1.4}$$

where the soliton amplitude E_{BS} is related to the pulse half-width S_o and the wave number shift λ_B by

$$S_o^{-1} = E_{BS}, \quad \lambda_B = -\frac{1}{2}E_{BS}^2. \tag{1.5}$$

In the normal dispersion region when $K_2 > 0$, Eq. (1.3) admits soliton solutions which represent envelope shocks and are called the dark solitons. They are given by

$$\psi_{DS}(\xi,S) = E_{DS}\tanh(S/\tilde{S}_o)\,\exp[i\lambda_D\xi], \tag{1.6}$$

where E_{DS}, $\tilde{S}_o$, and λ_D are related through

$$\tilde{S}_o^{-1} = E_{DS}, \quad \lambda_D = - E_{DS}^2. \tag{1.7}$$

Note that for dark solitons the nonlinear wave number shift is twice as large as the wave number shift in the case of bright solitons for the same amplitude.

Further, note that

$$\tilde{\psi}(\xi,S) = \psi(S - \frac{\xi}{u}) \exp[i(\frac{S}{u} - \frac{1}{2} \frac{\xi}{u^2})] \tag{1.8}$$

is also a solution which indicates that a frequency modulation at $\xi=0$ can produce a soliton which propagates with the speed u different from that without frequency modulation.

Let us now find out the expression for the peak power of the above mentioned solitons. Using Eq. (1.2a) and Eq. (1.5) the relationship between the peak electric field φ_o and the pulse width $2t_o$ (for which the pulse height drops to sech 1 = 0.65) can be written as [27]

$$\varphi_o = \frac{\sqrt{2}[\lambda^2(\partial^2 n/\partial\lambda^2)]^{1/2}}{\omega_o t_o (n_2)^{1/2}}, \tag{1.9}$$

where $\omega_o = 2\pi c/\lambda$ is the carrier angular frequency. The peak power P_o is related to φ_o by

$$P_o = \frac{1}{2} V_g \varphi_o^2 \varepsilon_o \tilde{S} n^2 \tag{1.10}$$

where $\tilde{S}$ is the cross-sectional area of the fibre, $V_g = c/n$ and the dielectric constant of the fibre $\varepsilon = \varepsilon_o n^2$ where $\varepsilon_o = 8.854 \times 10^{-12}$ F/m. Assuming $n_2 = 1.2 \times 10^{-22}$ $(m/V)^2$ and n = 1.5 we get

$$P_o = 3.32 \times 10^7 \; \tilde{S}(\mu m^2) \frac{\lambda^2(\partial^2 n/\partial\lambda^2)}{(\omega_o t_o)^2} \; (W). \tag{1.11}$$

This is the required expression for peak power. From this expression it follows that near the zero group dispersion the balancing power to produce a soliton can be minimum. However, at $\partial^2 n/\partial\lambda^2 = 0$ the soliton nature of the pulse is lost. Hence for realistic power one should choose λ in such a way that the higher order dispersions could be treated as perturbations [27].

For $\lambda = 1.3\ \mu m$ and $\bar{S} = 20\ \mu m^2$ one gets that

$$(2t_o)^2 . P_o = 1.6\ W(P\ sec)^2, \tag{1.12}$$

which shows that for 1 ps pulse one requires a minimum power of 1.6W. However, this minimum power can be changed by the choice of λ and some other factors.

At the end of this section we want to note that in the derivation of Eq. (1.1) the radial dependence of the field was treated by employing the radial eigenfunction of a linear dispersive fibre [1,37] which in general is not acceptable. Christodoulides and Joseph [29,30] have studied the propagation of bright solitons through optical fibres treating the radial dependence of the field in an exact way. For the weakly dispersive fibres they reproduce the old results. However, for highly dispersive fibres they conclude that higher order effects in the perturbation scheme for radial field dependence become important and may play a vital role in the formation of solitons. Using the same procedure they have obtained a new class of dark soliton solutions which exist in the normal dispersion region of the fibre.

II. EFFECT OF LOSSES

For $\gamma \neq 0$, Eq. (1.3) cannot be solved exactly. However, if $\Gamma\xi << 1$ the second term in Eq. (1.3) can be treated as perturbation and the one soliton solution can be written in a simple form [8,18,19]

$$\psi(\xi,S) = A\ \mathrm{sech}AS\ \exp(i\sigma), \tag{2.1}$$

where

$$A = \psi_o \exp[-2\Gamma\xi], \qquad \sigma = \frac{\psi_o^2}{8\Gamma}[1 - \exp(-4\Gamma S)]. \tag{2.2}$$

Thus the pulse amplitude decreases with the distance of propagation while the pulse width increases as $S_o \rightarrow S_o \exp[2\Gamma\xi]$. It puts some limitations on a large distance propagation of solitons in lossy fibres. In fact the condition for the perturbation to be valid is

$$\Gamma\xi = \frac{\tau^2 \gamma |K_2| x}{|K_2|\ \tau^2} = \gamma x << 1. \tag{2.3}$$

The characteristic distance determined by Eq. (2.3) is known as the period of soliton in lossy fibres [41].

Numerical simulation results [18] of Blow and Doran confirmed the perturbative results for small propagation distances. However, the results obtained for the evolution of higher order solitons in lossy fabric indicated that the complicated amplitude oscillations characteristic of higher order solitons gradually disappeared and the output shape at large distances is a single peak with a pulse width much less than in the corresponding linear case. These results demonstrate that for realistic optical fibre communication systems large amplitude solitons can be used for a significant reduction in pulse spreading over the linear regime, which is important for systems using long repeater spacing.

III. AMPLIFICATION AND RESHAPING OF SOLITONS

From the above discussions it is clear that in real fibres with losses it is not possible to have distortionless propagation for longer distances. For long distance communication systems periodic amplification is needed to retain the original pulse shape. If the soliton is periodically amplified at a distance determined by group dispersion and losses it should propagate without distortion for a large distance. To study this problem let us consider the following.

It has been shown [14-16] that if we include the acts of amplification the envelope amplitude $\psi(\xi,S)$ satisfies the following differential equation

$$i\frac{\partial\psi}{\partial\xi}+\frac{1}{2}\frac{\partial^2\psi}{\partial S^2}+i\Gamma\psi\quad -i\sum_{\ell=1}^{N}\alpha_\ell\quad\psi(S,\xi_\ell-0)\ \delta(\xi-\xi_\ell)+|\psi|^2\psi=0, \quad (3.1)$$

where $\xi_\ell = (\ell-1)\Delta\xi$, $\ell = 1,2,3,\ldots,N$ are the positions of the amplifiers, $\delta(x)$ is δ-function of x and α_ℓ is the exponential gain of the ℓ-th amplifier. Kodama and Hasegawa [14,15] showed that for $\Gamma \ll 1$ the soliton can propagate for a distance larger than $600.\Delta\xi$ provided that the amplifier gain is chosen constant, $\bar{\alpha}$, such that it exactly compensates for the loss rate

$$\bar{\alpha} = \exp[\Gamma.\Delta\xi] - 1. \quad (3.2)$$

However, if α_ℓ is variable one has to study the effect of random kicks, experienced by the soliton during amplification, on the stability and the detoriation of the soliton quality of the pulse.

Since due to the act of amplification only the amplitude of the pulse is increased to $(1+\alpha_\ell)$ while the pulse width is not affected the pulse deviates from the ideal soliton structure. The amplified pulse produces a new soliton with the amplitude $(1+\alpha_\ell)\psi_o$ where ψ_o is the original amplitude. The difference between the energy of the amplified pulse and the newly formed soliton is radiated out as linear dispersive wave. Kodama and Hasegawa [16] showed that the total energy $E(\xi)$:

$$E(\xi) = \int_{-\infty}^{+\infty} |\psi(\xi,S)|^2 dS \equiv F^2(\xi).E_o, \tag{3.3}$$

$$F(\xi) = \left(\prod_{\ell=1}^{n} (1+\alpha_\ell) \right) \exp(-\Gamma\xi); \quad \xi_n < \xi < \xi_{n+1} \tag{3.4}$$

diffuses as

$$\langle E(\xi) - E_o \rangle = [\exp(2D\xi) - 1]E_o \tag{3.5}$$

where D is defined by

$$\lim_{\Delta\xi \to 0} \frac{\langle (F_{n+1} - F_n)^2 \rangle}{\Delta\xi} = \lim_{\Delta\xi \to 0} \frac{\langle (\alpha_{n+1} - \bar{\alpha})^2 \rangle}{\Delta\xi} F_n^2 = 2DF_n^2 \tag{3.6}$$

under the assumption that the average value of α_ℓ is given by $\bar{\alpha}$ defined by equation (3.2).

Now, right after the n-th amplification $\xi = \xi_n + 0$ and the soliton is expected to have the form

$$\psi_{sol}(\xi_n + 0,S) = \psi_n \operatorname{sech}[\psi_n(S - \Theta_n)] \exp(i\sigma_n), \tag{3.7}$$

where Θ_n and σ_n are the centre and the phase of the soliton and

$$\psi_n = [1 + 2\alpha_n + 0(\alpha_n^2)] \exp[-2\Gamma\Delta\xi]\psi_{n-1}. \tag{3.8}$$

The deviation in ψ_n in the order α_n^2 is due to the interaction between the soliton and the radiation generated by the previous amplification and loss in the absence of which we would have got $\psi_n = (1 + 2\alpha_n)\psi_{n-1}$. From this equation we get that after the n-th amplification the energy of the soliton can be expressed as

$$E_n^{sol} = \int |\psi_{sol}|^2 dS = (1 + 2\alpha_n + \delta\alpha_n^2) \exp[-2\Gamma\Delta\xi].E_{n-1}^{sol}, \tag{3.9}$$

where $(1-\delta) \geq 0$ is the measure of the energy lost by the radiation of the dispersive wave. The ratio R_n of the soliton energy after n-th amplification and the total energy is given by

$$R_n = \frac{E_n^{sol}}{E_n} = \frac{1 + 2\alpha_n + \delta\alpha_n^2}{(1 + \alpha_n)^2} R_{n-1} . \tag{3.10}$$

It follows from this equation that the deviation of the soliton energy from the total energy is small after each amplification since $\alpha \approx \Gamma\Delta\xi$ is usually a small quantity. Thus the main problem in the study of soliton transmission with amplification is to find out the quantity δ since it determines whether there will be transfer of energy to the soliton from the dispersive waves ($0 < \delta < 1$) or from the soliton to the dispersive waves ($\delta < 0$) [16]. Besides that $\delta = 1$ indicates that no radiation appears and all the energy goes to the soliton while $\delta = 0$ indicates that there is no interaction between the soliton and the dispersive waves which means that the dispersive waves escape from the soliton quickly. For this purpose Kodama and Hasegawa numerically integrated equation (3.1) for various values of $\Delta\xi$ and the parameter 2D which represents the random error in the amplifier gain. Their results show that even for large errors in the amplifier gain an optical soliton can be stably transmitted through a glass fibre with low loss for thousands of kilometres.

Concluding the section we note that another adiabatic amplification scheme based on stimulated Raman gain has also been proposed by Hasegawa [22] which has the advantage that here the repeater spacing is decided only by fibre loss rate. In the first scheme discussed here the amplification has to be done every time before the dispersive wave leaves the soliton which means that the repeater spacing is decided by a distance controlled by group dispersion. The latter scheme utilizes the stimulated Raman process of the fibre itself in which the gain can be made small to make the amplification adiabatic (so that the change in the Raman gain due to fibre loss and pump depletion can be avoided). As a result the area of the soliton remains constant and the transfer of soliton energy to the dispersive waves is minimum.

IV. EFFECT OF FIFTH-ORDER NONLINEARITY

Let the index of refraction of the fibre be given by [24,25]

$$n = n_o(\omega) + n_2|\mathbf{E}|^2 + n_4|\mathbf{E}|^4, \tag{4.1}$$

where $n_2|\mathbf{E}|^2$ and $n_4|\mathbf{E}|^4$ are the intensity dependent nonlinear parts of the refractive index. In this case the propagation of the pulse envelope $\psi(x,\tau)$, where x is the longitudinal coordinate along the axis of the fibre, is governed by the following differential equation [24]

$$i\left(\frac{\partial\psi}{\partial x}+\gamma\psi\right) = \alpha\frac{\partial^2\psi}{\partial\tau^2}+ K_o|\psi|^2\psi + \lambda_o|\psi|^4\psi , \qquad (4.2)$$

where t is time, $\tau = t - k_o'x$ is a reduced time coordinate, $k_o' = (\partial k/\partial\omega)_{\omega=\omega_o}$ is the inverse of the group velocity at the group carrier frequency ω_o, $\alpha = \frac{1}{2}(\partial^2 K/\partial\omega^2)_{\omega=\omega_o}$, γ characterizes the losses in fibre and $K_o = -\omega_o n_2/2c$ and $\lambda_o = \omega_o n_4/2c$ express the magnitude of non-linearity in terms of n_2 and n_4. For $\psi = \varphi(x,\tau)e^{-\gamma x}$ Eq. (4.2) takes the form

$$i\frac{\partial\varphi}{\partial x} = \alpha\frac{\partial^2\varphi}{\partial\tau^2} + K_o e^{-2\gamma x}|\varphi|^2\varphi + \lambda_o e^{-4\gamma x}|\varphi|^4\varphi . \qquad (4.2a)$$

The pulse propagation through a fibre characterized by Eq. (4.2a) can be formulated as a variational problem in terms of the Lagrangian

$$L = \frac{i}{2}\left(\varphi\frac{\partial\varphi^*}{\partial x} - \varphi^*\frac{\partial\varphi}{\partial x}\right) - \alpha\left|\frac{\partial\varphi}{\partial\tau}\right|^2 + \frac{K(x)}{2}|\varphi|^4 + \frac{\lambda(x)}{3}|\varphi|^6, \qquad (4.3)$$

$$K(x) = K_o\exp[-2\gamma x], \quad \lambda(x) = \lambda_o\exp[-4\gamma x] , \qquad (4.4)$$

where the asterisk denotes a complex conjugate. Assume that φ initially had a gaussian form, i.e., that

$$\varphi(0,\tau) = A_o\exp[-\tau^2/2a_o^2], \qquad (4.5)$$

where A_o and a_o are the initial pulse amplitude and pulse width respectively. The further evolution of φ_o can be analyzed in terms of the trial function

$$\varphi(x,\tau) = A(x)\exp\left[-\frac{\tau^2}{2a^2(x)} + ib(x)\tau^2\right] , \qquad (4.6)$$

where the amplitude A(x), a(x) and the frequency chirp $2b(x)\tau$ all vary with the distance of propagation. Using the variational principle and performing the integration over τ we get four Euler differential equations for A(x), A*(x), a(x) and b(x) which after some algebra can be reduced to a single second order ordinary differential equation [24] for the normalized pulse width parameter $y(x) = a(x)/a_o$

$$\frac{d^2y}{dx^2} = \frac{2\mu}{y^3} - \frac{2\tilde{\lambda}e^{-4\gamma x}}{y^3} - \frac{\nu e^{-2\gamma x}}{y^2} , \qquad (4.7)$$

where

$$\mu = \frac{2\alpha^2}{a_o^4}, \quad \tilde{\lambda} = \frac{2\alpha\lambda_o E_o^2}{\sqrt{3}\, a_o^4}, \quad \nu = \frac{\sqrt{2}\alpha K_o E_o}{a_o^3}; \quad E_o = a(x)\ |A(x)|^2. \quad (4.8)$$

Kumar et al. integrated this equation numerically for $\mu = 1/8$ and $\gamma = 0.5$ dB/Km for various values of the parameters $f = (\nu/2\mu)^{\frac{1}{2}}$ and $\tilde{\lambda}$. The initial value for $y(x)$ was taken to be unity. The results of numerical integration are depicted in Figs. 1 and 2 and show that the pulse width initially oscillates and then starts increasing monotonically with the distance of propagation. While oscillating it passes through several minima whose positions vary with the relative strengths of dispersion and nonlinearity, which is in agreement with the character of solution obtained for damped NLS equation [13]. In the case of damped NLS equation for $f = 2$ the distance travelled by the pulse with consecutive compression and decompression before it disperses finally is the largest. In this case the final minimum of the pulse width occurs at a distance of about 8.6 km which in fact gives the spacing between the amplifiers to be used for long distance propagation. However if we increase f for nonzero $\tilde{\lambda}$ the position of the final minimum shifts towards larger distances. The best result is obtained for $\tilde{\lambda} = -0.035$ and $f = 2.5$ (curve

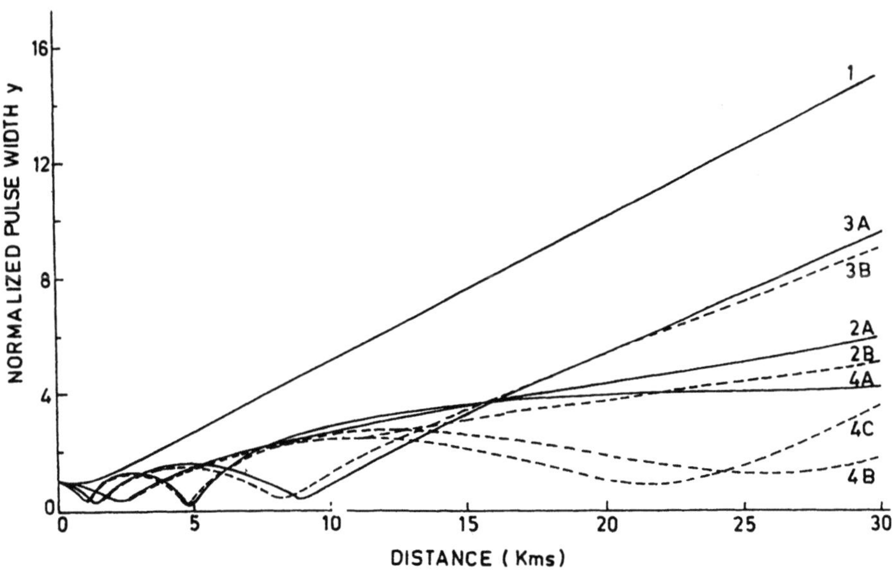

Fig. 1 Results of the numerical solution of Eq. (4.7) for different values of f and $\tilde{\lambda}$. Solid lines: 1, $f = 0.0$, $\tilde{\lambda} = 0.0$; 2A, $f = 1.5$, $\tilde{\lambda} = 0.0$; 3A, $f = 2.0$, $\tilde{\lambda} = 0.0$; 4A, $f = 2.5$, $\tilde{\lambda} = 0.0$. Dashed lines: 2B, $f = 1.5$, $\tilde{\lambda} = -0.025$; 3B, $f = 2.0$, $\tilde{\lambda} = -0.025$, 4B, $f = 2.5$, $\tilde{\lambda} = -0.025$; 3B, $f = 2.0$, $\tilde{\lambda} = -0.025$, 4B, $f = 2.5$, $\tilde{\lambda} = -0.025$; 4C, $f = 2.5$, $\tilde{\lambda} = -0.035$.

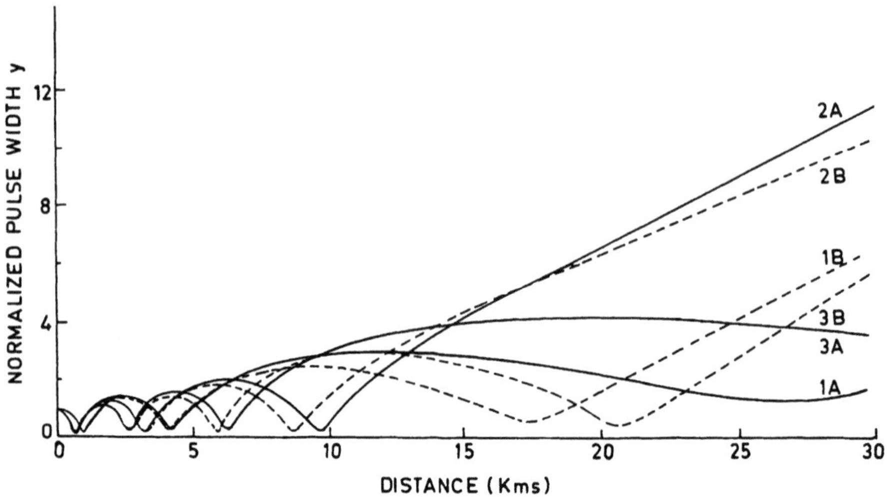

Fig. 2 Results of the numerical solution of Eq. (4.7) for different values of f and $\tilde{\lambda}$. Solid lines: 1A, f = 2.6, $\tilde{\lambda}$ = 0.0; 2A, f = 3.0, $\tilde{\lambda}$ = 0.0; 3A, f = 3.5, $\tilde{\lambda}$ = 0.0. Dashed lines: 1B, f = 2.6, $\tilde{\lambda}$ = -0.025; 2B, f = 3.0, $\tilde{\lambda}$ = -0.025; 3B, f = 3.4, $\tilde{\lambda}$ = -0.025.

4c of Fig. 1) when the pulse width passes through a minimum (before final dispersion takes place) at a distance of approximately 23 km. At this point the pulse width is about nine-tenth of the initial pulse width. For a glass fibre with effective core area of 20 μm^2 the minimum laser peak power, at $\lambda = 1.55\,\mu m$, required to materialise the above situation for a gaussian pulse of 1 psec width is of the order of 5W. Here the values of n_2 and n_4 used are $n_2 \approx 1.2 \times 10^{-22} (m/v)^2$ and $n_4 = 4.4 \times 10^{-37} (m/v)^4$.

Finally we note that since in our case the final minimum of the pulse occurs at a distance of about 23 km the frequency at which the amplification of the pulse should be done for long distance propagation is reduced by a factor of 3 (compared with the model based on damped NLS equation). This is a considerable advantage.

V. INDUCED MODULATIONAL INSTABILITY AND GENERATION OF SOLITONS

It was pointed out by Anderson and Lisak [21] that modulational instability in optical fibres makes a cw optical signal unstable if the carrier wavelength is more than 1.3 m, i.e., in the negative group dispersion region of the fibre leading to some difficulty for coherent signal transmission using amplitude or phase modulation (or both employed together).

For cubic nonlinearity the envelope amplitude satisfies the equation (see Eq. (4.2))

$$i\left(\frac{\partial\psi}{\partial x} + \gamma\psi\right) = \alpha \frac{\partial^2\psi}{\partial\tau^2} + K_o|\psi|^2\psi . \tag{5.1}$$

The steady state solution to it can be written as

$$\psi = \psi_o \exp\left[-i \int_o^x \Delta K(x)dx - \gamma x\right], \quad \psi_o^* = \psi_o \tag{5.2}$$

with $\Delta K(x) = K_o \psi_o^2 \exp[-2\gamma x]$. In the case of small modulation $\psi_1(x,\tau)$ representing $\psi(x,\tau)$ as

$$\psi(x,\tau) = [\psi_o + \psi_1(x,\tau)] \exp\left[-i \int_o^x \Delta K(x)dx - \gamma x\right], \quad |\psi_1| << \psi_o, \tag{5.3}$$

putting $\psi_1 = u + iv$ and seeking the solution to equations satisfied by u and v (obtained after the substitution of (5.3) into (5.1))in the form $u = u_o \cos[\omega x - \Omega\tau]$ and $v = v_o \sin[\omega x - \Omega\tau]$ we get that the local growth rate is given by [21]

$$I_m\, \omega(x) = \alpha\Omega^2\left[\frac{\Omega_c^2(x)}{\Omega^2} - 1\right]^{1/2} \tag{5.4}$$

where

$$\Omega^2 \leq \Omega_c^2(x) \equiv \Omega_{co}^2\, e^{-2\gamma x} \tag{5.5}$$

with $\Omega_{co}^2 = 2K_o|\psi_o|^2/\alpha$. Note that Ω_{co} is $\surd 2$ times the maximum growth rate frequency in the lossless case $\gamma = 0$. The total growth rate is determined by $\exp[K(x)]$, where

$$K(x) = \int_o^x I_m\omega(x')dx' = \frac{\alpha\Omega^2}{\gamma}\left[\left(\frac{\Omega_{co}^2}{\Omega^2} - 1\right)^{1/2} - \arctan\left(\frac{\Omega_{co}^2}{\Omega^2} - 1\right)^{1/2}\right] \tag{5.6}$$

Hence the maximum growth rate is given by $\exp[\max K(x)]$ where

$$\max(K(x)) = (1 - \pi/4)\frac{\alpha\Omega_{co}^2}{2\gamma} \tag{5.7}$$

and occurs at $\Omega = \Omega_{co}^2/2$. Thus if the product of the maximum growth rate and the damping length is larger than 1 the modulational instability becomes a limiting factor for coherent signal transmission because of the instability of the cw signal.

However, Hasegawa [20] has shown that if one uses an externally applied amplitude modulation which could induce modulational instability at the wavelength of the given modulation one would be able to produce a series of solitons with the pulse width in the range of 1-50 psec and repetition period ≈ 250 psec.

For this purpose Hasegawa solved Eq. (1.3) numerically for different wavelengths λ_M of the initial modulation and the depths of modulation A_M in a periodic boundary condition with a given period s = 48 which corresponds to 130 psec for the parameters used by him. In particular Eq. (1.3) was solved with the initial condition

$$\psi(0,S) = 1 + A_M \sin(2\pi S/S_M) \qquad (5.9)$$

with a fixed value of $\Gamma = 5.18 \times 10^{-2}$ corresponding to 0.3 dB/km. His calculations show that initially peak structures are formed which deform at a longer distance of propagation splitting into two or more peaks and a sequential pulse with a width $s_o \leq 1$ is produced at a repetition frequency given by the initial modulation τ_M independently of A_M. However, the distance of propagation needed to form the peaked structures varies as a function of A_M and τ_M (Table I of reference 20). This distance tends to increase as τ_M is increased and as A_M is reduced. It suggests that to produce pulses at a shorter distance it is better to use shorter modulation period and deeper depth of modulation.

VI. EXPERIMENTAL OBSERVATION OF BRIGHT SOLITONS

For the experimental verification of the existence of bright solitons one needs a fibre with low losses in the negative group dispersion region. The development of monomode silica glass fibres having low losses (0.2 dB/km) in the region of negative group dispersion [5,28] and mode-locked colour centre laser [39] tunable over that region made possible the experimental observation of bright solitons. Mollenauer, Stolen and Gordon [28] successfully demonstrated the distortionless propagation of a 7 psec pulse with peak power of ≈ 1.24 W at λ = 1.55 μm for a distance of about 700 meters using a monomode fibre of about 100 μm^2 cross section. For higher powers they observed substantial compression and well resolved splitting of psec pulses. These observations were in close agreement with the predictions of Hasegawa and Tappert. So far as the experimental observation of dark solitons is concerned we note that nothing has been reported so far.

VII THE SOLITON LASER

Using the basic properties of nonlinear pulse propagation through optical fibres and the mode-locked colour-centre laser Mollenauer and Stolen [17] created a device which can be called a soliton laser. It is basically a synchronously pumped mode-locked colour-centre laser with a certain length (which can be altered according to the required value of the pulse width) of a monomode polarisation preserving (highly birefrengent) fibre incorporated into its feedback loop. The operation of the laser is based on N = 2 soliton of NLS equation and the shape of the pulse obtained from this laser is $sech^2$ in intensity. Besides the ultra narrow pulses the soliton laser allows for longer precisely controlled pulses which are important for several experiments on pulse propagation in optical fibres.

The pulse narrowing and solitons are obtained using the combined effect of nonlinearity in the refractive index and negative group dispersion of the fibre as discussed earlier. The operation of the soliton laser can in short be explained as follows. Broad pulses initially obtained from the mode-locked colour-centre laser are narrowed by passage through the fibre. The narrowed pulses are reinjected into the cavity in such a way that they are coincident and in phase with those already existing in the cavity. As a result the laser itself starts producing narrower pulses. This process is continued until one obtains soliton pulses.

Concluding we wish to note that the polarisation preserving ability of the fibre is crucial for the operation of the laser for otherwise feedback into the mode-locked colour-centre laser, which is highly polarization sensitive, would fluctuate wildly with fibre length, wavelength, etc. [17].

REFERENCES

[1] A. Hasegawa and F. Tappert, Appl. Phys. Letts., **23** (1973) 142.
[2] A. Hasegawa and F. Tappert, Appl. Phys. Letts., **23** (1973) 171.
[3] B. Bendo, P. Gianino, N. Tzoar and M. Jain, J. Opt. Soc. Am., **70** (1980) 539.
[4] M. Jain and N. Tzoar, J. Appl. Phys., **49** (1978) 4649.
[5] T. Miya, Y. Terunuma, T. Hosaka and T. Miyashita, Elec. Letts., **15** (1979) 106.
[6] S. Shimada, T. Matsumaoto, T. Ito, S. Matsushita, K. Washio and K. Minemura, Elec. Letts., **15** (1979) 484.
[7] L. G. Cohen, C. Lin and W. G. French, Elec. Letts., **15** (1979) 334.

[8] R. Olshansky and G. W. Scherer, in **Rec. Optical Communication Conf.** (Amsterdam, The Netherlands) paper 12.5, Sept 17 (1979).
[9] R. H. Stolen, Proc. IEEE, **68** (1980) 1232.
[10] R. H. Stolen, in **Optical Fibre Telecommunications,** ed S. E. Miller and A. G. Chynoweth (Academic Press, New York, 1979).
[11] C. Gloge, Appl. Opt., **10** (1971) 2442.
[12] C. V. Shank, Science, **219** (1983) 1027.
[13] D. Anderson, IEE Proc., **132** (1985) 122.
[14] A. Hasegawa and Y. Kodama, Opt. Letts., **7** (1982) 285.
[15] Y. Kodama and A. Hasegawa, Opt. Letts., **7** (1982) 339.
[16] Y. Kodama and A. Hasegawa, Opt. Letts., **8** (1983) 342.
[17] L. F. Mollenauer and R. H. Stolen, Opt. Letts., **9** (1984) 13.
[18] K. J. Blow and N. J. Doran, Opt. Commun., **42** (1982) 403.
[19] D. Anderson, Opt. Commun., **48** (1983) 107.
[20] A. Hasegawa, Opt. Letts., **9** (1984) 288.
[21] D. Anderson and M. Lisak, Opt. Letts., **9** (1984) 468.
[22] A. Hasegawa, Opt. Letts., **8** (1983) 650.
[23] R. H. Stolen and E. P. Ippen, Appl. Phys. Letts.,**22** (1973) 276.
[24] Ajit Kumar, S. N. Sarkar and A. K. Ghatak, Opt. Letts.,**11**(1986)321.
[25] Ajit Kumar and M. S. Sodha, Elec. Letts., **23 (1987)** 275.
[26] L. F. Mollenauer, R. H. Stolen and M. N. Islam, Opt. Letts., **10** (1985) 229.
[27] A. Hasegawa and Y. Kodama, Proc. IEEE, **69** (1981) 1145.
[28] L. F. Mollenauer, R. H. Stolen and J. P. Gordon, Phys. Rev. Letts., **45** (1980) 1095.
[29] D. N. Christidoulides and R. I. Joseph, Opt. Letts.,**9** (1984) 229.
[30] D. N. Christodoulides and R. I. Joseph, Opt. Letts.,**9** (1984) 408.
[31] M. Jain and N. Tzoar, Opt. Letts., **3** (1978) 202.
[32] C. V. Shank, R. L. Fork, R. Yen and R. H. Stolen, Appl. Phys. Letts., **40** (1982) 761.
[33] B. P. Nelson, D. Cotter, K. J. Blow and N. J. Doran, Opt. Commun., **48** (1983) 292.
[34] D. Grischowsky and A. C. Balant, Appl. Phys. Letts.,**42** (1982) 1.
[35] W. J. Tomlinson, R. H. Stolen, and C. V. Shank, J. Opt. Soc. Am., **B1** (1984) 139.
[36] D. R. Nicholson and M. V. Goldman, Phys. Fluids. **19** (1976) 1621.
[37] V. I. Karpman and M. D. Kruskal, Soviet Phys. JETP **28** (1969) 277.
[38] G. R. Lamb, **Elements of Soliton Theory** (Wiley Interscience, New York, 1980).
[39] L. F. Mollenauer and D. M. Bloom, Opt. Letts., **4** (1979) 247.
[40] J. Satsuma and N. Yajima, Prog. Theor. Phys., Suppl. No. **55** (1974) 284.

Davydov's Soliton

A.C. Scott

Department of Electrical and Computer Engineering,
The University of Arizona, Tucson, AZ85721, USA

Davydov's theory for storage and transport of biological energy in protein is described and related to recent infrared absorption measurements in crystalline acetanilide. Some aspects of the quantum theory are considered in detail.

1. INTRODUCTION

In living organisms a fundamental mechanism for the transfer of energy into function proteins or enzymes is the hydrolysis of adenosine triphosphate (ATP) into adenosine diphosphate (ADP) according to the reaction

$$ATP^{-4} + 2\,H_2O \rightarrow ADP^{-3} + HPO_4^{-2} + H_3O^{+}. \tag{1.1}$$

Under normal physiological conditions about 10 kcal/mol or 0.422 eV of free energy is released by this reaction [1], leading to several interesting questions: How is this free energy transferred into protein? How is it stored there? How does it move inside a protein? How is it transformed into useful work?

To answer questions of this sort a theory was proposed by Davydov [2] which focused attention on the self-trapping of molecular vibrational energy in the amide-I (or CO stretch) vibration of the peptide unit (CONH), a basic structural element of all proteins. According to this theory, it was proposed that the localization of amide-I vibrational energy would alter the surrounding structure (primarily the hydrogen bonding) and that this local alteration would, in turn, lower the amide-I energy enough to prevent its dispersion.

At about the same time as the original paper by Davydov, Careri [3] published some unexpected spectral measurements in the amide-I region of crystalline acetanilide ($CH_3CONHC_6H_5$), or ACN. As the temperature was lowered from room temperature, he observed an anomalous amide-I band (at 1650 cm^{-1}) growing up on the red side of the normal amide-I band (at 1665 cm^{-1}). This 1650 cm^{-1} band was called anomalous because it could not be explained with accepted concepts of molecular spectroscopy (e.g.,Fermi resonance, Davydov splitting, etc.). At first Careri suspected some unusual one-dimensional phase transformation might provide an explanation, but no such evidence was found after several years of experimental work. Recently a self-trapping theory was proposed which is closely related to that of Davydov and explains the salient experimental facts [4-6].

The present situation, therefore, is that the 1650 cm^{-1} band in ACN seems to provide direct experimental evidence for a self-trapped state of molecular vibrational energy. The "red shift" of 15 cm^{-1} from the normal band can be considered as the binding energy of a Davydov-like soliton, and this interpretation leads to quantitative predictions of biological significance.

This paper is organized into three phases. The first is a review of Davydov's soliton theory and the experimental observations in crystalline acetanilide. The second phase is a detailed comparison of various attempts to provide a quantum mechanical explanation for self-trapping of molecular vibrations. Finally some questions of biological significance are briefly considered.

2. DAVYDOV'S SOLITON THEORY

This section is intended to provide a brief summary of Davydov's soliton theory for the convenience of the reader. Such a summary is helpful to appreciate the differences between the theory of self-trapping proposed for proteins and the theory proposed recently to explain experimental measurements on crystalline acetanilide. It is also necessary in order to see how the quantum theory developed by Davydov as a basis for self-trapping is related to other quantum analyses. Several surveys of this work are available for further reference [7,8].

Careful inspection of the α-helix structure of protein reveals three channels situated approximately in the longitudinal direction with the sequence

etc. H-N-C=O---H-NC=O---H-N-C=O---H-N-C=O etc.

where the dashed lines represent hydrogen bonds. For a detailed analysis it is necessary to consider the interaction of all three channels, but one is sufficient to lay out the basic ideas.

A single channel is governed by the energy operator

$$\hat{H} = \hat{H}_{CO} + \hat{H}_{ph} + \hat{H}_{int}. \tag{2.1}$$

Taking the components of $\hat{H}$ in order, $\hat{H}_{co}$ is an energy operator for the CO stretch (amide-I) vibration including the effects of nearest neighbor dipole-dipole interactions. Thus

$$\hat{H}_{CO} = \sum_n E_o \hat{b}_n^\dagger \hat{b}_n - J\hat{b}_{n+1}^\dagger \hat{b}_n + b_n b_{n+1}, \tag{2.2}$$

where E_o is the fundamental energy of the amide-I vibration, -J is the nearest neighbor dipole-dipole interaction energy, and $\hat{b}_n^\dagger(\hat{b}_n)$ are boson creation (annihilation) operators for amide-I quanta on the n-th molecule.

$\hat{H}_{ph}$ is the energy operator for longitudinal (acoustic) sound waves. Thus

$$\hat{H}_{ph} = \frac{1}{2} \sum_n [M^{-1}\hat{p}_n^2 + W(\hat{u}_n - \hat{u}_{n+1})^2], \tag{2.3}$$

where M is the mass of a molecule, W is the spring constant of a hydrogen bond, $\hat{P}_n$ is a longitudinal momentum operator for the n-th molecule, and $\hat{u}_n$ is the corresponding longitudinal position operator.

$$\hat{H}_{int} = \chi_a \sum_n (\hat{u}_n - \hat{u}_{n-1})\hat{b}_n^\dagger b_n, \tag{2.4}$$

where χ_a is the derivative of amide-I vibrational energy with respect to the length (R) of the adjacent hydrogen bond. Thus

$$\chi_a = dE_o/dR. \tag{2.5}$$

Davydov minimizes the average value of $\hat{H}$ with respect to the wave function

$$|\psi\rangle = \sum_n a_n(t)\exp[\hat{\sigma}(t)]\hat{b}_n^\dagger \; |0\rangle, \tag{2.6}$$

where

$$\hat{\sigma} \equiv - i \sum_n [\hat{\beta}_a(t)\hat{P}_n - \pi_n(t)\hat{u}_n]. \tag{2.7}$$

A straightforward calculation shows that

$$\beta_n(t) = \langle\psi|\hat{u}_n|\psi\rangle \tag{2.8}$$

and

$$\pi_n(t) = \langle\psi|\hat{p}_n|\psi\rangle \tag{2.9}$$

The wavefunction in (2.6) will be called Davydov's ansatz throughout this chapter. One of the aims here is to study the range of validity of this ansatz.

Assuming that Davydov's ansatz approximates the true wavefunction, (2.8) and (2.9) show that β_n and π_n are the average values of the position and momentum operators, respectively. Furthermore, a_n is the probability amplitude for finding a quantum of amide-I vibrational energy on the n-th molecule. The normalization condition $\langle\psi|\psi\rangle = 1$ implies that

$$\sum_n |a_n|^2 = 1. \tag{2.10}$$

Thus Davydov's ansatz describes the dynamics of a single quantum of amide-I vibrational energy.

Minimization of $\langle\psi|\hat{H}|\psi\rangle$ with respect to a_n, β_n and π_n leads to the differential-difference equations

$$(i \frac{d}{dt} - E_o)a_n + J(a_{n+1}) - \chi_a(\beta_n - \beta_{n+1})a_n = 0, \tag{2.11a}$$

$$M \frac{d^2\beta_n}{dt^2} - W(\beta_{n+1} - 2\beta_n + \beta_{n+1}) = \chi_a[|a_{n+1}|^2 - |a_n|^2]. \tag{2.11b}$$

Extensive numerical and theoretical analysis of (2.11) yields the following results [9-12]: (i) it is reasonable to expect soliton formation at the level of energy released by ATP hydrolysis (1.1), and (ii) such a soliton travels rather slowly with respect to the speed of longitudinal sound waves. This suggests neglecting the kinetic energy of longitudinal sound by assuming $\beta_n = 0$, whereupon

$$\beta_n - \beta_{n+1} = -\frac{\chi_a}{W} |a_n|^2 \tag{2.12}$$

and, in this "adiabatic approximation," (2.11) becomes

$$(i \frac{d}{dt} - E_o)a_n + J(a_{n+1}) + \gamma_a|a_n|^2 a_n = 0, \tag{2.13}$$

where

$$\gamma_a \equiv \chi_a^2/W. \tag{2.14}$$

Davydov has emphasized that a solitary wave solution of (2.11) cannot be created directly by absorption of a photon because of an unfavorable Franck-Condon factor. This is because the necessary intermolecular displacement in (2.11b) cannot occur in a time that is short enough for photon absorption. The Franck-Condon factor will be discussed in detail in the following section.

3. SELF-TRAPPING IN CRYSTALLINE ACETANILIDE

Just as in the α-helix, careful inspection of the crystal structure of acetanilide reveals channels situated in the b-direction with the sequence

etc. H-N-C=O---H-N-C=O---H-N-C=O---H-N-C=O etc.

Recent infrared absorption measurements on microcrystals of ACN show an unexpected band at 1650 cm^{-1} which rises with decreasing temperature to become the dominant spectral feature below 100 K. When this band was discovered, Careri suspected it to be caused by a subtle phase change along b-direction of the crystal [3], but careful studies over a period of several years failed to reveal any such evidence. The lack of a viable alternative eventually led to the suggestion that the 1650 cm^{-1} band might be caused by direct absorption of an infrared photon into a self-trapped state similar to that proposed by Davydov. The qualifier "similar" is important because, as was noted above, the Franck-Condon factor is unfavorable for direct photon absorption by a self-trapped solution of (2.11).

The corresponding theory proceeds, as in the previous section, by defining the energy operator

$$\hat{H} = \hat{H}_{CO} + \hat{H}_{ph} + \hat{H}_{int}, \tag{3.1}$$

where $\hat{H}_{CO}$ is again given by (2.2) but with [6]:

$$J = 3.96 \text{ cm}^{-1}. \tag{3.2}$$

In the present analysis, however, self-trapping is assumed to be caused by interaction with an optical phonon rather than an acoustic phonon. Thus

$$\hat{H}_{ph} = \frac{1}{2} \sum_n [m^{-1}\hat{p}_n^2 + w\hat{q}_n^2] \tag{3.3}$$

and

$$\hat{H}_{int} = \chi_o \Sigma\ \hat{q}_n \hat{b}_n^\dagger \hat{b}_n. \qquad (3.4)$$

Minimization of $<\psi|\hat{H}|\psi>$ with respect to the parameters of the Davydov ansatz wavefunction (2.6), where

$$\hat{\sigma} = -\frac{1}{h}\sum_n [q_n(t)\hat{P}_n - P_n(t)\hat{q}], \qquad (3.5)$$

leads to the dynamic equations

$$(i\frac{d}{dt} - E_o)a_n + J(a_{n+1} + a_{n-1}) - \chi_o q_n a_n = 0, \qquad (3.6a)$$

$$m\frac{d^2 q_n}{dt^2} - wq_n = \chi_o |a_n|^2. \qquad (3.6b)$$

As before

$$q_n(t) = <\psi|q_n|\psi>. \qquad (3.7)$$

The adiabatic approximation ($q_n = 0$) reduces (3.6) to

$$(i\frac{d}{dt} - E_o)a_n + J(a_{n+1} + a_{n-1}) - \gamma_o |q_n|^2 a_n = 0, \qquad (3.8)$$

where

$$\gamma_o \equiv \chi_o^2 a/w. \qquad (3.9)$$

A detailed numerical study of a system of equations similar to (3.8) but representing one hundred molecules of ACN in two coupled channels has recently been carried out by Eilbeck et al. [6]. This work shows that the red shift from the normal amide-I band at 1665 cm^{-1} to the 1650 cm^{-1} band is best fit by choosing

$$\gamma_o = 44.7\ cm^{-1}. \qquad (3.10)$$

We turn next to an estimate of the Franck-Condon factor for direct photon absorption by a self-trapped state of (3.6). Before absorption $|a_n|^2 = 0$, and after absorption $|a_n|^2 = 0$ over a localized region such that (2.10) is satisfied. Thus the ground state wavefunction of (3.6b) must shift from

$$\Phi_o = (\frac{w}{\pi\omega})^{\frac{1}{2}} \exp(-q_n^2 \frac{w}{2\omega}) \qquad (3.11)$$

before absorption to

$$\tilde{\Phi}_o = \left(\frac{w}{\pi\omega}\right)^{\frac{1}{2}} \exp[-(q_n + \gamma_o|a_n|^2)^2 \frac{w}{2\omega}] \qquad (3.12)$$

after absorption, where

$$\omega = (w/m)^{\frac{1}{2}} \qquad (3.13)$$

is the frequency of the optical mode that is mediating the self-trapping. The transition probability for soliton absorption is therefore reduced by the Franck-Condon factor

$$[\int\Phi_o\tilde{\Phi}_o^* \, dq_n]^2 \geq \exp(-\gamma_o/2\omega), \qquad (3.14)$$

which is close to unity for

$$\gamma_o \ll \omega. \qquad (3.15)$$

The frequency (ω) of the optical mode can be determined from the temperature dependence of the 1650 cm^{-1} line. Such temperature dependence is expected, because the probability of (3.6b) being in its ground state, and therefore able to participate in self-trapping, is $[1 - \exp(-\omega/kT)]$. Thus as the temperature is raised, the low temperature factor given in (3.14) should be reduced by the additional factor $[1 - \exp(-\omega/kT)]^2$. A least square fit to intensity data is obtained for $\omega = 131$ cm^{-1}. Together with (3.10) this implies $\exp(-\gamma_o/2h\omega) = 0.84$.

Further evidence tending to favor a self-trapping explanation for the 1650 cm^{-1} band is the recent observation of the overtone series shown in Table 1 [13]. Since the overtones $N \geq 2$ are self-trapped states involving more than one quantum of the amide-I vibration, it is interesting to consider states that avoid the constraint of (2.10).

Table 1 Overtone Series for the ACN Soliton

N	$\nu(N)$
1	1650 cm^{-1}
2	3250
3	4803
4	6304

4. THE QUANTUM THEORY OF SELF-TRAPPING

In this section we approach the problem from a classical perspective. Starting with the classical amide-I coordinates, P_n and Q_n, for which the Hamiltonian is $\Sigma_n[P_n^2 + Q_n^2]$, it is convenient to define the complex mode amplitudes

$$A_n = \omega_o^{\frac{1}{2}}(P_n + iQ_n). \qquad (4.1)$$

In terms of these complex mode amplitudes (including dipole-dipole interactions)

$$H_{CO} = \sum_n [\frac{E_o}{h}|A_n|^2 - J(A_{n+1}^*A_n + A_n^*A_{n+1})], \qquad (4.2)$$

where

$$\omega_o = E_o/h \qquad (4.3)$$

is the classical oscillation frequency of an amide-I vibration. (From here on we will assume $\hbar = 1$ and measure energy and frequency in the same units.)

With a classical interaction energy

$$H_{int} = \chi \sum_n q_n|A_n|^2, \qquad (4.4)$$

where q_n is the coordinate of some low-frequency phonon with adiabatic energy

$$H_{ph} = \frac{1}{2} w \sum_n q_n^2, \qquad (4.5)$$

one arrives at the total classical Hamiltonian

$$H = H_{CO} + H_{ph} + H_{int}. \qquad (4.6)$$

Minimizing (4.6) with respect to the q_n requires

$$q_n = -\frac{\chi}{w}|A_n|^2, \qquad (4.7)$$

whereupon (4.6) can be reduced to

$$H = \sum_n E_o|A_n|^2 - J(A_{n+1}^*A_n + A_n^*A_{n+1}) - \frac{1}{2}\gamma|A_n|^4, \qquad (4.8)$$

where

$$\gamma \equiv \chi^2/w. \tag{4.9}$$

The corresponding dynamical equation for A_n is

$$(i \frac{d}{dt} - E_o)A_n + J(A_{n+1} + A_{n-1}) + \gamma|A_n|^2 A_n = 0. \tag{4.10}$$

In additional to the energy, H, another constant of the motion along solutions of (4.10) is the number

$$N = \sum_n |A_n|^2. \tag{4.11}$$

To this point the discussion of the present section has been entirely classical. We now consider quantization in four special cases: (i) $J \ll \gamma$, (ii) $\gamma \ll J$, (iii) semiclassical quantization, and (iv) the Davydov ansatz. In each case it will be of particular interest to calculate an overtone series corresponding to that presented in Table 1 for crystalline acetanilide.

4.1 The Case $J \ll \gamma$

In this case we neglect the dipole-dipole interaction terms in (4.8) and (4.10), and write the energy

$$H = \sum_n h_n, \tag{4.12}$$

where

$$h_n = E_o|A_n|^2 - \tfrac{1}{2}\gamma|A_n|^4. \tag{4.13}$$

Under quantization, the terms in (4.12) become operators

$$h_n \rightarrow \hat{h}_n \tag{4.14}$$

through replacement of the complex mode amplitudes by creation and annihilation operators for bosons. Thus

$$A_n \rightarrow \hat{b}_n, \tag{4.15a}$$

$$A_n^* \rightarrow \hat{b}_n^\dagger. \tag{4.15b}$$

Since the ordering of these operators is not determined by (4.13), we take the averages

$$|A|^2 \rightarrow \frac{1}{2}(\hat{b}^\dagger\hat{b} + \hat{b}\hat{b}^\dagger), \tag{4.16}$$

$$|A|^4 \rightarrow \frac{1}{6}(\hat{b}^\dagger\hat{b}^\dagger\hat{b}\hat{b} + \hat{b}^\dagger\hat{b}\hat{b}^\dagger\hat{b} + \hat{b}^\dagger\hat{b}\hat{b}\hat{b}^\dagger + \hat{b}\hat{b}\hat{b}^\dagger\hat{b}^\dagger + \hat{b}\hat{b}^\dagger\hat{b}^\dagger\hat{b}). \tag{4.17}$$

where the subscripts have been dropped for typographical convenience. Noting that $\hat{b}^\dagger$ and $\hat{b}$ have the properties $\hat{b}^\dagger|N\rangle = \sqrt{N+1}\ |N+1\rangle$ and $\hat{b}|N\rangle = \sqrt{N}\ |N-1\rangle$ (where $|N\rangle$ is an harmonic oscillator eigenstate), it is straightforward to show that

$$\hat{h} \;=\; (E_o - \frac{1}{2}\gamma)(\hat{b}^\dagger\hat{b} + \frac{1}{2}) - \frac{1}{2}\gamma\ \hat{b}^\dagger\hat{b}\hat{b}^\dagger\hat{b}. \tag{4.18}$$

Thus

$$\hat{h}|N\rangle = E(N)|N\rangle, \tag{4.19}$$

where

$$E(N) \;=\; (E_o - \frac{1}{2}\gamma)(N + \frac{1}{2}) - \frac{1}{2}\gamma N^2. \tag{4.20}$$

In summary, eigenvectors of the operators defined through (4.13), (4.14), (4.16) and (4.17) are identical to those of an harmonic oscillator, but the corresponding eigenvalues are given (4.20).

The form of (4.20) is significant. It can be written

$$E(N) \;=\; E^C + E^L + E^{NL} \tag{4.21}$$

where E^C is the ground state ($N = 0$) energy, $E^L = N$ and

$$E^{NL} = -\frac{1}{2}\gamma N^2. \tag{4.22}$$

This "nonlinear" contribution is directly measured from the overtone series in Table 1.

4.2 The Case $\gamma \ll J$

In this case the classical equation (4.10) reduces to the nonlinear Schrödinger (NLS) equation of soliton theory. To see how this goes, assume the repeat distance between molecules of d and replace the discrete variable n by a continuous variable, $x = n$, which measures distance in units of d. Then (4.10) takes the form

$$(i\frac{\partial}{\partial t} - E_o + 2J)A + J\frac{\partial^2 A}{\partial x^2} + \gamma|A|^2 A \;=\; 0. \tag{4.23}$$

Quantization of this equation was originally performed using the Bethe ansatz method and recently it has been shown that such solutions can be efficiently constructed from a quantum version of inverse scattering theory [14,15].

Under quantization, the functions A and A* are replaced by annihilation and creation operators for boson fields, $\hat{\varphi}$ and $\hat{\varphi}^{\dagger}$. At equal times these have the commutation relations $[\hat{\Phi}(x), \hat{\Phi}^{*}(y)] = [\hat{\Phi}^{\dagger}(x), \hat{\Phi}^{\dagger}(y)] = 0$ and $[\hat{\Phi}(x), \hat{\Phi}^{\dagger}(y)] = \delta(x - y)$. In terms of the previous discussion, it is evident that $\hat{\Phi}(x)$ is equivalent (under scaling) to $\hat{b}_n$ in the continuous limit n = x. In effecting this limit two procedures are customary: (i) neglect consideration of the ground state energy which is unbounded in the limit, and (ii) "normal order" all operator expressions, i.e., move all creation operators to the left.

Since $\hat{b}\hat{b}^{\dagger} = \hat{b}^{\dagger}\hat{b} + 1$, normal ordering of (4.18) and neglect of the ground state energy imply

$$\hat{h}_n = (E_o - \gamma)\hat{b}_n^{\dagger}\hat{b}_n - \frac{1}{2}\gamma\, \hat{b}_n^{\dagger}\hat{b}_n^{\dagger}\hat{b}_n\hat{b}_n . \qquad (4.24)$$

Thus to put (4.23) in standard form for quantum analysis, let

$$A = \Phi \exp[-i(E_o - 2J - \gamma)t] \qquad (4.25)$$

and scale time as $t \to t/J$. Then (4.23) becomes

$$i\Phi_t + \Phi_{xx} + \frac{\gamma}{J}|\Phi|^2\Phi = 0, \qquad (4.26)$$

where a subscript notation is used for the partial derivatives. Under quantization $\Phi \to \hat{\Phi}$ and (4.26) becomes the operator equation

$$i\hat{\Phi}_t + \hat{\Phi}_{xx} + \frac{\gamma}{J}\hat{\Phi}^{\dagger}\hat{\Phi}\hat{\Phi} = 0, \qquad (4.27)$$

with energy operator

$$\hat{H} = \int dx\, \hat{\Phi}_x^{\dagger}\hat{\Phi}_x - \frac{1}{2}\frac{\gamma}{J}\int dx\, \hat{\Phi}^{\dagger}\hat{\Phi}^{\dagger}\hat{\Phi}\hat{\Phi} , \qquad (4.28)$$

number operator

$$N = \int dx\, \hat{\Phi}^{\dagger}\hat{\Phi} , \qquad (4.29)$$

and momentum operator

$$P = -i\int dx \hat{\Phi}^{\dagger}\hat{\Phi}_x . \qquad (4.30)$$

The quantum inverse scattering method provides exact wavefunctions, $|\psi>$, that diagonalize $\hat{N}$, $\hat{P}$ and $\hat{H}$ as follows.

$$\hat{N}|\psi> = N|\psi>, \tag{4.31}$$

where

$$N = \text{integer} \geq 0, \tag{4.32}$$

$$\hat{P}|\psi>= Np|\psi>, \tag{4.33}$$

where p is a real number, and

$$\hat{H}|\psi> = \left[Np^2 + \frac{\gamma^2}{48J^2}(N - N^3)\right]|\psi> . \tag{4.34}$$

Furthermore in the limit $\gamma/J \to 0$

$$|\psi> \to \int dx\, e^{ipx}\, \hat{\Phi}|0>. \tag{4.35}$$

Equations (4.32) and (4.34) imply an overtone series

$$E(N) = E^L + E^{NL}, \tag{4.36}$$

where $E^L \alpha$ N and

$$E^{NL} = - \frac{\gamma^2}{48\ J^2} N^3. \tag{4.37}$$

4.3 Semiclassical Quantization

In the parameter range $\gamma \simeq J$, it is possible to impose elementary quantum conditions on stationary solutions of (4.10). Writing such a solution in the form

$$A_n = (\frac{J}{\gamma})^{\frac{1}{2}}\ \alpha_n\ \exp[-i(\frac{E_o}{J} + \omega)t] \tag{4.38}$$

reduces (4.10) to the standard form

$$\omega\alpha_n + \alpha_{n+1} + \alpha_{n-1} + \alpha_n^3 = 0. \tag{4.39}$$

Using a shooting method [12] it is possible to find a family of numerical solutions for (4.39) with the following properties: (i) $\alpha_n = \alpha_{-n}$, (ii) for $n \geq 0$, $\alpha_n > \alpha_{n+1}$, and (iii) $\lim_{n\to\infty}\alpha_n = 0$. From such a solution the conserved quantities H and N defined in (4.8) and (4.11) are readily calculated as

$$N = \frac{J}{\gamma} \sum_n \alpha^2 \tag{4.40}$$

and

$$H(N) = E_o - J \frac{\sum_n \alpha_n \alpha_{n-1}}{\Sigma \alpha_n^2} N - \frac{1}{2}\gamma N^2 \frac{\sum_n \alpha_n^4}{\left(\Sigma \alpha_n^2\right)^2} . \tag{4.41}$$

Semiclassical quantization is effected by noting that stationary solutions are of the form

$$A_n(t) = A_{n0} \exp[-i\Theta(t)], \tag{4.42}$$

where $\dot{\Theta} = dH(N)/dN$. Thus N and Θ are conjugate variables and the quantum condition

$$\oint N \, d\Theta = 2\pi(\text{integer}) \tag{4.43}$$

together with the definition of N (4.11) imply

$$N = \text{integer} \geq 0. \tag{4.44}$$

Equation (4.41) has the form

$$E(N) = E^L + E^{NL}, \tag{4.45}$$

where

$$E^{NL} = -\frac{1}{2} \gamma N^2 \frac{\sum_n \alpha_n^4}{\left(\sum_n \alpha_n^2\right)^2} \tag{4.46}$$

In the limit $J \ll \gamma$, $\alpha_n \ll \alpha_0$ for $|n| \geq 1$ so (4.46) evidently reduces to (4.22). In the limit $\gamma \ll J$ it is straightforward to show that (4.46) reduces to (4.37). Thus (4.46) is expected to provide an accurate calculation of E^{NL} over the entire parameter range.

It is now possible to consider how data from the overtone series for the 1650 cm^{-1} band in acetanilide compare with these calculations. From (3.2) and (3.10),

$$\gamma_0/J = 11.3. \tag{4.47}$$

This lies in the range of (4.46) for which

$$E(N) \doteq E_0(N) - \frac{1}{2}\gamma N^2, \tag{4.48}$$

so the line at 1650 cm^{-1} implies

$$E_0 = 1672.3 \ cm^{-1}. \tag{4.49}$$

From the measured values of overtone frequency, $\nu(N)$ in Table 1, the non-linear contributions to the overtone spectrum can be calculated as

$$E^{NL} = \nu(N) - NE_0. \tag{4.50}$$

In Table 2 we compare these calculations with those computed from (4.48).

Table 2 Nonlinear Terms in ACN Overtone Series

N	$-E^{NL}(cm^{-1})$	$-\frac{1}{2}\gamma N^2(cm^{-1})$
1	22	22
2	95	89
3	214	201
4	385	357

4.4 Davydov's Ansatz

We are now in a position to evaluate Davydov's ansatz. In the context of an adiabatic approximation, the wavefunction introduced in (2.6) takes the form

$$|\psi\rangle = \sum_n a_n(t) \, \hat{b}_n^\dagger \, |0\rangle, \tag{4.51}$$

where the $a_n(t)$ are solutions of (2.13). This form of the Davydov ansatz has the following properties:

(i) In the limit $J \ll \gamma$, it reduces to the first eigenfunction, $|1\rangle$, in (4.19).

(ii) In the limit $\gamma \ll J$, it reduces to the asymptotic form of Bethe's ansatz in (4.35).

(iii) Between these two limits, Davydov's ansatz gives energies that agree with semiclassical calculations.

Thus one concludes that Davydov's ansatz is a useful approximation to the exact wavefunction over the entire parameter range $0 \leq \gamma/J < \infty$ with the constraint (2.10) which implies N = 1.

5. BIOLOGICAL SIGNIFICANCE OF SELF-TRAPPING

Measurements on crystalline acetanilide (ACN) confirm Davydov's theory of self-trapped states (solitons) in hydrogen bonded polypeptide chains. Furthermore, Table 1 shows that the "N = 2" state in ACN can absorb almost all (95%) of the free energy released in hydrolysis of adenosine triphosphate (ATP). It is reasonable to suppose that a corresponding state can form on the hydrogen bonded polypeptide chains of α-helix.

Over a decade ago McClare argued that the free energy released in ATP hydrolysis should transfer resonantly into a protein in order to avoid thermal degradation [16,17]. To store and transport this energy he posited an "excimer" state in protein which would be closely related to the amide-A band of α-helix at 3240 cm^{-1}. McClare's excimer is qualitatively similar to the "conformon" of Green and Ji [18] and the basic properties of both are provided by a Davydov soliton in the "N=2" state [2,7,19-22]. In the past such suggestions have been rejected or ignored by the biochemical community because a localized region of free energy within a protein was believed to be physically impossible. Since this view is no longer tenable, the early proposals of Davydov, McClare [16,17,22,23]and Green and Ji [18] must be reevaluated. A recent paper by Careri and Wyman [24], suggesting a soliton mechanism for cyclic enzyme activity, provides a first step in this direction.

REFERENCES

[1] R. F. Fox, **Biological Energy Transduction** (Wiley, New York, 1982), p. 216.
[2] A. S. Davydov, J. Theor. Biol. **38** (1973) 559.
[3] G. Careri, in **Cooperation Phenomena**, ed H. Haken and M. Wagner (Springer, Berlin, 1973), p. 391.
[4] G. Careri, U. Bountempo, R. Carta, E. Gratton and A. C. Scott, Phys. Rev. Lett. **51** (1983) 304.
[5] G. Careri, U. Buontempo, F. Galluzzi, A. C. Scott, E. Gratton and E. Shyamsunder, Phys. Rev. **B30** (1984) 4689.
[6] J. C. Eilbeck, P. S. Lomdahl and A. C. Scott, Phys. Rev. **B30** (1984) 4703.
[7] A. S. Davydov, Phys. Ser. **20** (1979) 387.

[8] A. S. Davydov, Soc. Phys. Usp. **25** (1982) 898 [Usp. Fiz. Nauk **138** (1982) 603].
[9] A. C. Scott, Phys. Rev. **A26** (1982) 578.
[10] A. C. Scott, Phys. Rev. **A27** (1983) 2767.
[11] A. C. Scott, Phys. Rev.**A29** (1984) 279.
[12] L. MacNeil and A. C. Scott, Phys. Scr. **29** (1984) 284.
[13] A. C. Scott, E. Gratton, E. Shyamsunder and G. Careri, Phys. Rev. **B32** (1985) 5551.
[14] E. K. Skylanin and L. D. Faddeev, Sov. Phys. Dokl. **23** (1978) 902 [Dokl. Acad. Nauk SSSR **243** (1978) 1430].
[15] H. B. Thacker and D. Wilkinson, Phys. Rev. **D19** (1979) 3660.
[16] C. W. F. McClare, Nature **240** (1972) 88.
[17] C. W. F. McClare, Ann. N. Y. Acad. Sci. **227** (1974) 74.
[18] D. E. Green and S. Ji, in **The Molecule Basis of Electron Transport** (Academic Press, New York, 1972), p. 1.
[19] A. S. Davydov, Biofizika **19** (1974) 670.
[20] A. S. Davydov, J. Theor. Biol. **66** (1977) 379.
[21] A. S. Davydov, Int. J. Quant. Chem. **16** (1979) 5.
[22] A. S. Davydov, **Biology and Quantum Mechanics** (Pergamon, New York, 1982).
[23] C. W. F. McClare, Ann. N. Y. Acad. Sci. **227** (1974) 74.
[24] G. Careri and J. Wyman, Proc. Natl. Acad. Sci., USA **81** (1984) 4386.

Generalized Nonlinear Schrödinger Equations in Quantum Fluid Dynamics

B.M. Deb and P.K. Chattaraj
Theoretical Chemistry Group, Department of Chemistry, Panjab University, Chandigarh 160014, India

This review article discusses several interesting generalized nonlinear Schrödinger equations (GNLSE) occurring in quantum fluid dynamics for both time-independent and time-dependent situations as well as the possibility of soliton or solitary wave solutions for such equations. It concludes with a brief report on a new GNLSE derived by the authors recently.

1. INTRODUCTION: THE NATURE OF QUANTUM FLUID DYNAMICS

Paradoxical though it may appear, generalized forms of nonlinear Schrödinger equation (GNLSE) do occur in the quantum theory of many-particle systems--and therefore in nuclear, atomic, molecular and solid state physics--provided one adopts Madelung's, rather than Schrödinger's, viewpoint. In Madelung's approach [1], the Schrödinger equation for a single particle is recast in the form of two classical fluid dynamical equations, a continuity equation and an Euler-type equation of motion (EOM) [2]. This is the origin of quantum fluid dynamics (QFD) where the system is described in terms of **two** real quantities, the charge density and the current density (both obeying the above two fluid dynamical equations), instead of the complex-valued wavefunction [3]. At a first glance, this Madelung fluid does not have a particle interpretation, unlike the Hamilton-Jacobi fluid [4] whose dynamics are governed by transformed Hamilton's EOMs. However, following Nelson [5], one may ascribe a particle interpretation to the Madelung fluid by considering the stochastic nature of the particle trajectories. In the semi-classical limit, the Bohm potential term [2] vanishes for the Madelung fluid which then reduces to the Hamilton-Jacobi fluid. Apart from employing a "classical" language within the quantum context, the main strength of the QFD viewpoint lies in its adoption of three-dimensional (3D) single-particle densities [3] as the basic variables. Even for a many-particle system, the continuity equation and the EOM can be brought to 3D space in terms of an orbital partitioning, e.g., natural orbitals [6] or Kohn-Sham orbitals [7], of the charge and the current density.

In this short review article, we discuss several interesting GNLSE's in both time-independent and time-dependent situations, the possibility of soliton or solitary wave solutions, along with a brief report on a new GNLSE derived by us recently [8]. The GNLSE's derived in our group have arisen in our attempts to deal with atomic and molecular phenomena using a blend of density functional theory (DFT) and QFD [2,3].

2. TIME-INDEPENDENT GNLSE

For a many-electron system, the current density vanishes in the ground state and therefore the QFD viewpoint reduces to the DFT viewpoint. The latter states that the ground state energy $E[\rho]$ is a unique functional of $\rho(\mathbf{r})$, the charge density, and attains its minimum value for the true ρ. Variational optimization of $E[\rho]$ with respect to a trial density, subject to the conservation of the total number of particles, leads to an Euler-Lagrange equation which is nothing but a time-independent GNLSE for the **direct** calculation of electron density, **by-passing** the wavefunction. One such equation, derived by Deb and Ghosh [9], has the form (atomic units employed throughout this paper)

$$[-\frac{1}{2}\nabla^2 + v_{eff}(\mathbf{r};\rho)]\Phi(\mathbf{r}) = \mu\Phi(\mathbf{r}), \tag{1}$$

$$\rho(\mathbf{r}) = |\Phi(\mathbf{r})|^2, \tag{2}$$

where μ is a Lagrange multiplier and v_{eff} is a one-body nonlocal, nonlinear effective potential consisting of kinetic, Coulomb and exchange-correlation terms. The nonlinearity in v_{eff} arises due to nonintegral powers of Φ as well as an integral. Similar equations have subsequently been derived, though not solved, by Levy et al. [10] and Hunter [11]. Equation (1) was numerically solved by Deb and Ghosh [9] in a model potential framework for noble gas atoms (Ne, Ar, Kr, Xe). For such systems, Eqs. (1) and (2) become one-dimensional in terms of the radial variable r. The calculated radial densities, energies and μ-values were quite satisfactory. However, an analytical solution was not attempted.

3. TIME-DEPENDENT GNLSE: THE POSSIBILITY OF SOLITON OR SOLITARY WAVE SOLUTIONS

The time-dependent GNLSE's are QFD EOMs. In order to examine the possibility of their soliton or solitary wave solutions, it is useful to fall back upon the familiar cubic NLSE (one spatial dimension),

$$i\,\frac{\partial\Phi}{\partial t} + \alpha\,\frac{\partial^2\Phi}{\partial x^2} + \beta\Phi|\Phi|^2 = 0, \tag{3}$$

whose envelope soliton solution exists when $\alpha\beta > 0$ and $\Phi \to 0$ as $|x| \to \infty$. This is the case of modulational instability of a train of modulated waves in a nonlinear, dispersive medium [12,13]. Note that, in general, a GNLSE may involve three spatial variables apart from time. For spherically symmetric systems, e.g., closed-shell atoms, such a GNLSE reduces from three to one spatial dimension. If one identifies x in Eq. (3) as the radial variable r, then Eq. (3) may be regarded as a spherically symmetric version of Eq. (1), apart from time, with an appropriate form for v_{eff} and suitable boundary conditions. Therefore, the solution of Eq. (3) can provide insight into its more general forms. Due to the difficulties associated with the analytical or numerical solution of higher dimensional GNLSE's, one frequently resorts to simpler forms of reduced dimensionality whose analytical/numerical solution is relatively easier to obtain. This is the approach adopted in Section 4. For the present, we continue our discussion on both one- and three-dimensional GNLSE's.

The following GNLSE [14] was helpful in the analysis of the Heisenberg spin chain with site-dependent exchange integral [15]:

$$i\,\frac{\partial\Phi}{\partial t} + \alpha\,\frac{\partial^2(f\Phi)}{\partial x^2} + \beta L\Phi = 0, \tag{4}$$

$$L = \int|\Phi|\,\frac{\partial}{\partial x}\,(f|\Phi|)dx, \tag{5}$$

where f is a real function of x. The corresponding Madelung fluid equations were obtained by Belic [16], by adopting the usual polar form for Φ, viz.,

$$\Phi = \rho^{1/2}e^{i\chi} \tag{6}$$

in terms of the density ρ and the velocity potential χ. Equation (4) has also been linked with the dynamics of a surface [16]. Earlier, Nonnenmacher and Nonnenmacher [17] had explicitly shown that the usual Madelung fluid (obtained from the Schrödinger equation) cannot have a soliton solution unless one adds a nonlinear term to it. Equation (3) is such a nonlinear equation and its corresponding Madelung fluid will have the expected solitonic behaviour. Such a Madelung fluid is equivalent to a nondissipative (ideal), isentropic and irrotational fluid [17].

Nassar [18] has suggested that the QFD viewpoint, using both the Madelung approach (i.e., transformation via the polar form of the wave-function) and the stochastic mechanical approach [4,5,19] which incorporates a stochastic force in Newtonian mechanics, may also be helpful in dealing with the quantum mechanical treatment of dissipative processes. Such processes occur, for example, in photochemistry, solid state physics, statistical physics, fission and heavy ion physics. Accordingly, a generalized nonlinear Schrödinger-Langevin equation (GNLSLE) of the following form has been derived [20]:

$$i\,\frac{\partial\Phi(x,t)}{\partial t} + \frac{1}{2}\,\frac{\partial^2\Phi\;(x,t)}{\partial x^2} - [\frac{\nu}{2i}\,\ln\,\frac{\Phi(x,t)}{\Phi^*(x,t)} + b\,\ln\rho(x,t) + V(x,t)]\;\Phi(x,t) \;=\; 0, \qquad (7)$$

$$\rho(x,t) \;=\; |\Phi(x,t)|^2, \qquad (8)$$

where ν is akin to a plasma collision frequency, b is a free parameter and $V(x,t)$ is an external potential which may contain a white-noise random force. The first term in square brackets in Eq. (7) accounts for dissipation. Nassar [21,22] has then employed the usual Madelung transform, Eq. (6), to obtain the QFD equations corresponding to Eq.(7). Note that Eq. (7) is not expected to have a solitary wave solution.

A different type of GNLSE (in three spatial dimensions) has been suggested by Himi and Fukushima [23] via QFD, in order to deal more effectively with quantum phenomena such as heavy ion collisions. The two QFD equations result on applying a generalized scaling approximation in the time-dependent Hartree-Fock framework. This GNLSE is, in effect, formally a time-dependent generalization of Eq. (1), viz.,

$$i\frac{\partial\Phi(\mathbf{r},t)}{\partial t} + \frac{1}{2}\,\nabla^2\,\Phi(\mathbf{r},t) + v_{eff}(\mathbf{r},t)\;\Phi(\mathbf{r},t) \;=\; 0, \qquad (9)$$

where v_{eff} is a nonlocal, nonlinear one-body potential given by

$$v_{eff}(\mathbf{r},t) \;=\; -\;[\frac{\delta E[\rho]}{\delta\rho} + v_{ext}(\mathbf{r},t) - \mu], \qquad (10)$$

with E as the total energy and v_{ext} as the external potential.

4. A NEW GNLSE BASED ON AN AMALGAMATION OF QFD AND DFT

We now return to our earlier theme in Section 1 of using a blend of QFD and DFT (called QFDFT). Since time occurs naturally in the QFD equations and a nonzero current density (in addition to the charge density) is obtained by solving them, through QFD DFT can go into both time-dependent situations [7,24] and excited states, two types of phenomena which have been **outside** the purview of traditional DFT. We have considered [8] the physical problem of high-energy proton-neon atom collisions and have derived a **single** equation (a GNLSE) for directly calculating the net time-dependent electron density and current density. The GNLSE has also been derived by us alternatively through stochastic mechanics, via a set of Fokker-Planck equations. The GNLSE has the following form, similar to Eq. (9), viz.,

$$i\frac{\partial\Phi(\mathbf{r},t)}{\partial t} + \frac{1}{2}\nabla^2 \Phi(\mathbf{r},t) + v_{eff}(\mathbf{r},t)\,\Phi(\mathbf{r},t) = 0. \qquad (11)$$

However, v_{eff} is of a different form from (10) and can be explicitly written as

$$v_{eff}(\mathbf{r},t) = -[\tfrac{5}{3}C_k\rho^{2/3} - \tfrac{4}{3}C_x\rho^{1/3} - \frac{a(N)}{r^2} - \frac{Q}{r} - \frac{1}{|\mathbf{R}-\mathbf{r}|} + \frac{f(R,N)}{N}], \qquad (12)$$

$$\rho(\mathbf{r},t) = |\Phi(\mathbf{r},t)|^2, \qquad (13)$$

where C_k and C_x are constants, a(N) depends on N, f(R,N) depends on both R and N, R being the internuclear distance and N being the number of electrons. Here Q refers to a screened nuclear charge. The nonlinearity in Eq. (11) enters through the one-body nonlocal potential v_{eff}, via nonintegral powers of Φ as well as an integral occurring in Q (cf. Section 1). In order to solve Eq. (11), for studying high-energy proton-atom collisions, we first consider a restricted case of spherically symmetric scattering system. For a spherically symmetric system (one spatial dimension), the GNLSE (11) may be transformed to

$$i\,\frac{\partial\psi(x,t)}{\partial t} + \frac{1}{\alpha x^2}\,\frac{\partial^2\psi(x,t)}{\partial x^2} - \{\frac{1}{\alpha x^3}\frac{\partial}{\partial x} + v_{eff}[\psi,t]\}\psi(x,t) = 0;\ \alpha>0, 0\leq x\leq\infty \qquad (14)$$

where $\psi = r\Phi$ and $x = \sqrt{r}$. ψ tends to zero as x goes to zero or infinity. We have numerically solved [8] Eq. (14) by applying a Crank-Nicolson-type finite difference scheme which is stable [25] and convergent. Figure 1 depicts the initial radial density profile and the corresponding time-evolved quantity after 61 time-steps (each time-step = 0.01 atomic unit). Since the initial profile is not a solution

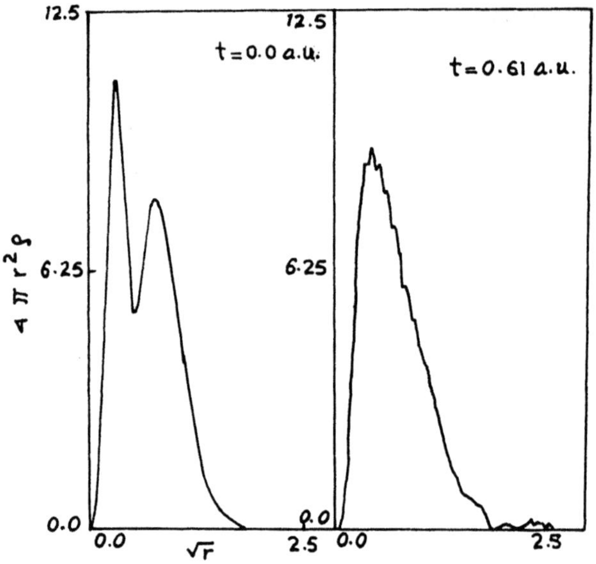

Fig. 1 Plot of radial densities against $\sqrt{r}$, in the spherically symmetric approximation, of a target neon atom colliding with a proton. The first figure at t = 0.0 corresponds to the internuclear distance R = 10.05 and clearly shows the two atomic shells (Hartree-Fock wavefunction [27] used). The second figure corresponds to R = 9.44 after 61 time-steps, each time-step being 0.01. All values are in atomic units.

of Eq. (14), it adjusts itself in course of time and from the thirtieth time-step onwards the envelope of the profile remains unchanged. In other words, these results indicate the possibility of solitary wave solutions. This requires further analysis. We also hope to look into the integrability properties of this type of equation, which might lead to an analytical solution. Incidentally, a spherically symmetric GNLSE also occurs in the study of higher dimensional Heisenberg spin chain [26].

The GNLSE (11) has also been solved numerically for a scattering system of cylindrical symmetry (two spatial dimensions). An alternating direction implicit-type finite difference technique has been used for this purpose. As the proton approaches the neon atom and the interaction is switched on, a nonzero current density is obtained which signifies that contributions from excited states are coming into picture [8]. We feel that, unlike the usual approaches in molecular reaction dynamics, our QFDFT approach [8] has the potential to monitor a time-dependent quantum process from start to finish, where the process occurs on a pulsating (time-dependent) potential surface given by $v_{eff}(\mathbf{r},t)$.

Acknowledgement

BMD and PKC would like to thank the CSIR, New Delhi, for a research grant and a Senior Research Fellowship respectively.

References

[1] E. Madelung, Z. Phys. **40** (1926) 332.
[2] S. K. Ghosh and B. M. Deb, Phys. Rep. **92** (1982) 1.
[3] A. S. Bamzai and B. M. Deb, Rev. Mod. Phys. **53** (1981) 96,593.
[4] F. Guerra, Phys. Rep. **77** (1981) 263.
[5] E. Nelson, **Quantum Fluctuations** (New Jersey: Princeton University, 1985).
[6] S. K. Ghosh and B. M. Deb, Int. J. Quant. Chem. **22** (1982) 871.
[7] B. M. Deb and S. K. Ghosh, J. Chem. Phys. **77** (1982) 342.
[8] B. M. Deb and P. K. Chattaraj, to appear.
[9] B. M. Deb and S. K. Ghosh, Int. J. Quant. Chem. **23** (1983) 1.
[10] M. Levy, J. P. Perdew and V. Sahni, Phys. Rev. **A30** (1984) 2745.
[11] G. Hunter, Int. J. Quant. Chem. **29** (1986) 197.
[12] A. Hasegawa, **Plasma Instabilities and Nonlinear Effects** (Berlin: Springer-Verlag, 1975), p. 201.
[13] R. K. Dodd, J. C. Eilbeck, J. D. Gibbon and H. C. Morris, **Solitons and Nonlinear Wave Equations** (London: Academic Press, 1982),p.495.
[14] M. Lakshmanan and R. K. Bullough, Phys. Lett. **80A** (1980) 287.
[15] R. Balakrishnan, J. Phys. **C15** (1982) L1305.
[16] M. R. Belic, J. Phys. **A18** (1985) L409.
[17] T. F. Nonnenmacher and J. D. F. Nonnenmacher, Lett. Nuovo Cim. **37** (1983) 241.
[18] A. B. Nassar, Lett. Nuovo Cim. **41** (1984) 476.
[19] G. C. Ghirardi, C. Omero, A. Rimini and T. Weber, Rev. Nuovo Cim. **1** (1978) 1.
[20] A. B. Nassar, Phys. Lett. **109A** (1985) 1.
[21] A. B. Nassar, J. Phys. **A18** (1985) L423.
[22] A. B. Nassar, J. Phys. **A18** (1985) L509.
[23] M. Himi and K. Fukushima, Nucl. Phys. **A431** (1984) 161.
[24] E. Runge and E. K. U. Gross, Phys. Rev. Lett. **52** (1984) 997.
[25] P. K. Chattaraj, S. Rao Koneru and B. M. Deb, J. Comp. Phys. (in press).
[26] M. Lakshmanan and M. Daniel, Phys. **A107** (1981) 533.
[27] E. Clementi and C. Roetti, At. Data Nucl. Data Tables **14**(1974)174.

Index of Contributors

Ablowitz, M.J. 44
Akutsu, Y. 282

Bullough, R.K. 7,250

Chattaraj, P.K. 359
Chowdhury, A. Roy 105

Dash, P.C. 196
Deb, B.M. 359
Dey, Bishwajyoti 188,233

Fokas, A.S. 66,118

Hasan, A. 184

Joseph, K. Babu 308

Kalra, M.S. 184
Kaushal, R.S. 226
Kumar, A. 328
Kundu, A. 86

Lakshmanan, M. 145

Papageorgiou, V. 66
Pilling, D.J. 250
Popowicz, Z. 212

Rajaraman, R. 240,321
Rangwala, A.A. 176

Rao, J.A. 176

Sachdev, P.L. 162
Santini, P.M. 118
Scott, A.C. 343
Sudarshan, E.C.G. 2

Tamizhmani, K.M. 145
Timonen, J. 250

Wadati, M. 282

MIX
Papier aus verantwortungsvollen Quellen
Paper from responsible sources
FSC® C105338

If you have any concerns about our products,
you can contact us on
ProductSafety@springernature.com

In case Publisher is established outside the EU,
the EU authorized representative is:
Springer Nature Customer Service Center GmbH
Europaplatz 3, 69115 Heidelberg, Germany

Printed by Libri Plureos GmbH
in Hamburg, Germany